THE
SOLAR
WATER HEATER
HANDBOOK

THE
SOLAR
WATER HEATER
HANDBOOK
A Guide to Residential Solar Water Heaters

RICHARD H. MONTGOMERY

JONATHAN L. LIVINGSTON

Illustrated by Wayne L. Lewis

JOHN WILEY & SONS

New York   Chichester   Brisbane   Toronto   Singapore

Library of Congress Cataloging in Publication Data:

Montgomery, Richard H.
  The solar water heater handbook.

  Includes index.
  1. Solar water heaters.   I. Livingston, Jonathan.
II. Title.
TH6561.7M66 1986   697'.78   85-17801
ISBN 0-471-86278-9

Printed in the United States of America

10 9 8 7 6 5 4 3 2 1

PREFACE

This book is a detailed guide to the maintenance and repair of America's solar water heaters. It teaches system identification, testing, and troubleshooting, tells how to evaluate the installation for structural integrity, sizing, safety, and health problems, and details the characteristics and performance of collectors, heat exchangers, heat transfer fluids, pumps, controllers, storage tanks, auxiliary heaters, piping, valves, and gauges.

It shows how to troubleshoot, remove, disassemble, repair, and replace collectors, heat exchangers, heat transfer fluids, pumps, controllers, valves, and tanks, and it details what instruments, tools, and supplies a serviceperson should carry to perform these tasks efficiently.

The book will be very valuable to students studying solar system maintenance and repair, plumbers, heating and air-conditioning contractors, solar contractors, and system owners who perform their own maintenance.

Most books on solar energy concentrate on system design and installation. To the best of our knowledge, no other textbook concentrates on maintaining and repairing the hundreds of thousands of existing systems.

In the past ten years, the solar water-heater manufacturing industry has proven conclusively that many different types of systems will perform very satisfactorily. No one system can be considered to be the best, and there are both low-tech and high-tech approaches. Each system has its place in the market, and each has its idiosyncrasies.

Maintenance and service are of prime interest to all industry participants. Therefore, the book serves as an integrated course for solar industry students, a field guide for servicepeople working on unfamiliar systems, and a reference to typical field problems for solar system designers.

In closing, we offer a few words of advice to new servicepeople.

If you disagree with the way the system was designed or installed, keep your opinion to yourself. Criticism of the system design or installation is also a criticism of the customer who chose that system. Maintain a positive, helpful, and tactful attitude.

Operate at a profit and do not run in the red to please your customers. The service department or business should be profitable in itself. Control your time and minimize your inventories. Charge rates that bring in a sound profit. Bill and collect for your services promptly. Use sound business practices and standard accounting procedures. Analyze your market and build a suitable marketing plan. You cannot substitute technical skills for sound management practices. You must know and practice both.

Troubleshoot consistently and systematically. You may elect to follow a different procedure than the one that we have shown for tracking down a malfunction, but that is not important. You must have a consistent, standard strategy for identifying and eliminating the problems that does not waste time.

When troubleshooting, always look for the underlying causes of the problem. Ask yourself if a sensor or pump is defective or if the system design or installation is at fault. It is important to uncover and correct the underlying problem. A novice might immediately replace a discharging pressure relief valve; an experienced serviceperson would recognize that insufficient fluid flow, not a faulty valve, could be causing the problem; and a master serviceperson would check out all the common causes of insufficient flow rate to determine if the problem is an incorrect valve setting, an obstructed pump, an air blockage, a faulty check valve, or a faulty pressure relief valve.

Recognize the inherent limitations of the individual system that you are servicing, the limitations of its components, and the limitations of your test equipment. Your function is to optimize the operation of that system as it exists, not to rebuild it into an optimized design. As a serviceperson, you are always limited by the test equipment, parts, and tools that you can carry conveniently on your truck.

Couple these guidelines with the technical information you are about to study, and you will conduct a sound service business.

Richard H. Montgomery

Jonathan L. Livingston

CONTENTS

TECHNICAL NOTES

Differences of opinion always exist as to how to cope with technical problems and design features in a book as broad in scope as this one. The following notes detail where such differences were experienced and how each was handled.

Differential temperature measurements were chosen as the most practical and easiest measurement technique for performance evaluation. It is impractical to ask a serviceperson to measure flow rates or solar insolation accurately during a service call.

The size of the differential temperature across the collectors that should be considered normal was the subject of much discussion. It was resolved using a 10 to 20°F differential as normal because, in the field, low flow rates that give a 30°F differential can result in draindown and drainback system performance problems.

Service-water tempering valves are not shown in Section 1, which teaches how to determine the type of system. This was done to simplify the designs, since the valve is usually located after the auxiliary heater and thus is not essential to describing the system type. Later in the book, readers are introduced to the concept of using a tempering valve in all systems.

A simplified bypass system is also used in Section 1 for the same reason. Bypass systems are described in detail later.

The pump strainer is shown on the supply side of the pump in all systems. Some designers always place the pump strainer on the discharge side of the pump. Either position seems to work satisfactorily in most systems. Many closed-system designs completely eliminate the pump strainer. In all cases, follow the manufacturer's recommendations.

THE
SOLAR
WATER HEATER
HANDBOOK

CHAPTER
1

Introduction

In this introductory chapter we discuss the nature of solar energy versus the nature of fossil fuels, show the challenge faced by solar engineers, suggest a set of basic solar engineering principles, show the limitations that solar system designers face, and discuss the sizing of small, residential, service-water solar heating systems.

In every scientific discipline there is room for technical disagreement. Some practitioners may adopt one set of standards and others may adopt another. However, as disciplines mature, shared experiences and technical compromises bring opposing viewpoints together.

Although the American Society For Testing Materials (ASTM) has provided a forum for setting standards, the solar water-heating industry is still undergoing the maturation process. Thus there are legitimate opposing viewpoints about some areas of solar technology.

In this chapter we set the stage for the technical decisions shown throughout the book by defining our technical position.

The Nature of Solar Energy

As a fuel for heating residential service water, solar energy has eight important characteristics.

- It is not available on demand.
- It is available only 4 to 10 hours per day.
- It is low in intensity and cannot readily be collected at temperatures in excess of 300°F in economically priced residential heating equipment.
- It cannot be turned off when the system overheats.
- It cannot be turned on when the system cools down.
- It attacks a number of common construction materials.
- It must be converted to thermal energy outdoors.
- It costs nothing.

If the energy were not free it would not be commonly used in our modern society, because these fuel characteristics do not fit society's technical requirements.

The Nature of Fossil Fuels

In comparison to solar energy, fossil fuels have these characteristics:

- They are available on demand.
- They are available 24 hours per day.
- They are high in intensity and can provide very high temperatures in economically priced equipment.
- They can be turned off when the system overheats.
- They can be turned on when the system cools down.
- The by-products of combustion are corrosive to some common construction materials.
- The energy can be converted to thermal energy indoors in a protected environment.
- Each unit of fuel costs money and the price per unit is increasing.

Fossil fuels fit society's technical requirements much better.

A Comparison

Table 1.1 summarizes these important comparative fuel characteristics. These intrinsic fuel characteristic differences are important to understand, because they explain why the engineering approach to designing effective solar water-heating systems must be different from the approach to designing gas-fired (or electric) water-heating systems.

The Solar Engineering Challenge

The nature of the fuel presents the solar design engineer with the following five challenges.

Design a Fail-safe System

A fail-safe system is one that fails in a nondestructive mode so that it is not physically damaged. The fuel cannot be turned off and on at will, so solar systems can overheat and/or be exposed to freezing temperatures many times during their design life. These incidents must be anticipated and must not cause *catastrophic failure*. Catastrophic failure is failure that results in physical damage to the system.

Solar system designers differ widely in defining what constitutes a fail-safe system, because it is expensive to design true fail-safe characteristics. In this book we define a fail-safe system as one that will fail-safe

Table 1.1

A COMPARISON OF FUEL CHARACTERISTICS

CHARACTERISTIC	SOLAR FUEL	FOSSIL FUEL
Availability	4–10 hr/day	24 hr/day
Top available temperature	Under 300°F	Very high
Hourly energy intensity	250–300 Btu/ft^2	High
On/off controllability	None	At will
Degradation mode	Ultraviolet rays	Combustion by-products
Location	Usually outdoors	Controlled environment
Cost	Free	Expensive

anywhere in the United States under any ordinary circumstances if the system malfunctions or external power fails.

Utilize Nondegradable Materials

The part of the system that is located outdoors must not be damaged by ultraviolet rays or by mildly acid or alkaline atmospheric conditions, and the fluid passages of the system must not be damaged by any chemicals used as heat transport fluids.

Many solar system designers still fail to recognize that the outdoor environment in which solar systems must operate is very aggressive and unforgiving. Premium-grade materials designed to survive for many years are required.

Capture the Maximum Energy

The thermal energy provided by the sun in a square foot of area is very limited, so the system designer must spread the energy-collection section of the system over the widest practical area so that the amount of energy that falls on the system is maximized.

The fossil-fuel system designer takes just the opposite approach and designs the fossil-fueled system into the least possible amount of space.

Eliminate All Possible Standby Losses

Because solar energy must be collected when it is available and cannot be called up on demand, longer storage periods and larger storage volumes are required. Therefore, standby losses assume much more importance than they do in fossil-fueled systems, where energy is available on demand.

Provide a 10- to 15-Year Minimum Design Life

Obviously, meeting the design criteria for solar water-heating systems will be more costly than meeting fossil-fueled system design criteria. The difference in cost is only justifed when the system is designed to last for a long enough time to allow fuel savings to pay for the additional capital investment.

The Basic Solar Engineering Princples

The previous discussion leads to establishing six basic engineering principles that must be followed to meet our definition for a successful solar design.

- Build an efficient and long-lived solar collector that will survive for *at least* 10 years outdoors and will convert the sun's radiant energy to usable thermal energy.
- Build an efficient storage unit that will hold the collected energy with minimal losses for 24 to 48 hours.
- Provide a positive mechanism that will prevent catastrophic failure when the system overheats.
- Provide a positive mechanism that will prevent catastrophic failure when the system is exposed to freezing temperatures.
- Build an efficient and long-lived transport system to move the collected energy from the solar collector to the storage unit.

- Provide an automatic control mechanism that will collect as much solar energy as possible.

Design Limitations

The need to meet these six solar design principles places five major constraints on the system designer.

- Solar collection must be *separated* from solar storage. It is not possible to maximize energy collection and minimize storage losses in one device at an economical cost with currently available technology.
- The system must contain devices or materials that prevent *overheating* or *freezing* of the system. These materials or devices must be positive acting when all external power is removed from the system; they cannot depend on the owner to take special action to protect the system.
- No materials can be included that do not have a design life of less than 10 years outdoors, assuming that the investment in the system will be paid for in a 10-year period or less. Periodic maintenance of the system to meet this criteron is acceptable.
- No materials of construction can be used that would be degraded by the chemicals chosen for the transport system or that would degrade those chemicals. Periodic maintenance of the chemicals to meet this criteron is acceptable.
- The cost of the external energy required to operate the system must be a very small fraction of the value of the energy collected by the system.
- The economics must be favorable; that is, the energy savings must exceed the investment cost during the design life of the system.

The Uncertainty

Solar systems are like all products engineered by human beings; there is a degree of uncertainty in every design that is only removed when enough experience has been gathered under actual operating conditions to prove statistically that the design criteria were correct.

Not all solar designs will meet the previously mentioned criteria. Throughout the book, you will read about the limitations of each design as it is shown. You will read about the operating experience that has been gathered for the older designs, and you will read our opinions of some of the newer designs for which no extended operating experience exists.

These opinions are based on the discussion you have just read. They may or may not agree with the opinions of others who hold a differing view on solar design.

Solar System Sizing

Sizing solar systems for residential service-water heating is complex, because the required size of the system depends on four major variables: *geographical location, system design, water-heating load,* and *available capital.*

It is difficult to quantify all four of these variables properly in any one installation. Available capital can be readily defined, but the amount of solar energy available, the size and nature of the load, and the system performance will vary widely.

Predicting Available Solar Energy

The amount of sunshine available in any given location on any given day cannot be predicted accurately in advance. However, the approximate amount of sunshine available in any given month can be predicted by averaging historical data. The amount of sunshine available in any given year can be predicted within narrower limits, and the amount of sunshine available over the design life of the system can be predicted with a great deal of confidence.

The solar designer therefore must work with average solar data in sizing systems. He or she cannot predict how much solar energy will be available on Tuesday, but a monthly, yearly, or lifetime estimate can be made.

Predicting Load

The demand for hot water in any given time period on any given day is also unpredictable but, in contrast to solar data, historical use data are generally unavailable for any given family and life-style. Hot water demand varies tremendously from one installation to another.

Initially, U.S. Department of Housing and Urban Development's (HUD's) average hot water usage standards were used by designers and installers to predict load. These standards were based on number of occupants and types of appliances in the residence. When actual load measurements were taken for several hundred solar installations in the Northeast, the HUD standards were found to be useless for predicting demand.

Predicting System Efficiency

The efficiency of a solar system is both design dependent and use dependent. The designer can control the collection, transport, and storage efficiency, but he or she has no control over the use patterns. Therefore, the true operating efficiency of the system cannot be determined in advance of the system's use.

The Law of Diminishing Returns

The widely varying nature of the energy source, the system load, and the system efficiency subject solar installations to a law of diminishing returns that can be stated as follows: "After a certain optimum size has been achieved, the cost of collecting solar energy will rise if a larger system is installed." Let us examine why this is so.

It is economically impossible to satisfy 100% of the hot water requirements in all installations. To do so would require the building of an enormous system, at a huge cost, that would provide storage through long, nonsunshine periods.

To satisfy all energy needs, a combination of solar energy plus a fossil-fueled or electrical backup system is required, so a complete solar system uses two energy sources that exhibit different costs.

Therefore, the cost of the system expressed as cost per unit of solar energy collected must be compared to the cost of purchasing conventional fuel to satisfy the balance of the energy demand.

F-Chart, a Theoretical Approach to System Sizing

During the early 1970s, W. A. Beckman, S. A. Klein, and J. A. Duffie, working at the University of Wisconsin under the sponsorship of the National Science Foundation and the Energy Research and Development Administration, conducted a number of experiments and developed a sizing theory now known as *F-chart*. The details of this approach can be found in W. A. Beckman, S. A. Klein, and J. A. Duffie, *Solar Heating Design*, John Wiley & Sons, Inc., New York, 1977.

F-chart uses the average monthly solar energy, the average daily load, the system's component efficiencies, the installed cost, and the cost of conventional fuel to calculate the optimum size.

An economic analysis of cost can also be carried out for the design life of the system by incorporating financial data relating to tax savings, interest on investment, fuel inflation, and discount rates.

F-chart outputs the *optimum size based on financial considerations*. In effect, the analysis says:

> To get the lowest cost per unit of energy, build a system containing *X* ft² of collector and *Y* gal of storage. This system will satisfy *Z*% of your total energy needs.

If the design life economic analysis is incorporated in the calculations, the system also returns the cost of providing hot water through a solar + backup system and calculates the annual dollar savings over providing all the energy with a fossil-fuel or electrically fueled system.

F-Chart Has Been Badly Abused

Many solar marketing and sales companies have used F-chart calculations as a major sales tool for selling individual installations and have, in numerous instances, presented the results as indicative of that individual installation's performance.

Given the variable nature of the input data, such use of this technique is highly improper, because the results of the calculations are not valid except when considered as average values for a statistical number of installations.

Proper F-Chart Use

The Solar Decision Book (Richard H. Montgomery and Jim Budnick, John Wiley & Sons, Inc., New York, 1978) contains a set of tables that was built by using F-chart to run several hundred simulations using model data. These tables were then used to build simple graphs that showed the number of square feet of collector needed to satisfy a certain percentage of the total load with solar energy. A typical chart is shown in Figure 1.1, where the horizontal axis is the square feet of collector, the vertical axis is the fraction of the load satisfied by solar energy, and the three curves represent three different daily hot water loads.

On this particular chart, 50 ft² of collector would satisfy 40% of a 120-gal daily load, and 90 ft² would satisfy 60% of the load.

These data are only valid when considered for 1 year or more and when the load patterns conform to those used to define the parameters used in the F-chart model. Therefore, these data should only be presented as typical data, not as specific data relating to an individual installation.

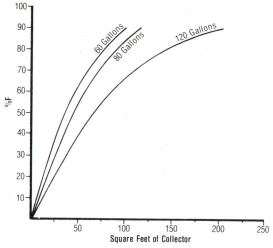
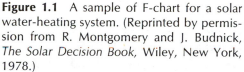

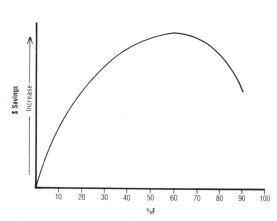

Figure 1.1 A sample of F-chart for a solar water-heating system. (Reprinted by permission from R. Montgomery and J. Budnick, *The Solar Decision Book,* Wiley, New York, 1978.)

Figure 1.2 A typical F-chart cash flow analysis for a solar system. (Reprinted by permission from R. Montgomery and J. Budnick, *The Solar Decision Book,* Wiley, New York, 1978.)

The Optimum Size

If the cost of energy is to be used to determine the optimum system size, which is not always the case, then F-chart can be used to run a cash flow analysis. Various researchers have run thousands of these simulations. Figure 1.2 shows the general results. In this figure the horizontal axis represents the fraction of the load provided by solar energy, and the vertical axis represents the dollars saved.

The shape of this curve shows that savings optimize at about a 60% load fraction and rapidly drop off after a 70% fraction.

Most researchers have found that the optimum size of the average residential service-water solar heating system—assuming lowest-cost energy is the optimizing criterion—lies between 40 and 70%.

General Sizing Guidelines

The tremendous variation in actual conditions leaves F-chart as a designer's tool to be used with a number of varying installations in order to arrive at some empirical sizing guidelines that can be used for individual installations.

We have devised such sizing guidelines for this book, but we caution readers that because of the large geographical differences across the United States, care should be used in adopting these in areas that are high in sunshine, such as Miami, Florida or Phoenix, Arizona, and in areas that are notoriously low in sunshine, such as Portland, Oregon, and Seattle, Washington.

These guidelines were devised using the following considerations.

- A single person uses a certain amount of hot water.
- A family can afford an economic investment of a certain number of dollars.
- The physical location will only support so many square feet of collecting surface.
- Fractions of a collector cannot be utilized.

Based on these considerations, the following generalized empirical decisions were made.

- For most systems designed with relatively good efficiency, the system should contain 20 to 25 ft² per occupant to satisfy 40 to 60% of the total load.
- If 20- to 25-ft² collectors are being used, at least two collectors should be used. Installation of one small collector is generally a poor economic decision.
- When the system contains more than 100 ft² of collector area, other design considerations exist that lie beyond the range of this type of rule-of-thumb sizing.
- Once the square feet of collecting surface has been determined, the rest of the system should be designed to match the collector size. Storage tank, pump, and pipe sizing are discussed later.

The final sizing decision comes down to deciding whether to install one, two, three, or four collectors.

Although this may seem like an unscientific or casual approach, it is not. Such designs—when checked back through the F-chart program—agree with F-charted sizing. Only by extensive study of an individual site and load can more precise sizing be accomplished.

The Scope of This Book

This book has been written specifically about small, residential, solar water-heating systems and for servicepeople engaged in the maintenance and repair of installed systems.

It is not intended as a solar designer's or solar installer's manual, even though it does delve into elements of both design and installation in order to help servicepeople identify and repair solar heating system faults.

SECTION
1

Identify the System

There are eight major types of solar water heaters: thermosiphon systems, integral collector storage systems, pumped service-water systems, draindown systems, drainback systems, liquid heat-exchanged systems, phase-change heat-exchanged systems, and air heat-exchanged systems. The first task of the serviceperson is to identify the system being serviced.

Chapters 2 to 8 are about the various systems. The system's name, end use, component parts, component functions, operating cycle, expected performance, and sizing parameters are discussed.

Each chapter includes a discussion of how to test performance and what can go wrong. A troubleshooting diagnostic chart and a checklist of what the serviceperson should have learned will complete the chapter.

At the end of these seven chapters, readers should be able to:

- Identify any solar water-heater system.
- Identify each major component in the system.
- Understand the operating cycle of each system.
- Understand the expected performance of each system.
- Know the advantages and disadvantages of each system.
- Understand the sizing of each system.
- Known the common problems associated with each system.
- Test system performance.
- Perform general troubleshooting.

CHAPTER

2

Thermosiphon and ICS Systems

Thermosiphon: The convective circulation of fluid occurring in a closed system in which the dense, cold fluid sinks and displaces the less dense, warm fluid upwards. The solar industry calls thermosiphon and ICS systems *passive systems*.

When water is heated, it expands. That expansion results in water that weighs less per gallon than cooler water. This event can be observed by observing a hot water heater in operation (Figure 2.1). Although the gas flame heating the water is at the bottom of the tank, the water rises to the top of the tank as it is warmed, and the cooler water sinks to the bottom of the tank.

If a second tank were located above the first tank (Figure. 2.2) and were connected to the top and the bottom of the first tank, the hot water would collect in the top tank.

This convective flow event can be put to work to create a simple **thermosiphon solar water heater.** A thermosiphon solar water-heater system is a system in which the water is circulated by the difference in density between hot and cold water and no pump is used.

In Figure 2.3 the gas water heater has been replaced by a solar collector. The bottom and the top of the solar collector are connected to the tank. The water in the solar collector is warmed by the sun and rises from the top of the collector to the top of the tank. Denser, cooler water from the bottom of the tank sinks to the bottom of the collector to be warmed. The warming process continues as long as there is a difference in temperature. The water in the system just gets hotter and hotter as it recirculates.

For thermosiphoning to occur, the following construction details must be observed.

- The solar collector must be lower than the storage tank.
- The connections between the tank and the collector must be made as shown in Figure 2.3.
- The piping must be continuously sloped so that there are no *traps* in the lines. Traps are intermediate high points that can impede or stop the flow of fluid. Figure. 2.4 shows a trap where hot water or air bubbles can become lodged. To pass through the trap, the lighter hot water or air would have to move down through a heavier

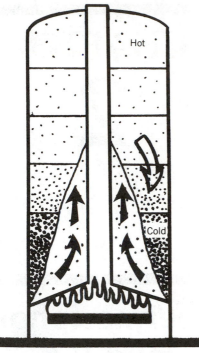

Figure 2.1 Hot water rises because it weighs less.

fluid. This violates the law of gravity and cannot occur in the absence of a mechanical pump. Therefore, the system in Figure 2.4 will not thermosiphon properly.

The rate at which thermosiphoning takes place depends on three factors.

- The height of the tank above the collector.
- The difference in temperature between the tank and the collector.
- The amount of resistance to flow caused by the friction generated in the piping.

The greater the difference in height, the faster the flow. The greater the difference in temperature, the faster the flow. The less the resistance to flow in the piping, the faster the flow. Obviously, good thermosiphoning systems must have good height differentials, good heat fluxes, and large piping. Specific guidelines for each type of thermosiphon system will be provided as the systems are explained.

The Vertical Tank Thermosiphon System

Figure 2.5 shows a typical **vertical tank thermosiphon water heater.** The vertical tank configuration first came into use in the early 1900s, but systems more than 25 years old are rare in most areas of the United States. The serviceperson is more likely to encounter custom-made *retrofit* vertical tank systems built since the mid-1970s. A retrofit solar heating system is one installed some time after the house was constructed. The system in Figure 2.5 contains two solar collectors with a combined square footage of 40 to 50 ft².

The storage tank holds 60 to 100 gal. The collectors are mounted below the tank at an angle to the horizontal that generally lies between 15 and 45°, although angles between 10 and 90° are satisfactory in some systems. They are *connected in parallel*. Parallel-connected collectors are

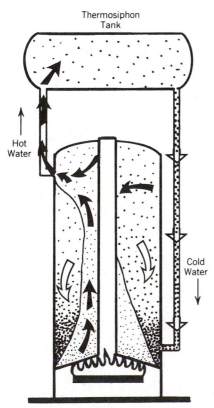

Figure 2.2 A convective loop that thermosiphons.

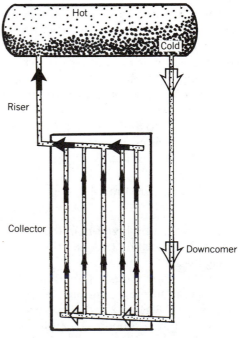

Figure 2.3 A solar convective loop.

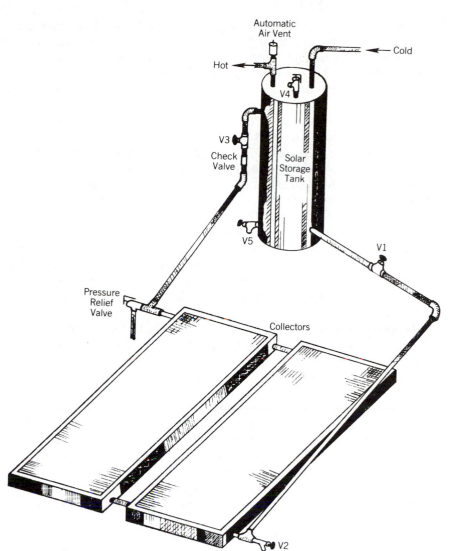

Automatic
Air Vent

Cold

Hot

V4

V3
Check
Valve

Solar
Storage
Tank

V5

V1

Pressure
Relief
Valve

Collectors

V2

Figure 2.5 A vertical tank thermosiphon solar water heater.

connected so that the flow is divided across them. The bottom collector manifolds are connected to the *cold downcomer,* and the top collector manifolds are connected to the *hot riser.*

Refer to Figure 2.5. The system may be built with or without a **check valve.** A check valve is a valve that allows fluid flow in only one direction. In systems built without a check valve, VI is a **backflush valve** that can be closed to backflush the collectors through the tank; V2 is a flush and drain valve for the system; and V3 is a manual valve that can be closed along with V1 so that the collectors can be worked on without shutting down the water system.

In systems containing a check valve, VI cannot be used for backflushing and serves only as an isolation valve to isolate the tank from the collectors.

Both V1 and V3 must be nonrestricting-flow gate or ball valves. V1, V2, and V3 must never all be closed at the same time. This would lock the water into the collectors. As the fluid expanded when warmed by the sun, pressure would be built up in the collectors. This pressure could become high enough to deform or burst the collectors. A pressure relief valve has been installed at the top collector manifold to prevent this.

Valve V4 is a **temperature and pressure relief valve** that prevents tank

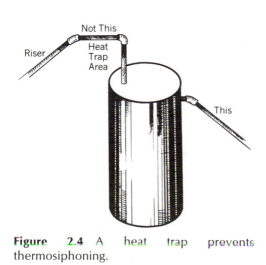

Not This

Riser

Heat
Trap
Area

This

Figure 2.4 A heat trap prevents thermosiphoning.

overpressurization and overheating; V5 is a tank sediment drain valve located below the incoming cold water supply pipe. The **downcomer** and **riser** piping are fabricated from $\frac{3}{4}$ in. or larger copper pipe to minimize friction losses. The entire system is under service-water pressure, which provides the driving force to supply hot water from the system to the residence.

Cold water from the service line enters into the bottom of the tank through a dip tube located at the tank top and fills the system. During the initial filling process, an automatic air vent valve located on the hot water supply pipe opens to remove the air from the system. This automatic air vent valve also can admit air to allow full drainage of the system at the onset of the freezing season. Once the system is filled and the hot water supply valve is closed, the system is at equilibrium and under service-water pressure. If the sun is shining, the water in the collectors is warmed and rises into the tank, and cooler, denser water from the bottom of the tank sinks down the downcomer into the bottom of the collectors to be warmed. This thermosiphoning process continues as long as the water in the collectors is hotter than the water in the bottom of the tank.

Reverse thermosiphoning can occur at night unless a check valve is installed in the riser or the riser is bonded to the tank. Reverse thermosiphoning is the passing of the heat collected in the tank back down to the collectors, where it is radiated back out to the sky.

Reverse thermosiphoning occurs in this system when the riser tube cools faster than the tank. As the water in the riser cools, it sinks into the collector, and hot water from the tank is forced into the riser.

Solar collectors can freeze at night, even when air temperatures are above 32°F because of **night sky radiation.** Night sky radiation is the tendency for a warm collector to radiate heat to a cold night sky. On a very clear night the sky can have a radiant temperature as low as −50°F. This radiant effect can cause standing water in a solar collector to freeze and damage the collector tubes.

The check valve used to prevent reverse thermosiphoning should be a rising/sinking ball check valve, because it offers low resistance to forward flow. Rising/sinking ball check valves must be installed in a vertical pipe. They will not operate properly in a sloping or horizontal pipe.

The System Connected as a Preheater

Figure 2.6 shows the vertical tank thermosiphon system connected as a **solar water preheater** to a fossil fuel-fired or electric hot water heater. A solar water preheater is a solar water heater that is connected in series with the regular or auxiliary heater so that the service water that would normally flow into the auxiliary heater is prewarmed in the solar heater prior to going to the auxiliary heater.

In most retrofit installations the thermosiphon preheater is located outdoors, and the auxiliary heater is located in the residence. This configuration requires that the system be installed so that it can be fully drained and removed from operation in freezing weather.

When the preheater is operating, V7 and V8 are open, and V6 is closed. The cold service water is routed to the bottom of the solar water preheater tank through the dip tube and is returned, warmed, to the cold water supply of the auxiliary heater.

When the preheater is out of service, which would normally occur in freezing weather, V7 and V8 are closed, and V6 is open. The solar water

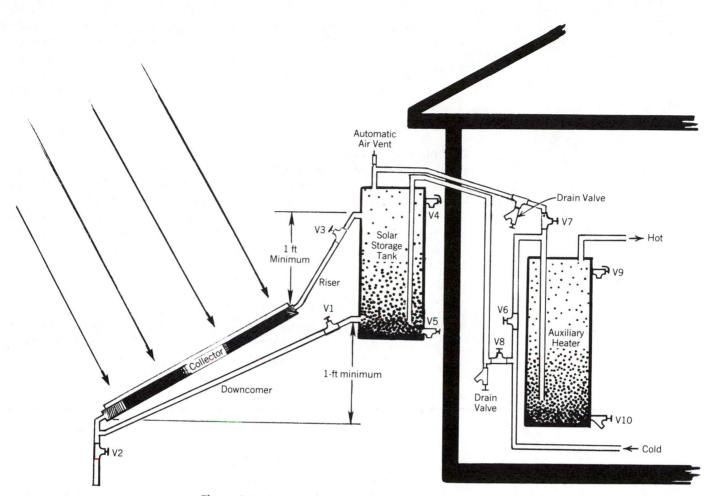

Figure 2.6 A vertical tank system connected as a preheater.

preheater system is bypassed, and the cold water supply goes directly to the auxiliary heater. Drain valves, located next to V7 and V8, assure complete pipe drainage.

The piping that is connected to the preheater must be pitched so that it drains fully. It is best if all the piping can be pitched to be drained to the indoors. If it cannot be pitched down to the interior of the residence, it should be pitched back to the tank. It is extremely important that no undrainable pockets be created in the piping system. V6, V7, and V8 must be nonrestrictive, full-flow gate or ball valves. Figure 2.6 also shows the height differentials that are normally considered as appropriate minimums in a vertical tank thermosiphon solar water-heater system. Both the hot riser and the cold downcomer should pitch 1 ft or more in their runs from the tank to the collector.

Note, also, that the tank drain V5 is located at the very bottom of the preheater tank. Typically, solar collectors contain riser tubes that vary from $\frac{1}{4}$ to $\frac{5}{8}$ in.[1] in outside diameter. If sediment is not drained periodically from the bottom of the tank, the collector riser tubes may, in time, become clogged. It is also good practice to provide a drain leg at the bottom of the collectors where the downcomer connects to the collec-

---

[1]Standard solar and plumbing practice is to refer to copper tubing or pipe by its nominal size or, less frequently, by its outside diameter. The nominal tubing size is equal to the tubing outside diameter minus $\frac{1}{8}$ in. Unless otherwise indicated, all tubing sizes used in this book are nominal sizes.

tor. This will allow sediment to drop out into the drain leg. If the service water is dirty enough to present a serious problem, it is advisable to place a filter in the service line ahead of all the fixtures in the residence instead of in the auxiliary heater or the solar preheater system. In the case of hard water, the system owner should consider installing water-softening equipment for the household water supply.

The auxiliary heater must have its own separate temperature and pressure relief valve and drain valve, as indicated by V9 and V10. The local plumbing codes will dictate the specifications for these valves.

Reverse thermosiphoning between the auxiliary heater and the solar water heater usually does not occur, since there is no full circulation path between the series-connected heaters. In rare instances it might be necessary to consider using a flapper or spring-type check valve in the connection between the top of the solar preheater tank and the auxiliary heater to eliminate auxiliary heat loss caused by reverse thermosiphoning.

Expected Performance

Solar water-heater performance is both site dependent and system dependent; that is, solar water heaters installed in Southern California or Arizona will see more sunshine and perform better than comparable heaters installed in northern Minnesota or even in the Middle Atlantic states, so statements of performance must be generalized. In discussing relative system performance in this book, we will assume that

- The system is installed in Nebraska.
- The hot water usage averages 20 gal per person per day.
- The water is delivered at 120°F.
- The use pattern throughout the day is that of an energy-conscious user but is not dictated by the existence of a solar system, which would require that the user temper demand to suit availability.

The collectors in a vertical tank thermosiphon system can be expected to operate at higher temperatures than the collectors in a pumped system because the circulation of the water is dictated by the difference in density between the hot and cold water instead of by the design parameters of the pump. Generally, the water flow will be slower, the collectors will run hotter, and the system will collect energy at a lower efficiency. The differences should not be dramatic and can be compensated for by increasing the ratio of storage water to collector footage. A storage tank of at least 1.5 gal/ft² of collector should be used, but no further increases in efficiency should be expected once the system contains 2 gal of storage per square foot of collector.

The year-round use of vertical tank thermosiphon systems is not practical except in areas where there is no freezing weather. In Nebraska, freezing weather usually occurs from the beginning of October until the end of March, with some low temperatures occurring from the middle of September until the middle of April. This indicates that the solar heater can only be used for $6\frac{1}{2}$ to 7 months of the year, and performance is zero during the winter months. However, the system can be engineered for maximum performance during its operating season.

A thermosiphon system in Nebraska that is faced due south at a tilt angle of 20 to 30° and that contains 1.5 to 2 gal of storage tank per square foot of collector will give excellent performance over the nonfreezing months of the year.

Sizing Parameters

Based on the assumptions shown in the last section, vertical tank thermosiphon systems should be sized at 20 to 25 ft^2 per resident. Under most circumstances, this will produce 40 to 60% of the required hot water. Servicepeople encountering systems designed outside these parameters should query the owners as to unusually high or low water usage. Like all generalized statements of solar performance, this rule of thumb will have a number of exceptions and should only be used as a general guideline. However, by understanding it, if the serviceperson were to run into a 400-ft collector array with an 82-gal storage tank for a family of two in Vermont, he or she would immediately understand that the nonperformance was a design problem, not a system fault.

Other Considerations

The glass-lined solar storage tank, which is generally available from almost all water-heater manufacturers, is a very satisfactory tank. Although it is not shown on any of the drawings in this chapter, the glass-lined storage tank in a vertical tank thermosiphon system should contain a magnesium anode rod to protect against tank corrosion.

The piping and fittings should all be copper and brass to minimize corrosion problems. Special high-temperature plastics may also be used in some situations. Large-diameter piping should be used for the riser and downcomer. We recommend $\frac{3}{4}$-in. minimum diameter piping for up to two collectors and 1-in. minimum diameter piping for three or four collectors.

The tank should be insulated with R-16 or higher-value insulation so that the collected heat will not be lost to the surrounding air. Most pipe insulation runs below R-10 but, if R-10 pipe insulation can be found, the pipes also should be insulated to R-10. All parts of the system that are outdoors must be very carefully protected against adverse weather. The tank and the collectors must be mounted securely on well-designed supports that will carry the weight and not be affected by frost and wet weather.

Advantages

The vertical tank thermosiphon system has these advantages.

- The system is inexpensive to install because no pump, controller, special heat transfer fluid, or heat exchanger is required.
- The system is simple and noncomplex.
- The seasonal performance is excellent.
- Very little maintenance or service is required.

Disadvantages

The vertical tank thermosiphon system has these disadvantages.

- The system cannot be operated year round except where there is no freezing weather.
- The system must be cared for by filling and draining it at the start and the close of the season.
- The system is constantly recirculating fresh water. If the water is not properly conditioned, corrosion, scale, and sediment buildup can cause premature system or component failure.
- The storage tank must be placed higher than the collectors. This will pose structural or aesthetic problems in most retrofit installations.

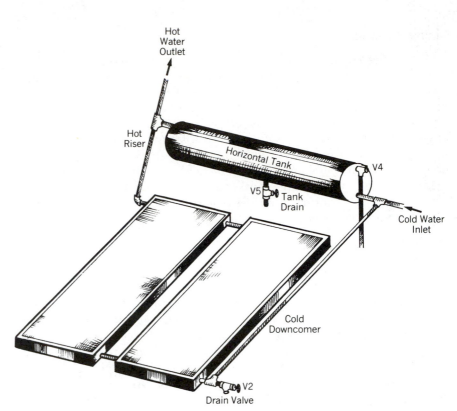

Figure 2.7 A horizontal tank thermosiphon solar water heater.

The Horizontal Tank Thermosiphon System

Figure 2.7 shows a horizontal tank thermosiphon system. In this system the vertical tank has been replaced with a horizontal tank located just above the collectors. The collector area may range from 20 to 65 ft², and the tank should be sized at 30 to 80 gal.

Unlike vertical tank systems, horizontal tank thermosiphon systems usually are manufactured commercially. Horizontal tank systems were rarely installed in the United States before 1975.

In order to provide for **stratification** of the water in the tank, the tank should be at least 18 in. in diameter. Stratification is the layering of the water in the tank into hot and cold sections. Figure 2.3 depicts this layering. Table 2.1 gives some common tank sizes and capacities.

For convenience and ease of construction, the length of the tank is

Table 2.1

COMMON TANK SIZES (WITHOUT INSULATION)

LENGTH (in.)	DIAMETER (in.)	CAPACITY (gal)
36	18	40
48	18	53
54	18	60
60	18	66
66	18	73
72	18	79
36	24	71
48	24	94
54	24	106
60	24	118

usually less than or equal to the width of the collector array. The tank must be mounted so that the bottom of the tank is no lower than a line drawn parallel to the back of the collectors, as illustrated in Figure 2.8.

Horizontal tanks must have at least one end with two openings, one near the top and one near the bottom, and one end with one opening at the top in order to plumb the tank properly for thermosiphoning. There is usually also an opening at or near the bottom of the tank for a drain valve. Refer to Figure 2.7 to see why the placement of these tank openings is so critical.

We recommend that a minimum angle of 20° from the horizontal be used for the collector tilt angle in order to assure that the system will thermosiphon at a reasonable rate. Again, just as in vertical tank systems, the riser and downcomer must be pitched a minimum of 1 ft for proper thermosiphoning.

The valving is similar to the valving for the vertical tank system, except that V1 and V3 can usually be eliminated. The valves in Figure 2.7 have been numbered to correspond to the valve functions in Figures 2.4 and 2.5. Figure 2.9 shows the details of the piping layout.

The components, their functions, the operating cycle, the expected performance, and the sizing of the horizontal tank thermosiphon system are similar to those of the vertical tank thermosiphon system, with one exception. Virtually all horizontal tank thermosiphon systems manufactured since 1975 contain an automatic freeze protection system.

Advantages

The horizontal tank system has these advantages.

- A low-profile system (for yard or roof mounting).
- A compact arrangement that requires minimum structural bracing.
- A horizontal distribution of storage tank load over several roof support members.
- A configuration that allows modular assembly instead of on-site fabrication.
- A system that can be given automatic freeze protection in mild climates.

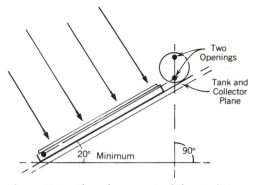

Figure 2.8 The placement of the tank in a horizontal tank system.

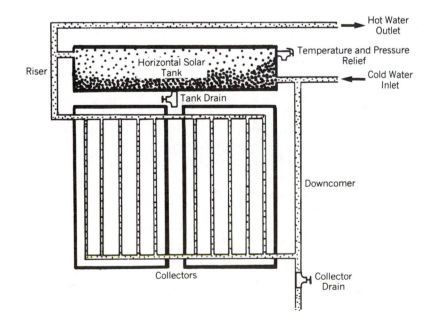

Figure 2.9 The horizontal tank system piping layout.

Disadvantages

The horizontal tank system has these disadvantages.

- Less stratification within the tank because of the low profile.
- A tendency for the collectors to run hotter because of the lower stratification.
- A standard magnesium anticorrosion rod cannot be properly suspended in the tank.
- Many roofs are not engineered to withstand the weight.
- Not suitable for year-round operation in many areas of the United States.

Solar Collector Freeze Protection in Thermosiphon Systems

Solar collectors are very susceptible to freezing. When water freezes, it expands. If this expansion occurs inside a solar collector riser tube, either the tube will burst or a soldered or brazed joint will rupture.

Repairing freeze-damaged collectors is complex and expensive. Damage at multiple points can even necessitate replacing an entire collector. It is essential that **positive freeze protection** be provided. Positive freeze protection is protection that occurs without human assistance and when external power fails.

Even though solar system piping will also freeze in cold weather, collectors are virtually always the first system component to freeze; they contain very small pipes and readily radiate energy to the night sky, while the remainder of the system consists of insulated larger-diameter pipes.

Freeze Protection Strategies

Five different strategies can be used to prevent solar collectors from freezing.

- The collectors can be fabricated from nonmetallic materials that can withstand freezing without bursting.
- The collectors can be drained.
- Warm water from the storage tank can be circulated through the collectors.
- Auxiliary power can be used to heat the collectors.
- A nonfreezing liquid or gas can replace the water in the collectors.

The first strategy is not technically practical at present. No long-lasting material has been developed that can withstand freezing, exposure to ultraviolet light, and high temperature and, at the same time, be economical and collect solar energy efficiently.

Vertical tank themosiphon systems, which are mostly site fabricated, very rarely incorporate any automatic freeze protection. They must be drained manually and taken out of service in freezing weather.

Horizontal thermosiphon systems may or may not incorporate automatic freeze protection. Most manufacturers incorporate one of the last three strategies listed above into their designs in order to extend the geographic range in which their products can be used year round.

Circulation of warm water from the storage tank through the collectors can be induced by providing a mechanical *dribble valve*. A dribble valve is a valve that opens when exposed to near-freezing water. Water pressure forces warm water from the tank to travel through the collectors and out the valve. However, dribble valves do not work when the water pressure is lost and are only effective in mild climates where hard freezes do not occur. For example, one manufacturer only warrantees its dribble valve-protected thermosiphon system for installation at altitudes below 2000 ft.

Thermosiphon solar collectors can incorporate an electric heating element inside the lower collector header, but the operation of this element is expensive. In cold climates it can consume more energy than the solar collector produces. During power failures, there is no freeze protection. Electrically heated, freeze-protected collectors used to be warranteed down to 20°F. They are no longer marketed.

One system, the Solahart thermosiphon solar water heater, separates the collector fluid from the storage water with a heat exchanger that surrounds the storage tank. An antifreeze is circulated through the collectors and the heat exchanger. Solahart warrantees this system down to 5°F—provided that the collectors are not covered with snow for more than 24 continuous hours.

Although the Solahart system storage tank contains water, it is regarded as freeze protected because it is insulated and because of the long time required to cool a large mass of water enough to freeze it.

Even the best of the thermosiphon systems mentioned cannot positively protect against freezing except in temperate climates. In Chapters 4 to 8, you will study systems that are capable of operating year round in more severe climates.

The Integral Collector/Storage Solar Water Heater

Integral collector/storage solar water heaters (ICS) are rudimentary thermosiphon heaters. They are batch-heater systems that combine solar collection and solar storage into one unit. They do not require pumps or controllers, so they have been included here.

Batch heaters do not meet the specifications we proposed in Chapter 1 for efficient solar water heating, but they are in general use in Florida and California and thus deserve to be discussed.

Figures 2.10 and 2.11 show a simple configuration that has been in use since the 1890s. This configuration is called a **breadbox solar water heater.** The breadbox solar heater places the storage tank inside the collector. However, the collector consists only of an insulated box with a transparent cover, so the efficiency of the breadbox is low compared to the efficiency of a standard solar collector. Generally, the tank is painted flat black, and the interior of the box is painted white or silver to provide high reflectance from the box to the tank surface.

Modern manufacturers' ICS systems can be more sophisticated. Among the design improvements introduced to increase ICS system efficiency and reduce nighttime heat loss are the following.

- Double or triple glazing sheets.
- Selective transmission glazing coatings.
- Selective surface tank coatings.[2]
- High-gain external and internal reflectors.

---

[2] A selective surface is an absorber panel coating that absorbs most of the sunlight hitting it and emits very little thermal energy.

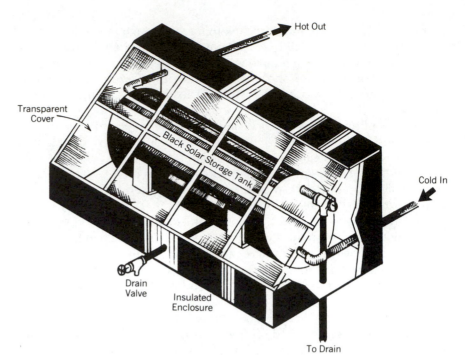

Figure 2.10 A breadbox solar water heater.

Figures 2.12 and 2.13 show one example of a higher-efficiency system. Here the breadbox design has been replaced by a more conventional configuration that increases the amount of energy that the system receives by returning to a collector-like shape. This flat ICS system heats up faster during the day but tends to lose more heat at night than a more compact cylindrical tank.

Figure 2.12 shows a cutaway of the collector, which contains four tubes connected in series. As the water in the system heats up, it rises through the collector. The use of four separate tubes connected in series insures that the hottest available water will be delivered to the user. These tubes are the solar storage tank as well as the energy-collecting surface.

Figure 2.13 shows the collector/storage system connected to an auxiliary heater. Cold water flows up to the collector/storage hardware, is warmed by the sun, and passes down to the auxiliary heater, where conventional gas or electric fuel augments solar energy to complete the heating process.

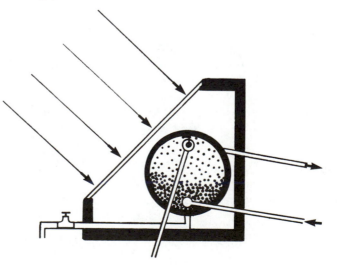

Figure 2.11 End view of a breadbox heater.

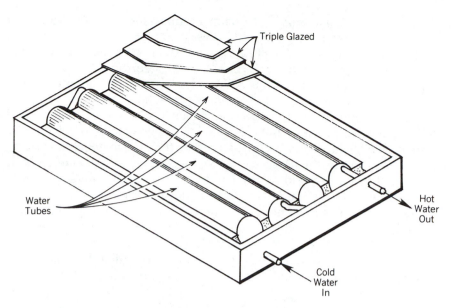

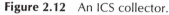

Figure 2.12 An ICS collector.

Expected Performance and Sizing

Like the thermosiphon and other solar heating systems, ICS systems are subject to performance variations, depending on installation site and local climate, system design specifics, and owner use patterns. Until recently, no standards for systematically sizing or predicting performance of ICS systems existed. The combination of collector and storage into a single unit could not be modeled by the F-Chart computer program or described by a collector efficiency curve.

Now a new national rating program (see Chapter 9) has eliminated some of the subjective, hit-or-miss ICS performance and sizing guidelines. Normal ICS system sizing calls for 20 to 25 ft² of glazing per person and 1.6 to 2 gal of storage for every square foot of glazing.

Because of the inherent collection inefficiencies of ICS systems and because of their greater nighttime heat loss, they can only collect up to 75% as much heat per square foot of surface area as a separated collection/storage system. The following restrictions on the preceding guidelines should also be observed, or system performance will be curtailed.

- ICS systems must be insulated to R-16 to R-20 on all sides except where the solar glazing aperture is located.
- The full aperture must be exposed to sunlight during all daylight hours. An aperture that is half exposed to sunlight is losing almost as much heat as it is collecting.
- Reflector surfaces are less effective at capturing solar energy than glazing and storage surfaces.
- Reflectors located inside the ICS box are more effective than external reflectors that are exposed to the elements.

Freeze Protection and Other Considerations

The absence of small-diameter collector tubes makes ICS systems inherently more freeze resistant than other solar heating systems. One popular product, the Cornell Energy Model 360, is warranteed down to −10°F. It features R-16 insulation and triple glazing, together with a selective surface tank coating that absorbs sunlight more readily than it reradiates heat. Similar systems can be expected to perform well and avoid freezing

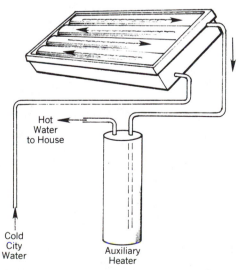

Figure 2.13 An ICS system connected to an auxiliary heater.

even in severe climates. A high-quality thermostatic "dribble" valve can add extra freeze resistance to such a system.

ICS systems tanks are usually glass-lined steel or stainless steel. In the first case, anodic protection is required to arrest corrosion. A special, rigidized magnesium anode rod is needed for horizontally mounted tanks. Stainless steel normally needs no anodic protection.

Except for the glazed aperture, the ICS enclosure should be insulated to R-16 or better. Because the ICS system is designed to be piped directly into the service hot water system as a preheater, pipe materials and sizing can be the same as the house supply piping, usually $\frac{3}{4}$-in. steel or copper. The supply piping should be insulated to R-10 or better for several feet upstream from the ICS and continuously downstream.

The ICS system should be roof mounted only when the roof members are engineered to support the concentrated weight of the system. An 80-gal ICS system can weigh close to 800 lb and may require a bearing wall instead of reinforced roof rafters for support. Supporting structures must not be subject to weakening due to frost, wind, or wet weather.

Advantages and Disadvantages

ICS systems have earned a place in the market as a low-cost approach to solar water heating. They are most often seen in the southern states or in vacation homes. Their advantage is that they are inexpensive and simple. However, they do have high standby heat loss characteristics, offer limited freeze protection, and give low-efficiency collection.

ICS systems have been badly abused by some companies desiring to make a fast killing in the solar market, and many improperly designed products have proven to be close to worthless.

The serviceperson can repair physical damage to ICS systems, such as pipe leaks, broken glazing, or damaged insulation. Performance problems usually are the result of poor design and can be remedied only by rebuilding or replacing the entire ICS unit.

Connecting the Solar Preheater to the Auxiliary Heater

There are many ways to pipe solar preheaters to auxiliary heaters. In this book, a simple connection allowing either the use of both heaters or just the auxiliary heater will be used. This connection is shown in Figure 2.14.

When A is closed and B and C are open, the solar heater is in operation and the service water is routed through it. When A is open and B and C are closed, the solar water heater is bypassed.

Figure 2.15 shows a different connection, where the solar water heater is connected so that either heater may be used while bypassing the other. In this figure, A and B are bypass valves that allow the use of either heater alone or the two together.

As a serviceperson, you will encounter several different piping configurations. The owner's manual usually will show you how these are laid out. When it does not, sketch them out so that you understand the function of each valve.

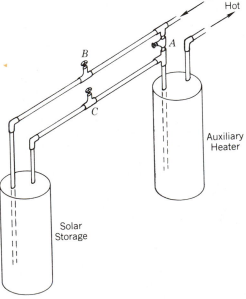

Figure 2.14 The heater connection used in this book. The backup heater cannot be bypassed.

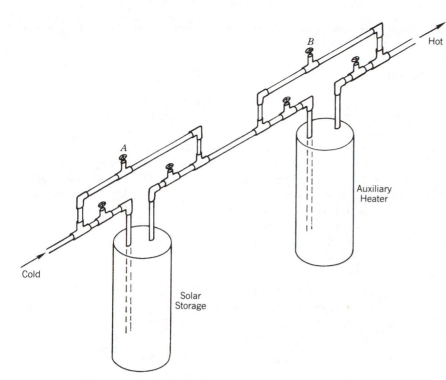

Figure 2.15 A heater connection that allows the use of either heater to be bypassed.

How to Test Performance

Performance testing of thermosiphon heaters is simple. The serviceperson needs two temperature gauges and a clear day. Testing proceeds as follows:

- These tests must be performed between 10 A.M. and 2 P.M. on a clear day in direct sunlight.
- The hot water service must be cut off so that no water can be drawn from the system.
- A temperature gauge is placed on the downcomer where it enters the collector (preferred) or where it exits the tank.
- A second temperature gauge is placed on the riser where it enters the tank (preferred) or where it exits the collector.
- The temperature gauges are allowed to come to equilibirum, which should take 3 to 4 minutes.
- The gauges are read and the readings recorded on a chart like the one in Table 2.2.
- Thirty minutes to 1 hour is allowed to pass.

Table 2.2

TEMPERATURE CHART—THERMOSIPHON PERFORMANCE TEST

TIME READ	TEMPERATURE DOWNCOMER (°F)	TEMPERATURE RISER (°F)	TEMPERATURE DIFFERENCE (°F)	TEMPERATURE TANK (°F)
°F Rise				

- The gauges are read again and the readings are recorded in the chart like the one in Table 2.2.

Filling Out the Chart

1. Subtract the downcomer temperature reading from the riser temperature reading and place the result under temperature difference. Do this for both readings.
2. Add the downcomer and the riser temperature readings together. Divide the sum by 2 and place the result under tank temperature. This is the average tank temperature. Do this for both readings.
3. Subtract the first downcomer temperature reading from the second downcomer reading and record the difference under °F rise in the downcomer column.
4. Subtract the first riser temperature reading from the second riser temperature reading and record the difference under °F rise in the riser column.
5. Subtract the second tank temperature from the first tank temperature and record the difference under °F rise in the tank temperature column.

The Meaning of the Readings and Calculations

Step 1 provides the difference in temperature between the hot and the cold sides of the collection loop. If the loop is not circulating, there will be little or no difference, usually less than 5 to 10°F.

Step 2 provides an estimated average tank temperature. Some of the water in the tank is at the temperature of the riser, some of the water is at the temperature of the downcomer, and the balance lies in between.

Step 3 provides the temperature rise with time at the downcomer. The system is closed, with no water being drawn, so the water in the system constantly recirculates and should get hotter. If the loop is not circulating, there will be little or no temperature rise in 30 minutes to 1 hour.

Step 4 provides the temperature rise with time at the riser. This temperature should also increase with time if the system is circulating.

Step 5 provides the average tank water temperature rise with time. Again, if the loop is not circulating, there will be little or no temperature rise in the tank water with time.

Interpreting the Results

A properly operating system should show a difference between the downcomer and the riser temperature of 10 to 30°F after 30 minutes. If the difference in temperature is 0°F or close to it, the fluid is not circulating in the system. If the difference lies between 5 and 10°F, the system is extremely efficient.

If the difference in temperature is over 30°F, the fluid is circulating too slowly. Usually the system should be running with a 15 to 25°F difference between hot and cold.

Since the system is depending on the water's density to circulate it, the temperature difference will vary from one tank temperature to another. As the tank gets hotter, the temperature difference will decrease because of higher heat losses from the collectors and piping.

Do not interpret an extremely high temperature difference as a sign of good performance. A high temperature differential lowers the efficiency of the collection process.

The system should experience a good temperature rise in 30 minutes to 1 hour. It is hard to give a definite range, because the sizing of the system affects the rate of temperature increase. Certainly, when the ambient temperature of the air is 65°F or higher, you should experience at least 9°F temperature rise in the downcomer, the riser, and the average tank temperature in 1 hour. Obviously, the higher the rise, the better the system is performing. If the tank temperature does not rise by 9°F in 1 hour in direct sunlight, the system is not performing. When the ambient temperature is below 65°F, the temperature rise may fall back to 5°F if the system is not well insulated. This is due to excess system and piping heat losses.

WHAT CAN GO WRONG

Assuming that the system has operated successfully prior to the service call, five things can go wrong.

1. The owner can change the valve settings.
2. The owner can plumb a heat trap into the system.
3. The system can become air bound.
4. The system can become clogged with scale and/or sediment.
5. The system can suffer physical damage or develop a leak.

The serviceperson can determine which of these defects exists with a troubleshooting approach.

TROUBLESHOOTING CHART

1. Look for physical damage.

Item	Yes	No
Crushed pipes	_____	_____
Leaky valves or joints	_____	_____
Broken collector cover	_____	_____
Freeze damage	_____	_____
Structural movement	_____	_____

2. Check the valve settings.

Valve Number	Solar System Mode		
	Operative	Nonoperative	Start-up
V1	Open	Closed	Open
V2	Closed	Open	Closed
V3	Open	Closed	Open
V4	Closed	Closed	Closed
V5	Closed	Open	Closed
V6	Closed	Open	Closed
V7	Open	Closed	Open
V8	Open	Closed	Open
V9	Closed	Closed	Closed
V10	Closed	Closed	Closed

3. Check for air bubbles.

- Check the automatic air vent's operation.
- Open the hot water faucets in the residence and exhaust any air in the piping.

4. Drain the scale and sediment.

- Open V10 and run until clear.
- Open V5 and run until clear.
- Open V2 and run until clear.

5a. Check the riser flow (if there is a check valve in the riser).

- Close V1.
- Open V2.
- If the water runs out V2, the check valve is frozen open.

5b. Check the riser flow (if there is no check valve in the riser).

- Close V1.
- Open V2.
- Run until clear (backflushes the collectors).

6. Check the downcomer flow.

- Close V3.
- Open V2.
- Run until clear (flushes downcomer).

7. Run a system performance test.

- Recheck the valve settings.
- Close V7.
- Install temperature gauges.
- Check for clear sky and time of day (see text).
- Run test and record temperatures in Table 2.2.
- Evaluate the test results.

    1. Did the temperature rise at least 9°F in 1 hour (5°F on a cold day with poor insulation)?
    2. Does a 10 to 30°F temperature difference exist between the gauges?
       If the answers to these two questions are both yes, the system is performing satisfactorily.

8. Make the necessary repairs.

- Repairs are covered in Section 3. Turn to the appropriate chapter for guidance.

9. Visually check over the system.

- Return V7 and other operating valves to their proper positions.
- Are any structural repairs needed?
- Is any painting needed?
- Is there any loose or missing insulation?
- Are there any leaks?

10. Finish the job.

- Clean up the area.
- Remove all tools.
- Secure the area.

- Complete the service report for the owner.
- Present a bill for services.

ON THE JOB

Be certain that you have replacement boiler drain valves with you in case V2, V5, or V10 fail to reseat.

If the owner has specifically pointed out leaks or structural damage to the system, repair these first and visually recheck them during step 9.

Once any damage is repaired, start the performance test outlined in step 7.

Perform steps 2, 3, and 4 before step 1, because step 1 may involve going outdoors on a ladder and following the pipe runs.

Follow step 1 with steps 5 and 6.

By now, the problem should have been corrected, and step 7 can be performed again to insure that the system is operational while you secure the area and fill out your service reports.

WHAT YOU SHOULD HAVE LEARNED

1. A thermosiphon system operates on the difference in density between hot and cold water. It does not contain a pump or a controller.
2. The collector must be lower than the tank, and there must be no heat traps in the system.
3. The height of the tank over the collector, the difference in temperature between the tank and the collectors, and the friction in the pipes determine the flow rate in the system.
4. The two types of thermosiphon systems are vertical tank and horizontal tank.
5. Both systems can be used alone or can be used as preheaters for auxiliary heaters.
6. Thermosiphon systems must be drained in freezing weather. They cannot be operated in the winter, except in mild climates.
7. The performance depends on the site and the system. For good performance, the system should

   - Be sized at 20 to 25 ft^2 per occupant.
   - Contain $1\frac{1}{2}$ to 2 gal of storage for each square foot of collector area.
   - Be tilted at an angle of not less than 10° to the horizontal for a vertical tank and 20° for a horizontal tank system.

8. Cooper and brass piping and fittings should be used throughout the system. Glass-lined tanks with a magnesium anode rod work well for storage.
9. The system is operating correctly when

   - The tank temperature rises at least 9°F in 1 hour on a warm, clear day between 10 A.M. and 2 P.M. (5°F rise on a cold day with a poorly insulated tank).
   - A temperature differential of 10 to 30°F between the downcomer and the riser is maintained under testing.

10. ICS systems are rudimentary thermosiphon heaters. Their low cost is balanced by their relatively poor performance.

CHAPTER
3

Pumped Service-Water Systems

Pumped Service-Water System   A solar water heater system which has the collectors mounted higher than the storage tank and which pumps the storage water through the collectors to heat it.

A system cannot thermosiphon unless the storage tank is mounted higher than the collectors. In many applications such a configuration is not practical, and the collectors must be mounted higher than the tank. In these instances the water must be pumped around the loop, and a pump controller must be present to turn the pump off and on at the proper times. Any system requiring mechanical or electrical energy input is known as an *active system*. The simplest active system is the *pumped service-water system*. This system configuration is shown in Figures 3.1 and 3.2. The system operates as follows.

When the collectors are a few degrees hotter than the storage tank, the pump is turned on by the controller. Water from the lower part of the solar storage tank flows through the strainer, past the drain leg into the pump, and is pumped up to the bottom of the collectors through V3, which is a **throttle valve** that can be used to control the rate of flow. A throttle valve is a ball or butterfly valve that can be partially closed to restrict the flow.

The water is distributed across the bottom of the collectors by the lower manifold (see Figure 3.2), flows up through the risers, is collected by the upper manifold, and returns through the check valve and V1 to the top of the solar storage tank.

When the solar tank reaches a temperature close to the temperature of the collectors, the controller turns the pump off. The check valve prevents reverse thermosiphoning from occurring. Without the check valve, the hot water in the tank would rise to the collectors, which would then lose heat to the environment.

The solar storage tank can be connected in series to an auxiliary heater in the same manner as the thermosiphon system. It may be used alone,

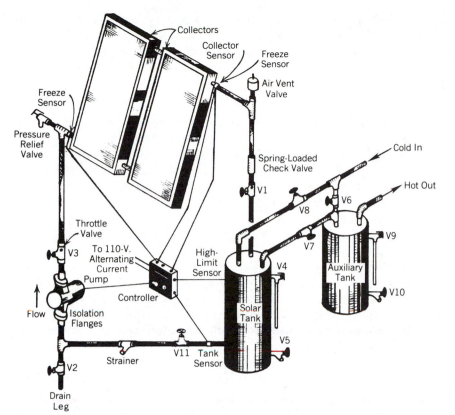

Figure 3.1 A pumped service-water solar water heater.

without an auxiliary heater, or it may be connected with bypass valves (see Figure 2.15).

In the system illustrated in Figure 3.1, the control system contains four different types of sensors: a collector sensor, a tank sensor, a high-limit sensor, and a freeze sensor. When the pumped service-water system contains freeze sensors, it is also known as a *recirculation freeze-protec-*

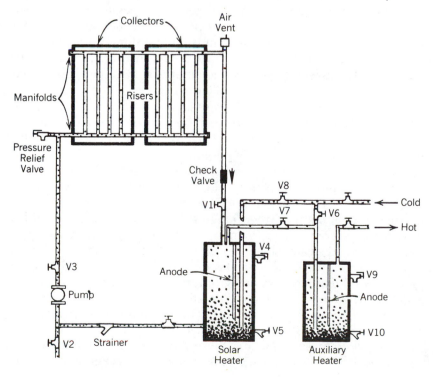

Figure 3.2 The piping layout of a pumped service-water solar water heater.

tion system. A recirculation freeze-protection system is a system that recirculates hot water from the storage tank through the collectors in freezing weather. Recirculation freeze protection is similar in its range of applications and limitations to the thermostatic dribble valve described in Chapter 2.

Each of these sensors serves a very specific purpose.

- The collector sensor and the storage tank sensor constantly change in electrical resistance as the collector and tank temperature change, allowing the controller to compare the two temperatures. When the collector temperature rises to a preset threshold differential of 8 to 18°F hotter than the storage tank, the controller turns the pump on. When the difference between the collector and tank temperatures falls below a 2.5 to 6°F threshold, the controller turns the pump off, because continued circulation would result in no meaningful heat gain.
- The high-limit sensor lets the controller override these signals and shut down the pump when the tank temperature reaches a predetermined temperature, usually between 160 and 180°F. This prevents the solar tank from overheating.
- The freeze sensors measure the temperature of the collectors and the outdoor piping. When either of these drop to 38 to 42°F, the sensors signal the controller to override the tank and collector sensors and turn the pump on to circulate warm water from the tank through the collectors. This action prevents a freeze-up from an unexpected cold snap. However, it does not provide the positive freeze control that is needed to run the system in the winter months.

The high-limit sensor is an important safety feature. It prevents the solar hot water from exceeding scald prevention limits established by safety standards and building codes.

The thermosiphon and ICS systems described in Chapter 2 were protected from boiling only by an emergency temperature and pressure relief valve. Because they do not use a pump and controller, they cannot include an antiscald, high-limit function. Some systems are sized conservatively to prevent them from reaching dangerously high temperatures, but most of the newer, high-efficiency thermosiphon and pumped service-water systems can generate boiling water.

This is most likely to occur during summer vacation periods, when daily hot water use may drop to zero. If the solar heat does not leave the system as hot water, it will add to the temperature of the storage tank until the water reaches the boiling point.

Temperature and pressure relief valves have three shortcomings that make them unsatisfactory as antiscald, high-limit devices. First, they are designed to protect system components from excessive stresses and do not operate until the hot water temperature has risen far above human safety limits. Second, they produce a sudden unsignaled discharge of boiling water and steam when they operate. This discharge may be hazardous to anyone in the immediate area. Third, they are engineered to *open* on a high-temperatures or high-pressure condition, but not to *close* again fully when this condition is corrected, so it is common to find leakage from a temperature and pressure relief valve that has operated only once. To eliminate this leakage, the valve must be replaced.

The high-limit sensor eliminates these shortcomings. By shutting down the pumps, it keeps the storage tank temperature within human safety limits and well below the temperature and pressure relief valve operating range. The collectors continue to heat up, but the energy is not transferred to the storage tank.

Instead, the collector pressure relief valve is triggered, and hot water is vented from the collectors as steam. Any condensate will discharge into a roof drain or gutter, where danger to humans is slight. In most cases, as the collector refills with cooler water, the pressure relief valve will reseat with no leakage. You will learn more about this stagnation-pressure relief cycle in Chapter 4.

Recirculation freeze-protected systems have these major limitations.

- Recirculation uses up the heat in the solar storage tank. This lowers the value of the system.
- In severely cold weather recirculation can eventually freeze up both the tank and collectors.
- If the power to the pump fails, no recirculation occurs and the collectors may freeze.

In summary, recirculation freeze protection is unsuitable for use in cold climates.

The following construction details must be observed in the pumped service-water system.

- Assuming that the piping running from V3 through the collectors to V1 is located outdoors, all piping must be pitched to drain fully when valves V1 and V3, which should be located indoors, are closed. These must be nonrestrictive flow gate valves with drain plugs.
- The collectors must be pitched to drain fully. They should be pitched to V3.
- An automatic air vent must be located at the highest point in the collector array so that during start-up, the system will not become airbound.
- V1 and V3 must be located at different levels, or a vacuum relief valve must be added at the high point of the system so that the system can be fully drained at the onset of the freezing season.

The rate at which the system operates depends on three factors.

- The operating characteristics of the pump.
- The throttle valve setting.
- The amount of resistance to flow caused by the friction generated in the piping and the collectors.

Expected Performance

Again, the performance of the system is both site dependent and system dependent.

A properly designed and sized pump service-water system is extremely efficient. The flow rate can be tailored to make the collectors operate at maximum efficiency. No heat exchanger is located in the system to cause efficiency losses, and water is a highly efficient heat transfer fluid.

However, the system can be run only in nonfreezing weather; therefore, it is a seasonal system that cannot be operated during the winter months except in the extreme southern latitudes. Because the short operating season allows the collectors to be tilted with less compromise than a year-round system, it is usually the most efficient system available during the operating period.

Sizing Parameters

The pumped service-water system should also be sized at 20 to 25 ft² per resident. This will usually produce 40 to 60% of the required hot water unless the family has an unusually high or low water use rate.

There should be 1.2 to 2 gal of storage in the solar tank for each square foot of collector.

The pump should be sized to provide a flow rate of up to 0.03 gpm/ft² of collector, and a strainer should be placed upstream of the suction end. Generally, $\frac{1}{40}$- to $\frac{1}{12}$-hp pumps are found in most residential-size pumped service-water systems.

A controller should be selected that will turn the pump on when the collectors are 8 to 20°F hotter than the tank and will turn the pump off when the differential drops to about 2.5 to 5°F. The high-limit switch should activate at or below 180°F, and the freeze sensor should turn the pump on at 38 to 42°F and off at 50°F. The controller should contain one or more controlled outlets that operate at 120 V to run the pump.

CAUTION In a large system, if the pump is $\frac{1}{3}$ hp or larger, check if the controller's pump relay is large enough for both start-up and continuous running.

Other Considerations

A glass-lined storage tank is very satisfactory. The tank must contain a magnesium anode to protect against premature corrosion failure.

Aluminum water passages in the collectors should be avoided because the use of fresh, unconditioned water in contact with aluminum can cause severe corrosion problems.

The piping and fittings should all be copper and brass to minimize any corrosion problems. The size of the piping will depend on the length of the runs, the size of the pump, and the number of collectors. In general, assuming that less than 100 ft total of piping separates the tank from the collectors, $\frac{1}{2}$-in. nominal piping is satisfactory for two collector systems but may be marginal for three collector systems; $\frac{3}{4}$-in. pipe is generally satisfactory for three to four collectors. If there are more than four collectors, 1-in. pipe should be used. Be certain that the collector mainfolds are at least as large as the supply piping. Systems containing more than four collectors need careful design, and the serviceperson should check with the designer if any flow problems are encountered.

The tank and, if possible the associated piping, should be insulated with R-10 or better insulation. Outdoor insulation must be carefully chosen and protected against adverse weather as described in Chapter 9.

The tank and the collectors must have adequate fittings and be mounted securely on well-designed supports that will carry their weight and not be affected by frost and wet weather.

Advantages

The pumped service-water system has these advantages.

- The system is relatively inexpensive to install when compared to

positive freeze-protected systems but is more expensive than a thermosiphon system.
- The system is more complex than thermosiphon systems but less complex than positive freeze-protected systems.
- The seasonal performance is outstanding.
- Only pump and controller maintenance is required.

Disadvantages

The pumped service-water system has these disadvantages.

- The system cannot be operated year round except where there is no freezing weather.
- The system must be cared for by filling and draining it at the start and the close of the season.
- The system is constantly circulating fresh water. If the water is not properly conditioned, corrosion or scale and sediment buildup can cause premature system or component failure.
- A power outage will cause freeze protection to be lost.
- Stored heat is lost when recirculation occurs.

Recirculation freeze-protection systems are not recommended where hard-freeze conditions are encountered.

How to Test Performance

Performance testing of pumped service-water systems proceeds in two stages. First, a test similar to the thermosiphon system test is carried out. If this test shows that the system is operating correctly, further testing is not necessary. If the test shows that the system is operating incorrectly, the pump and controller must be further tested to determine where the problem lies. Testing proceeds as follows:

- The tests should be carried out between 10 A.M. and 2 P.M. on a clear day in direct sunlight.
- The hot water service must be cut off so that no water can be drawn from the system.
- The controller is turned off until the test gauges are placed and read.
- A temperature gauge is placed on the hot downcomer as close to the top of the entrance of the tank as possible.
- A temperature gauge is placed on the cold riser where it exits the tank as close to the tank as possible.
- The temperature gauges are allowed to come to equilibrium, which should take 3 to 4 minutes.
- The gauges are read and the readings are recorded on a chart like the one shown in Table 3.1.
- The pump controller is turned to automatic.
- If the pump starts up within 2 to 3 minutes, proceed with the test.
- If the pump does not start within 2 to 3 minutes, turn the controller switch to manual.
- If the pump starts up on manual but will not run on automatic, four possibilities exist.

    1. The controller is faulty.
    2. One or more of the sensors is faulty.
    3. The temperature differential between the collectors and the tank is not high enough to activate the controller.

Table 3.1

TEMPERATURE CHART—PUMPED SERVICE-WATER PERFORMANCE TEST

TIME READ	TEMPERATURE RISER (°F)	TEMPERATURE DOWNCOMER (°F)	TEMPERATURE DIFFERENCE (°F)	TEMPERATURE TANK (°F)
_____	_____	_____	_____	_____
_____	_____	_____	_____	_____
°F rise	_____	_____		_____

4. The tank temperature is high enough to activate the high-limit sensor.

- If the pump will run on manual, proceed with this test and check for temperature rise with time. When the test is completed, go to the pump/controller test procedure.
- If the pump will not run on manual, terminate the test and go to the pump/collecter test procedure.

Proceeding with the Test

Assuming that the pump will run, continue the temperature testing with the pump on manual as the preceding procedure indicates. If the pump will not run, you will return to this point in the test procedure after you have corrected the pump or controller fault.

- Allow 30 minutes to 1 hour to pass with the pump operating.
- Record the readings of the temperature gauges in a chart like the one in Table 3.1.

Filling Out the Chart

1. Subtract the riser temperature from the downcomer temperature reading and place the result under temperature difference. Do this for both readings.
2. Add the riser and the downcomer temperature readings together. Divide the sum by 2 and place the result under tank temperature. This is the average tank temperature.
3. Subtract the first riser temperature from the second riser temperature and record the difference under °F rise in the riser column.
4. Subtract the first downcomer temperature from the second downcomer temperature and record the difference under °F rise in the downcomer column.
5. Subtract the first tank temperature from the second tank temperature and record the difference under °F rise in the tank temperature column.

The Meaning of the Readings and the Calculations

Step 1 provides the difference in temperature between the hot and cold sides of the storage tank. When the loop is not circulating, the difference can be either minimal or high, depending on the conditions that existed when the test was started. Any temperatures between 55 and 180°F may be seen and should be considered normal. After the loop has been running, the downcomer should be at a higher temperature than the riser if the system is operating correctly.

Step 2 provides an estimated average tank temperature because some of the water in the tank is at the downcomer temperature, some is at the riser temperature, and the balance lies in between.

Step 3 provides the temperature rise with time for the riser. The system is closed, with no water being drawn, so the water in the tank should get hotter and hotter with time. If the loop is not circulating, there will be little or no temperature rise in 30 minutes to 1 hour.

Step 4 provides the temperature rise with time for the downcomer. This temperature will also increase with time if the system is circulating.

Step 5 provides an estimated average tank temperature with time. Again, if the loop is not circulating, there will be little or no temperature rise in the tank with time.

Interpreting the Results

A properly operating system should show a difference between the riser and the downcomer temperature of 10 to 20°F after 1 hour.

The flow rate of the fluid determines the temperature difference between the riser and the downcomer. One of the following five conditions can exist:

1. The temperature difference can be between 10 and 20°F. This system would have a satisfactory flow rate.
2. The temperature difference can be greater than 20°F. This would indicate too slow a flow rate.
3. The temperature difference can be less than 10°F but large enough to keep the pump running. (Most controllers turn the pump off at more than a 2.5°F differential but at less than a 6°F differential.) This would indicate too fast a flow rate.
4. The temperature difference can fluctuate from a very low differential to a very high differential. This also indicates too fast a flow rate. The pump will cycle on and off over and over again as the temperature increases and decreases.
5. No difference in temperature can exist. This indicates that no flow is occurring.

A no-flow condition would result in no energy being collected. Too slow a flow would result in decreased collector efficiency. Too fast a flow would cause the pump to cycle off and on too often, which will prematurely wear out the pump and can result in inefficient energy collection.

The throttle valve, V3, is used to control the flow rate. If the flow rate is too fast, close down the valve slightly and adjust the flow rate to fall within the 10 to 20°F range. In making the adjustments, give the system about 10 minutes to come to equilibrium between adjustments. If the flow rate is too slow, check the throttle valve to see if it can be opened to increase the flow. Also check V1; it may not be wide open. V1 can be used for rough throttling if V3 cannot control the rate. However, the best procedure is to leave V1 wide open and control the rate of flow with V3.

The system should experience a temperature rise of 9 to 15°F in 1 hour. Obviously, the higher the temperature rise, the better the system is performing. Do not be concerned about higher temperature rises. The 9 to 15°F range was chosen assuming average sun, weather, and system conditions. If the system does not experience at least a 9 to 15°F rise in 1 hour, it is most likely not operating at good efficiency and needs adjusting or repairs.

The Pump/Controller Test

If the pump will not run when the controller is turned to manual, proceed as follows:

- Check the circuit to the controller at the fuse box for a blown fuse or a circuit breaker that has tripped.
- If the pump is the plug-in instead of the hard-wired type, remove the plug from the controller and plug the pump into an extension cord that has been tested with a trouble lamp to assure that it has electricity flowing to it. If the fault lies in the controller, the pump should now operate. In the case of a hard-wired pump, see the test procedures outlined in Chapter 16.
- If you have determined from the preceding tests that the pump is faulty, turn to Chapter 16 and repair the pump.
- If the pump tests good, the fault lies in the control system. In most controllers the manual controller switch overrides all the sensors. Therefore, chances are that the sensors or the controller's internal circuits are at fault if the pump will run on manual.
- If the pump will not run on the automatic controller setting, the control system needs troubleshooting. Turn to Chapter 17 and follow the control system troubleshooting procedures.
- Once the pump or controller problem has been solved, return to the performance test and complete it.

WHAT CAN GO WRONG

Assuming that the system has operated successfully prior to the service call, six things can go wrong.

1. The owner can change the valve settings.
2. The system can become airbound.
3. The system can become clogged with scale and/or sediment.
4. The system can suffer physical damage or develop a leak.
5. The pump can fail.
6. The controller or sensors can fail.

The serviceperson can determine which of these defects exists with a troubleshooting approach.

TROUBLESHOOTING CHART

1. Look for physical damage.

Item	Yes	No
Crushed pipes	_____	_____
Leaky valves or joints	_____	_____
Broken collector cover	_____	_____
Freeze damage	_____	_____
Structural movement	_____	_____
Broken sensor wire	_____	_____

Damaged or missing sensor   _____ _____

Electric service interruption   _____ _____

Sticking air vent valve   _____ _____

Sticking check valve   _____ _____

2. Check the valve settings.

Valve Number	Solar System Mode		
	Operative	Nonoperative	Start-up
V1	Open	Closed	Open
V2	Closed	Open	Closed
V3	Varies	Varies	Open
V4	Closed	Closed	Closed
V5	Closed	Open	Closed
V6	Closed	Open	Closed
V7	Open	Closed	Open
V8	Open	Closed	Open
V9	Closed	Closed	Closed
V10	Closed	Closed	Closed
Air vent	Operating	Operating	Operating
Check valve	Operating	Operating	Operating
Strainer	Unclogged	Unclogged	Unclogged

3. Drain the scale and sediment.

    • Turn the controller off.
    • Open V10 and run until clear.
    • Open V5 and run until clear.
    • Open V2 and run until clear.

4. Clean the strainer and check the tank outlet flow.

    • Close V11.
    • Open V2 to relieve pressure from the system.
    • If flow continues, the check valve is faulty.
    • Repair or replace the check valve and continue.
    • Remove and clean the strainer screen with a stiff brush.
    • Open V11 and flush the strainer case.
    • Close V11. Replace the strainer screen and plug.
    • Open V11.
    • Open V2 to flush the drain leg again.

5. Backflush the collectors, the pump, and the riser.

    • Close V11.
    • Open V2.
    • Allow the collector array to drain.

6. If the collectors do not drain,

    • Remove the pressure relief valve.
    • If the collectors drain, the riser or the pump is clogged. Check V3. If V3 is open, unclog the pump.
    • If the collectors do not drain, the air vent valve is not operating. Repair the air vent valve.

7. Run the system performance test.

- Recheck the valve settings.
- Check that the controller is off.
- Close V7.
- Install temperature gauges.
- Check for clear sky and time of day (see text).
- Run test and record temperatures in Table 3.1.
- If the pump will not run, skip to step 8.
- Evaluate the test results

1. Does the pump run on automatic?
2. Did temperatures rise at least 9 to 12°F?
3. Does a 10 to 20°F temperature difference exist between the gauges?

If the answer to these three questions are yes, the system is performing satisfactorily. However, you have not checked one thing. Does the pump shut off when the temperature of the collectors drops down to 2.5 to 5°F of the tank (controller's pump shutoff differential)? Chances are that the owner would have told you that the pump did not shut off when he or she called for service. To check shutoff operation, see controller troubleshooting in Chapter 17.

Skip steps 8 and 9 if the pump is functioning correctly on the automatic controller setting.

8. If the pump will not run on manual,

- Check for a blown fuse or faulty circuit breaker.
- Check the pump on a separate circuit if it is the plug-in type.
- Check for electricity at the pump power cord with a trouble lamp or voltmeter if the pump is hard wired.
- Determine whether the pump or controller is faulty.
- Proceed to troubleshoot the pump as outlined in Chapter 16.

9. If the pump will not run on automatic,

- Recheck the controller sensors and sensor wiring for damage.
- Proceed to troubleshoot the controller as outlined in Chapter 17.

Complete the system performance test after repairing the pump and controller.

10. Visually check over the system.

- Return V7 and other operating valves to their proper positions.
- Is the controller placed in the automatic mode?
- Are any structural repairs needed?
- Is any painting needed?
- Is there any loose or missing insulation?
- Are there any leaks?

11. Finish up the job.

- Clean up the area.
- Remove all tools.
- Secure the area.
- Complete the service report for the owner.
- Present a bill for services.

ON THE JOB

First, be certain that you have replacement boiler drain valves with you, that you are carrying replacement control sensors, a voltmeter, and your sensor substitution box. A long extension cord and a trouble lamp or circuit tester should also be in your toolbox.

Second, if the owner has pointed out or if you see leaks or structural damage as you arrive on the job, repair these first and then recheck them in step 10. Once any damage is repaired, see if the pump will run. If it runs on both manual and automatic, proceed with the performance test outlined in step 7.

If the pump will not run on either manual or automatic, perform step 8.

If the pump runs on manual but not automatic, proceed to troubleshoot the control system (step 9). Once the pump is repaired, start the performance test outlined in step 7.

While performance is under test, perform steps 1 and 2.

If performance is unsatisfactory, perform steps 3 to 6 and then recheck the performance. Any problems should now be corrected.

(In California, recirculation freeze protection is not recommended at elevations greater than 1000 ft. Only use these systems in areas that have less than 20 freezing days per year. In northern coastal California, we would only expect to see them in San Francisco.)

Flow rates below 0.02 gpm/ft^2 are low in our experience. They produce too high a temperature differential across the collector. Think of 0.01 to 0.019 as only being satisfactory flow rates when a major repair job would be needed to increase them.

WHAT YOU SHOULD HAVE LEARNED

1. The pumped service-water system is used when the collectors must be located higher than the storage tank.
2. The pumped service-water system contains a pump and a controller.
3. The pumped service-water system cannot be used in freezing weather but must be drained.
4. The control system contains up to four types of sensors.
5. Two of these sensors measure the difference in temperature between the collector and the tank. They turn the pump on when the collectors are a few degrees hotter than the tank and turn the pump off when the tank has risen to close to the collector temperature.
6. One sensor is a high-limit switch that prevents the pump from turning on when the tank reaches a predetermined temperature, usually around 160 to 180°F.
7. One or more freeze sensors turn the pump on at 38 to 42°F to circulate warm water through the collectors and to give some limited protection against freeze-ups. The pump turns back off at 50°F.
8. The pumped service-water system can be used as a preheater for an auxiliary heater or it can be used alone.
9. An automatic air vent is needed at the highest point in the system to prevent the system from becoming airbound.
10. Flow rate depends on the pump chosen and the amount of resis-

tance to flow caused by the friction generated in the piping and the rest of the system.

11. The performance of the system is both site and system dependent. A pumped service-water system is extremely efficient because the flow rate can be fine-tuned, it contains no heat exchanger, and water is a highly efficient heat transfer medium.

12. An average system is sized at 20 to 25 ft^2 per occupant, would typically produce 40 to 60% of the hot water needed, contains 1.2 to 2 gal of storage per square foot of collector, and has a flow rate that generally ranges from 0.020 to 0.030 gpm/ft^2 of collector.

13. Glass-lined tanks may be used. Aluminum water passages should be avoided. Piping and fittings should be copper and brass.

14. The system is operating correctly when

- The tank temperature rises at least 9 to 15°F in 1 hour on a clear day between 10 A.M. and 2 P.M.
- A temperature difference of 10 to 20°F is maintained across the collectors under test.
- The pump is being turned off and on at the correct times by the controller when set on automatic.

Draindown Systems

Draindown System A solar water heater that is designed so that the water in the collectors and the outdoor piping is automatically drained when the temperature drops to 40 to 42°F or when a power failure occurs.

The thermosiphon, the ISC, and the pumped service-water solar water-heater systems do not offer positive freeze protection and cannot be operated safely in winter weather. The portions of these systems that are exposed to freezing weather must be drained at the onset of cold weather. These systems cannot be operated year round in most North American climates.

The draindown system has been devised to extend the use of solar water heaters into the winter months. Early draindown systems had a poor record of reliability because of valve failures. In recent years valves manufactured specifically for draindown solar systems have been marketed. These valves have eliminated most of the problems.

The Three-Way Valve Draindown System

In Figure 4.1, two *three-way electrically actuated valves* have been placed in the piping. An electrically actuated valve is a valve that is actuated by either an electric motor or a solenoid. One valve is located in the riser between the throttle valve and the collectors. The other is located in the downcomer above the check valve. Both valves must be located indoors, where they are not subject to freezing. All the piping and parts of the system that are located on the collector side of the two valves are pitched to drain to one of these two valves.

Figure 4.2 shows a three-way electrically actuated valve. Note the direction of the three arrows, A, B, and C, located on the illustration. Opening A can be connected to either opening B or opening C, depending on whether the valve is energized.

In Figure 4.1 assume that opening A is connected to the collector piping, opening B is connected to the piping adjacent to the check valve, opening C is connected to a drain pipe and, when the valve is not energized, openings A and C are connected and opening B is shut.

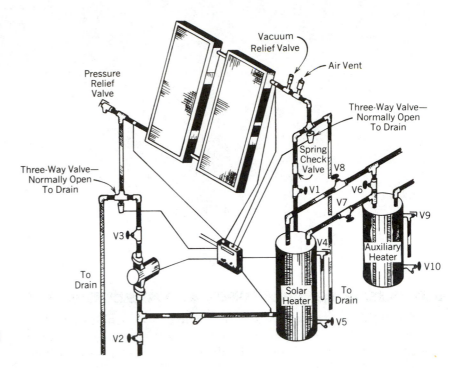

Figure 4.1 A three-way electrically actuated valve draindown system.

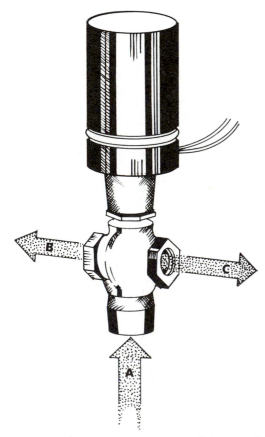

Figure 4.2 A three-way electrically actuated valve.

Under this set of conditions, the collectors would be drained whenever the valves were not receiving power. When electric power was supplied the valves would be energized, and openings A and B would be connected and opening C would be shut. This would connect the collectors to the system, and the system would operate just like a pumped water system.

To assure draindown and to allow the system to start up, a **vacuum relief valve** and an **air vent valve** must be installed in the system. The vacuum relief valve is an automatic valve that opens when the draining water creates a vacuum in the collectors. It admits outside air. The air vent valve allows the air to escape when the collectors are refilled. Although the two functions can be incorporated in one valve, the use of two separate valves is recommended. By opening during draining and filling, the valves allow air to enter and leave the collectors so that they will drain and fill fully.

Considerable attention must be paid to pitching the pipes to the drain valves in a draindown system because the system will be operating during the winter. The collectors and the associated piping must drain fully, without any attention from the owner. A $\frac{1}{8}$- to $\frac{1}{4}$-in. vertical drop per running foot is required.

Note that the collectors are *reverse return manifolded*. In reverse return manifolding, the manifolds are connected so that the flow from the riser enters at one end of the collector array and exits to the downcomer at the other end of the array. Reverse return manifolding is used to equalize the flow across the collectors by equalizing the pressure drop across them.

The Two-Way Valve Draindown System

Figures 4.3, 4.4, and 4.5 illustrate a *two-way electrically actuated valve* and its use in the draindown system. A two-way electrically actuated valve is a valve that has only an off and on position; it is either open or

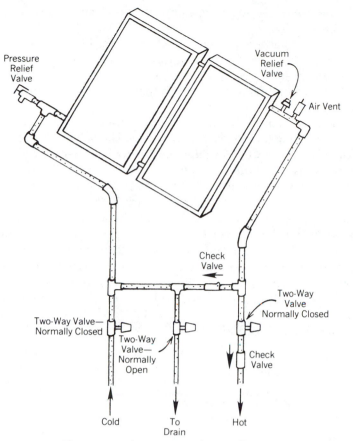

Figure 4.3 A two-way electrically actuated valve draindown collector array and piping.

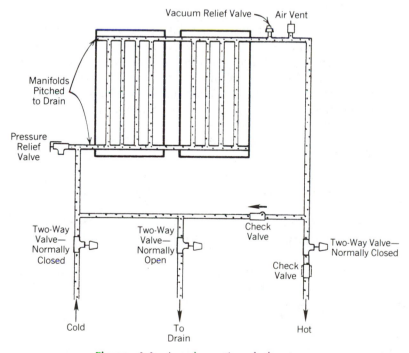

Figure 4.4 A schematic of the two-way electrically actuated valve draindown system's collector array and piping.

Figure 4.5 A two-way electrically actuated valve.

closed. The use of two-way valves in draindown systems is illustrated in Figures 4.3 and 4.4.

One two-way valve is placed between the throttle valve and the collectors in the riser. Another two-way valve is placed in the downcomer between the check valve and the collectors. These two valves are normally closed when the power is off so that water cannot flow to the collector array.

Two pitched drainpipes are run from the riser and the downcomer to a tee or wye fitting that leads to a third two-way valve. This valve is normally open. A check valve installed in the downcomer leg prevents flow between the riser and the downcomer when the system is operating.

When the system is energized, the drain valve closes, the riser and downcomer valves open, and the pressure in the storage tank fills the collector array with water. The system is then operational. Some air may be bound in the drainpipes. This will flow gradually to the top of the collector array and escape from the system through the air vent valve.

Without the check valve, the water in the riser would flow directly to the downcomer and bypass the collectors, since there is less resistance to flow in the drainpipes than there is in the collector array.

Using Two Two-Way Valves

Systems have been installed in which the electrically actuated valve in the downcomer has been eliminated. In this case the check valve in the downcomer must operate flawlessly, or water from the tank will continuously leak back into the drained system. In this type of application a spring check valve is much more reliable than a swing- or flapper-type check valve, despite the greater resistance to flow created by the spring check valve.

The Sunspool System

As stated earlier, the electrically actuated valve problems experienced in early draindown systems have been largely overcome with a new type of valve. One product in particular, the Sunspool valve, has a good service record backed by strong manufacturer support. Figure 4.6 shows a draindown system built using the Sunspool valve. One additional piping change that solves a long-standing, irritating noise problem has also been made.

The Sunspool valve contains both the riser and the downcomer dump valves plus an integral downcomer check valve. The valves are operated by a very low-current-drain heat motor. This motor moves at very slow speed and eliminates water hammer. It also aids in air control during filling.

The closed Sunspool valve is spring loaded. When the valve is deenergized, the spring forces it open immediately, and the collector array drains.

Sunspool's manufacturer also recommends the use of an auxiliary vacuum relief valve located just above the Sunspool valve in case the collector vacuum relief valve fails to operate and the use of an auxiliary air vent valve at the top of the solar storage tank to purge any accumulated air from the storage tank.

Servicepeople encountering systems that are prone to freezing up should seriously consider installing a spool-type valve, such as the Sunspool valve, to replace the valves currently in the system.

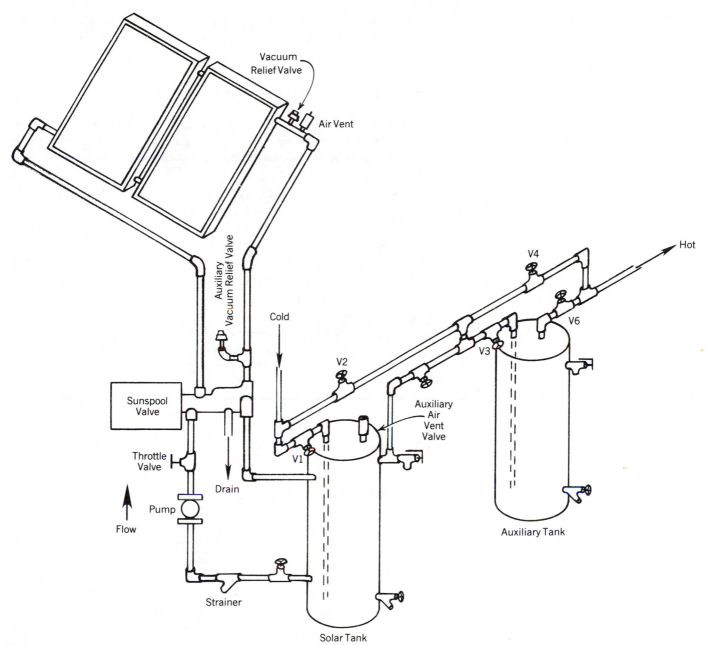

Figure 4.6 A Sunspool valve system.

Solving the Draindown System Noise Problem

When a draindown valve system refills the collectors, air is introduced into the storage tank; it passes on first to the auxiliary water heater and then to the hot water lines in the house.

This problem can be solved very simply. Repipe the hot water line so that it connects to a tee fitting at the temperature and pressure relief valve outlet (see Figure 4.6). Replace the relief valve with a deep-probe relief valve that meets the local plumbing code.

Then plumb an air vent valve to the hot water outlet at the top of the solar storage tank. Now that the air vent and the hot water line are separated, the noise will be eliminated.

If for some reason you cannot replumb the hot water line to the side port of the tank, Sunspool also manufactures an automatic air vent that combines the air vent and the hot water outlet tee fitting with a 6-in. dip tube. This unit may be installed in the hot water outlet port at the top of the tank to eliminate the noise.

Controlling the Draindown System

The draindown system requires a different controller than the pumped water system because the controller must operate the electrically actuated valves as well as the pump. The controller must perform these functions.

- Measure the difference in temperature between the collectors and the tank so that it can turn the pump on and off at the proper times.
- Shut down the system when the tank reaches a preset high limit.
- Energize the electrically actuated valves to fill the system.
- Deenergize the electrically actuated valve with the onset of freezing temperatures and, in some cases, when the tank's high limit is reached.

These controller requirements make it necessary to use a controller that contains a separate **freeze dump circuit.** A freeze dump circuit is a circuit that deenergizes the valving when the temperature reaches 40 to 42°F and reenergizes the valves when the temperatures rises to about 50°F. The freeze dump circuit also must override the pump on/off circuit so that the pump cannot turn on when the valves are deenergized.

The freeze dump circuit can be controlled by the collector sensor, by a separate freeze sensor placed elsewhere in the system, or by a combination of sensors, any one of which can trigger the freeze dump circuit. Sensor choice and placement are critical but vary from controller manufacturer to controller manufacturer and from solar system designer to solar system designer. The designer's and the manufacturer's recommended placement should be followed unless the serviceperson finds that placement is inadequate and the system is freezing up.

The following construction details must be followed in draindown systems.

- All water passages in the collector array and its associated piping must drop vertically at least $\frac{1}{8}$ to $\frac{1}{4}$ in./ft to drain properly.
- The system must contain either a combined vacuum relief/air vent valve or a separate vacuum relief valve and an air vent valve. The separate valves are recommended. These valves must be located at the highest point in the system.
- A controller that contains a separate freeze dump circuit must be used in the system.
- The electrically actuated valves must be installed in a way that leaves them in the dump position when the power is off.
- In a two-way valve system a check valve must be installed in the drain leg leading to the downcomer, or the collectors will be bypassed during system operation.

Control Strategies

Because of the draindown system's freeze dump capability, three high-limit control strategies are possible.

- The **boil option,** where the circulator pump is shut down but the collectors are not drained.

- The **bake option,** where the circulator pump is shut down and the collectors are drained.
- The **heat rejection option,** where the solar heating system is engineered to dissipate heat actively at high temperatures.

The boil option results in intermittent triggering of the storage tank temperature and pressure relief valve, or of the pressure relief installed at the collectors when the water in the hottest part of the collectors expands and exceeds the valve's maximum pressure limit. This is followed by the refilling of the collectors by cooler water. This cycle will repeat as long as the solar energy input is sufficient to cause the collector pressure to rise over the relief valve limit.

Using the boil option has two disadvantages. First, the pressure throughout the house is raised to the relief valve set point, which places high stress on all the pipes and valves as much as 20 or more times per day. Second, the boiling water can build up mineral deposits in the collectors. This will lower the collectors' efficiency with time.

The bake option results in raising the empty collectors to very high temperatures (250 to 400°F). If hot water is drawn from the system, the tank temperature will eventually be lowered to where the collectors refill with cool water. This can generate superheated steam and it can also expose the collectors to thermal shock, a much discussed but little documented factor in reducing the life of the collectors. The generation of superheated steam can result in high pipe temperatures; this can seriously degrade pipe insulation, because most solar pipe insulations are rated for no more than 220°F service. This steam can also damage temperature or freeze sensors and melt sensor wire insulation where it contacts the heated pipes.

Heat rejecting collectors, collectors that are engineered to dump heat actively at about 205°F, are now rarely used in small residential systems.

When snap switches are used as sensors to initiate freeze dump action, placement of the switches is critical. Typical locations are on the inlet and outlet headers where the collector array attaches to the riser and downcomer pipes. Some experimenters have observed temperatures as much as 10°F lower than the header temperatures in the center of the collectors, so the snap switch should drain the collector at 44°F or higher to preclude freezing. It is possible to cut into the back of the collector and mount a snap switch on the collector absorber plate fin, but this is usually not practical in a system that has already been installed.

One manufacturer's controller design prevents the draindown system from refilling until the collector array temperatures rises to 65°F. This would keep the system from potentially filling and draining down two or more times in one night because of weather conditions that cause a more than 10° but less than a 20° shift in temperature. This multiple drain cycle can occur because typical snap switches have only a 10°F open-to-close operating differential.

Expected Performance

The performance of the system is both site dependent and system dependent, so the assumptions used in Chapter 1 still apply to the performance comments.

A properly designed draindown system operates at the same high efficiency as a pumped service-water system. The flow rate is adjustable, there is no heat exchanger in the system, and water is used as the heat transfer fluid.

The system can be used in freezing weather because it has positive freeze protection. This extends the seasonal service beyond the seasonal service of the thermosiphon and pumped service-water systems and results in higher annual energy collection.

However, it is debatable whether the draindown system will operate properly year round in extremely cold climates. Servicepeople working in the northern states, where there is a long subzero season, should think seriously about recommending that the system be drained and shut down during January and February.

Sizing Parameters

Sizing parameters are similar to the sizing parameters used for pumped service-water systems. The collector array should be sized at 20 to 25 ft^2 per occupant to provide 40 to 60% of the family's hot water needs.

There should be 1.2 to 2 gal of storage per square foot of collector array, and the pump should be sized to provide a flow rate of 0.015 to 0.03 gpm/ft^2 of collector.

The controller must contain a separate freeze dump system to run the electrically actuated valves, and the controller should start up the pump when the collectors are 8 to 20°F hotter than the tank, *provided that the freeze dump system is not overriding the pump on/off circuit.* The controller should turn the pump off when the collectors drop to only 2.5 to 5°F hotter than the tank. The high-limit switch should operated at a preselected point somewhere below 180°F. The freeze dump circuit should deenergize the valves at about 40 to 44°F and reenergize them at about 50°F.

The controller freeze dump voltage output must correspond to the operating voltages of the electrically actuated freeze protection valves, or a transformer must be placed between the valves and the controller to change the voltage. Many electrically actuated valves use 12 to 24 V to run the valve, but most controllers use 120 V as their operating voltage. A mismatch would quickly burn out the valve. Check amperage requirements of the valves versus the rating for the controller's freeze dump circuit carefully. It may be necessary to use a relay and a separate source of current to run the valves if their current draw exceeds the rating of the controller. Keep in mind that the valves are draining current whenever the system is running, so the controller must be rated for *continuous service* at the required current draw.

The controller must contain one or more outlets that are controlled by the pump off/on circuit. These outlets will supply 120 V and up to 10 A of current. Motors $\frac{1}{3}$ hp or higher should be checked for start-up current requirements that may exceed the rating of the controller.

Other Considerations

A glass-lined tank is satisfactory. It should contain a magnesium anode to protect against corrosion. Aluminum water passages in the system should be avoided.

The operating piping in the system should all be copper and brass, but the drain piping may be plastic if the plastic will withstand the higher temperatures that may be encountered. The piping should be large enough to facilitate easy draindown.

Assuming a total of 100 running feet of piping between the collectors and the tank, excluding the drain piping, $\frac{3}{4}$-in. nominal piping is generally satisfactory for up to four collectors; 1-in. piping should be used for more than four collectors. Be certain that the manifolds are at least the size of the piping system. Check with the designer when there are more than four collectors in the system if problems in flow are encountered.

The vacuum relief valve and the air vent valve can each create distinct problems in draindown systems. The successful operation of the system is highly dependent on the proper operation of these valves. *These valves are subject to freezing in extremely cold weather.* Servicepeople encountering recurring problems with a suspected freezing collector vacuum relief valve in this location should add an auxiliary valve, as shown in Figure 4.6. If the array does not refill fully after dumping in freezing weather, the air vent valve may be freezing and need relocating or additional insulating.

The tank and, if possible, the associated piping should be insulated to R-10 or better, and the insulation must be carefully chosen and protected against adverse weather.

The tank and the collectors must have adequate fittings and be mounted securely on well-designed supports that will carry their weight and not be affected by frost and wet weather.

All freeze-protection valves should be mounted in a protected, non-freezing area to assure reliable operation. They cannot be placed in a freezing environment.

Advantages

The draindown system has these advantages.

- The seasonal performance is outstanding.
- The season is extended over the season of a thermosiphon or pumped service-water system, which results in higher annual energy collection.
- The system offers positive freeze protection.

Disadvantages

The draindown system has the following disadvantages.

- The system may not be able to be operated year round in extremely cold climates.
- The system is more complex than either a thermosiphon or pumped service-water system.
- The system is constantly circulating fresh water. If the water is not properly conditioned, corrosion or scale and sediment buildup can cause premature system or component failure.
- If the system fails to drain through the failure of the vacuum relief valve or the sticking of an electrically actuated valve, freezing can cause mechanical damage.

How to Test Performance

Performance testing of draindown systems proceeds in three stages.

- The system draindown operation is tested.
- The system heat collection operation is tested.
- The pump and controller operation is tested.

System Draindown Testing

The system should drain down under two (or three) conditions.

- When power failure occurs.
- When the freeze dump sensor senses a preset ambient temperature that is normally about 40 to 42°F.
- When the high-limit control trips (bake control strategy only).

A power failure may be simulated by cutting off the power to the controller. The freeze dump sensor can be tested by simulating a 40 to 42°F temperature at the controller location with a sensor substitution box (see Chapter 17) or at the sensor location with a Freon aerosol spray. To test the system, proceed as follows.

- Place a pail or pails under the freeze dump drainpipes.
- Calculate the approximate amount of water that should be contained in the system using the following guidelines.
  For each collector—1 gal.
  For 100 ft of $\frac{1}{2}$-in. pipe—1 gal. ($\frac{1}{2}$-in. pipe is not recommended).
  For 100 ft of $\frac{3}{4}$-in. pipe—$2\frac{1}{2}$ gal.
  For 100 ft of 1-in. pipe—4 gal.
  Total the amount of water contained in the system.
- Cut off the power to the controller by either switching it off, unplugging it from the circuit, or defusing the circuit if the controller is hard wired. The collector array should dump its contents into the pails.
- Examine the contents of the pails. The water should be clean and free of foreign particles.
- Weigh or measure the amount of water collected. If weighed, 1 gal = 8 lb, 5 oz. The amount of water should be about equal to your calculations.
- Dump the water and replace the pails under the drainpipes.
- Remove the vacuum relief valve.
- Examine the pails to see if any additional water drained from the system. If the valve is working correctly, no appreciable volume of water should be collected. If the valve is not working correctly, more water will drain from the system.
- Replace the vacuum relief valve. If the valve is working correctly, use the original. If it is not, install a new valve and retest the system.
- Reenergize the controller and monitor the air vent valve. If the air vent valve is working properly, the serviceperson will hear the air escaping while the system fills. If the system does not fill, the air vent valve must be replaced.
- Locate the freeze dump sensor or sensors.
- Carefully free the sensor or sensors from the insulation, taking extreme care not to damage the sensors or their wiring.
- Place a thermometer against the back side of the sensor.
- Using a Freon aerosol spray such as those sold in electronic supply stores, spray the sensor housing to lower its temperature to below 40 to 44°F.
- If the freeze dump cycle operates correctly, watch the sensor and the thermometer and note at what temperature the freeze dump valves reclose. It should be at about 50°F. (A limited number of controls close at 65°F.)
- If the valves operate properly, reinstall the sensor.
- If the valves fail to operate, troubleshoot the control system according to the procedures shown in Chapter 17.

Testing Heat Collection

The heat collection test procedure is similar to the procedure used in pumped service-water systems. Turn to Chapter 3 for instructions.

Testing the Pump and Controller

The pump and controller tests are carried out in the same manner as the pump and controller tests in pumped water systems.

Interpreting the Results

Use Table 3.1 and the instructions in Chapter 3 to interpret the results of the performance testing.

WHAT CAN GO WRONG

Assuming that the system has operated successfully prior to the service call, seven things can go wrong.

1. The owner can change the valve settings.
2. The system can become airbound.
3. The system can become clogged with scale or sediment.
4. The system can suffer physical damage or develop a leak.
5. The pump can fail.
6. The controller can fail.
7. The freeze protection valves can fail.

The serviceperson can determine which of these defects exists with a troubleshooting approach.

TROUBLESHOOTING CHART

1. Look for physical damage.

Item	Yes	No
Crushed pipes	_____	_____
Leaky valves or joints	_____	_____
Broken collector cover	_____	_____
Freeze damage	_____	_____
Structural movement	_____	_____
Broken sensor wire	_____	_____
Damaged or missing sensor	_____	_____
Electrical service interruption	_____	_____
Sticking air vent valve	_____	_____
Clogged vacuum relief valve	_____	_____

Sticking check valve   _____ _____

Sticking electrically actuated valve   _____ _____

Broken electrically actuated valve wires   _____ _____

2.   Check the valve settings (see Figure 3.1).

	Solar System Mode		
Valve Number	**Operative**	**Freeze Dump**	**Start-up**
V1	Open	Open	Open
V2	Closed	Closed	Closed
V3	Varies	Varies	Varies
V4	Closed	Closed	Closed
V5	Closed	Closed	Closed
V6	Closed	Closed	Closed
V7	Open	Open	Open
V8	Open	Open	Open
V9	Closed	Closed	Closed
V10	Closed	Closed	Closed
V11	Open	Open	Open
Air vent	Operating	Operating	Operating
Vacuum relief	Operating	Operating	Operating
Strainer	Unclogged	Unclogged	Unclogged
Check valves	Operating	Operating	Operating
Three-way riser dump valve	Open to collector	Open to drain	Open to collector
Three-way downcomer dump valve	Open to collector	Open to drain	Open to collector
Two-way riser valve	Open	Closed	Open
Two-way downcomer valve	Open	Closed	Open
Two-way dump valve	Closed	Open	Closed

3.   Drain the scale and sediment.

   - Turn the controller off.
   - Open V10 and run until clear.
   - Open V5 and run until clear.
   - Open V2 and run until clear.

4.   Clean the strainer and check the tank outlet flow.

   - Close V11.
   - Open V2 to relieve the pressure.
   - Remove and clean the strainer screen with a stiff brush.
   - Open V11 and flush the strainer case.
   - Close V11. Replace the strainer screen and the plug.
   - Open V2 to flush the drain leg again.

5.   Backflush the collectors and check drain leg flow.

   - Turn the controller on to the manual setting.
   - Wait for the collectors to fill.
   - If the collectors do not fill, the air vent valve is faulty or the electrically actuated valves are not closing.
   - Check the electrically actuated valve drainpipes. No water should be flowing.

- Unplug or deenergize the controller. Valves should actuate, and the collector array should drain.

6. If the collectors do not drain,

   - Remove the pressure relief valve.
   - If the collectors drain, the vacuum relief valve is faulty.
   - If the collectors still do not drain, the electrically actuated valve on the riser is not opening.

7. Run the system draindown test.

   - Put pails under the drains.
   - Calculate the amount of water in the system.
   - Cut off controller power.
   - Measure the amount of water dumped.
   - Throw out the water dumped and replace the pails.
   - Remove the vacuum relief valve.
   - Check for more water dumped.
   - Replace the original valve or install a new vacuum relief valve.
   - Restore power to the controller.
   - Let the system refill while checking the air vent valve operation.
   - Locate and free up the freeze dump sensors.
   - Check to see if freeze dump occurs.
   - Check the temperature at which the system refills.
   - Troubleshoot the controller and sensors if the system is not operating correctly.
   - Retest the repaired system.

8. Run a system performance test.

   - Recheck the valve settings.
   - Check that the controller is off.
   - Close V7.
   - Install temperature gauges.
   - Check for clear sky and time of day.
   - Run the test and record temperatures in Table 3.1.
   - If the pump will not run or the valves do not operate, skip to step 9.
   - Evaluate the results.

     1. Does the pump run on automatic?
     2. Did temperatures rise at least 9 to 15°F in 1 hour?
     3. Does a 10 to 20°F temperature difference exist between the gauges?

   If the answers to these three questions are yes, the system is performing satisfactorily. *However,* you have not checked the pump shutoff action. See Chapter 17 to check this function. Skip steps 9 and 10 if the pump is functioning correctly on the automatic controller setting.

9. If the pump will not run on manual or the system will not fill,

   - Check for a blown fuse or faulty circuit relief.
   - Check the pump on a separate circuit if it is the plug-in type
   - Check for electricity using the procedure in Chapter 16 if the pump is hard wired.
   - Isolate whether the pump or the controller is faulty.
   - Proceed to troubleshoot the pump as outlined in Chapter 16.

- Check for electricity at the valves with a voltmeter.
- Check for stuck or frozen valves.
- Check for a malfunctioning air vent valve.

10. If the pump will not run on automatic,

   - Recheck the sensors and sensor wiring for damage.
   - Proceed to troubleshoot the controller as outlined in Chapter 17.

   Complete the system performance test after repairing the pump and controller.

11. Visually check over the system.

   - Return V7 and other operating valves to their proper positions.
   - Is the controller placed in the automatic mode?
   - Are any structural repairs needed?
   - Is any painting needed?
   - Is there any loose or missing insulation?
   - Are there any leaks?

12. Finish up the job.

   - Clean up the area.
   - Remove all tools.
   - Secure the area.
   - Complete the service report for the owner.
   - Present a bill for services.

ON THE JOB

Your toolbox should include a Sunspool valve, boiler drain valves, replacement control sensors, a voltmeter, the sensor substitution box, a can of Freon freeze spray, a long extension cord, and a trouble lamp or circuit tester. You may also want to carry 12-V and 24-V replacement transformers and coils for electrically actuated valves.

Draindown systems are more complicated than the thermosiphon and pumped water systems that you have studied. Unless you carefully quiz the owner on the problems he or she is experiencing, you may spend an excess of time checking out the entire system. Gear your troubleshooting strategy to the problems that the owner is encountering.

Repair any known physical damage first. Once it is repaired, switch the controller to automatic and see if the pump runs, the valves close, and the system fills.

If any problems are encountered in automatic operation, switch to manual and see if the pump runs, the valves close, and the system fills. If the system works on manual but not automatic, troubleshoot the control system (step 10).

If the pump will not run, the electrically actuated valves will not close or the system will not fill, troubleshoot the pump, the electrically actuated valves, or the air vent valve (step 9).

Shut off the controller and see if the system drains. If the system does not drain, check the electrically actuated valves and the vacuum relief valves.

If there is any question in your mind about whether the entire system is draining, run through a complete step 7.

Run the performance test outlined in step 8. While performance is under test, perform steps 1 and 2.

If performance is unsatisfactory, perform steps 3 to 6. Recheck performance when complete.

Any problems should now be corrected. Proceed to finish up the job.

Flow rates below 0.03 gpm/ft^2 are low in our experience. If there is a large difference between the collector inlet and the collector outlet temperature, try to get the flow rate up unless a major repair job would be needed to increase it.

Continual freeze-ups on older systems are almost always either valve related or freeze sensor location related. Carefully check out the freeze dump control strategy and sensor locations and give strong consideration to replumbing the system to use the Sunspool valve.

WHAT YOU SHOULD HAVE LEARNED

1. A draindown system is similar to a pumped water system except that electrically activated freeze protection valves have been placed in the system that drain the collector array in freezing weather. These valves must be protected against freezing.

2. Draindown systems provide positive freeze protection and extend the season in which the solar water heater can be operated.

3. Two three-way valves or three two-way valves are needed to pipe the draindown system. In a two-way-valved system, an additional check valve is also needed.

4. The draindown system is designed so that the system "fails safe." That is, the system returns to a freeze dump mode when power is removed from the system.

5. A vacuum relief valve must be placed in the system so that the water can drain safely. An air vent valve is still needed so that the system can refill.

6. Proper pipe pitch is extremely important. All piping on the collector side of the electrically actuated valves must be pitched to drain. The vertical drop must be at least $\frac{1}{8}$ to $\frac{1}{4}$ in./ft.

7. A controller with a separate freeze dump circuit is needed in a draindown system to shut down and drain the system at 40 to 44°F. A controller designed for a recirculation freeze-protection system cannot be used in a draindown system.

8. Freeze dump sensor placement is highly critical. Usually, more than one freeze dump sensor should be used.

9. Three control strategies can be used: the boil, the bake, and the heat rejection strategies.

10. Draindown systems are just as efficient as pumped service-water systems, and they have longer operating seasons. However, in hard-freeze climates, the serviceperson should consider shutting off the system in January and February.

11. Draindown systems are sized the same way as pumped service-water systems. Use 20 to 25 ft^2 per occupant to obtain 40 to 60% of the hot water needed. Provide 1.2 to 2.0 gal of storage per square foot of collector. Use pump flow rates of 0.015 to 0.03 gpm/ft^2 of collector.

12. Electrically actuated valves sometimes use different voltages than the controller circuit. A transformer or a relay may be needed to prevent burning out the valves if the voltages are mismatched.

The controller freeze dump circuit must be rated for continuous operation because the valves are energized all the time that the system is operating.

13. Glass-lined storage tanks may be used. The tank should contain a magnesium anode. Aluminum water passages should be avoided. Piping should be copper and brass except for the drain-down piping, which may be copper or high-temperature plastic.

14. It may be necessary to place auxiliary vacuum relief and air vent valves indoors in very cold climates or to eliminate annoying fill and/or drain problems in older systems.

15. There are three separate tests that the serviceperson should perform on a draindown system.

    - Draindown operation.
    - Performance testing.
    - Pump and controller testing.

    The troubleshooting strategy should be tailored to the problem encountered in order to keep the service call short.

CHAPTER

5

Drainback Systems

Drainback System A solar water heater system that is designed with air space in the storage tank or *drainback tank* so that the water drains out of the collector array into the tank below when the system is not collecting solar energy.

Chapter 4 discussed draindown solar water heaters, which offer a system of positive freeze protection. The draindown system was cited as not being able to be operated year round in extremely cold climates, being fairly complex in design, and being subject to freezing because of mechanical failure.

The drainback system has been designed to overcome these problems and still allow the use of water as the heat transfer fluid. A number of different drainback systems have been designed, but they all have one common feature. There is room in either the solar storage tank or in a special drainback tank to hold the water from the collector array. Therefore, when the controller turns the collector pump off, the water in the collectors drains back into the tank. Thus, gravity provides the freeze protection.

The proper design of the heat exchanger and the storage tank has been the biggest problem in obtaining effective drainback systems. Several different designs will be shown in this chapter.

These designs are presented in the chronological order in which they were developed. Notice that the newer designs tend to have fewer potential failure modes and greater efficiency than the older ones.

The Compressed Air Drainback System (Not Recommended)

Figure 5.1 shows an early drainback system that uses a blanket of compressed air in the solar storage tank to create room for the water from the collector array. The air compressor is located close to the top of the storage tank. A water-level sensor is placed at an appropriate level on the side of the tank. When the sensor detects the presence of water at that tank level, it turns on the air compressor, which pressurizes the tank enough to overcome the service-water pressure and drive the water

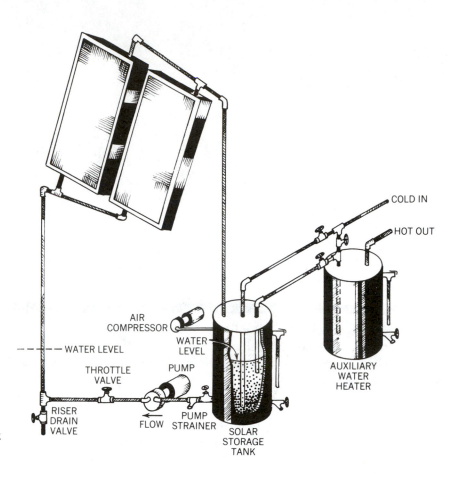

COLD IN

HOT OUT

AIR
COMPRESSOR

WATER
LEVEL

- - - WATER LEVEL

THROTTLE
VALVE

PUMP

AUXILIARY
WATER
HEATER

RISER
DRAIN
VALVE

FLOW

PUMP
STRAINER

SOLAR
STORAGE
TANK

Figure 5.1 The compressed-air drainback system.

level down to below the sensor. At that point, the air compressor turns off and an air void is held at the top of the solar storage tank.

Note the pipe that carries the hot water back to the auxiliary heater or to the house hot water lines in the case of a single-tank system. This pipe must extend down inside the tank to below the lowest level to which the water will be driven.

When the collector pump is not operating, the top of the tank and the collector array are filled with air, as indicated by the level shown in the tank and the line labeled water level across the riser.

The compressed air drainback system is now rarely encountered because the system is subject to catastrophic freezing failure. A power failure plus a slow leak in the collector array will result in the loss of the air pressure. If the air pressure is lost, the collector array will fill with water even though the collector pump is not operating. In cold weather, collector freeze damage is almost certain.

Note the location of the pump. In all drainback systems the pump must be located well below the lowest possible water level. Pumps located higher than the lowest water level will lose their prime and start pumping air. This can lead to overheating that will destroy most pumps.

The pump in a drainback system is always pumping against a vertical pump head. The pump must be more powerful than the pump in a drain-down or a pumped service-water system, and a pump must be chosen that will pump efficiently against a high head.

The Load-Side Exchanger Drainback System

The design problems encountered in a compressed air drainback system can be eliminated by building a tank that contains a **load-side heat-exchanger coil.** A *heat exchanger* is a device, usually made of metal, that transfers heat across an internal wall that separates two fluids, two gases, or a fluid and a gas. A load-side heat-exchanger coil is a large coil of tubing, usually made of copper, that is inserted in the drainback tank. The tank serves as a reservoir for the collector fluid. Heat is transferred from the large tank into the coil.

Figure 5.2 shows such a tank. In this system the service water for the home is isolated from the water that is circulated through the collectors by the heat-exchanger coil. This allows the use of a nonpressurized tank that is vented to the outside air. The water from the bottom of the tank is pumped up to the collector array and returns back into the top of the tank. The hot water in the drainback tank warms the water in the heat-exchanger coil. When the hot water tap is turned on in the home, cold water from the service lines passes into the heat-exchanger coil, and the warm water in the coil passes into the auxiliary heater or directly to the house when no auxiliary heater is used.

The drainback tank is filled by placing a hose on the tank drain valve and opening the drainback fill valve, which is shown in an upside-down

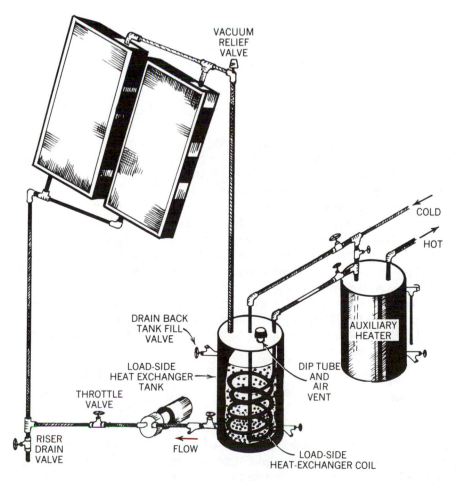

Figure 5.2 A load-side heat-exchanger coil system connected to an auxiliary heater.

position on the side of the tank. The tank is filled until water drains from the drainback fill valve. At this point, the two valves are closed and the system is operational.

The drainback fill valve is shown in an upside-down position for a specific reason. Now that the service water that will be used in the home is separated from the collector water, it is possible to condition the collector water to prevent scale and sediment buildup. To condition the water, detach the fill hose and place a pail under the drain valve.

Open both the drainback fill valve and the drain valve to fill the pail with water. Close off the drain valve, add the water-conditioning chemicals to the water in the pail, and pour this water back into the drainback fill valve.

CAUTION The choice of conditioning chemicals is critical. They must be both nontoxic and compatible with metals used in that particular system. The details can be found in Chapter 10.

The placement of the pump relative to the drainback fill valve is important. The pump must be located 3 ft below the lowest tank water level to avoid loss of prime and resulting damage. Ample drainback tank height above the pump and tank water volume during operation are crucial.

Although such a system solves the catastrophic failure possibilities shown in the air pressurized drainback system, it creates another design problem: the efficient transfer of the energy from the drainback tank to the auxiliary heater.

The amount of water that can be heated at one time is limited by the volume of the heat-exchanger coil. The amount of water that is moved into the auxiliary tank or the service line is limited to the amount of water drawn by the users. For example, suppose that the coil was 100 ft of $\frac{1}{2}$-in. copper tubing. This coil would only hold 1 gal of water; 100 ft of $\frac{3}{4}$-in copper tubing only holds $2\frac{1}{2}$ gal of water; and 100 ft of 1-in. tubing only holds 4 gal of water. The problem is easily visualized: it takes a very large exchanger to heat a lot of water.

Also, the amount of heat actually transferred through the heat-exchanger walls to the line water inside depends on three factors: the heat-exchanger surface area, the temperature difference between the water in the drainback tank and the water in the heat-exchanger coil, and the length of time that the colder water is in the coil.

Although the coil surface area is fairly large and the water in the drain back tank is considerably hotter, the length of time for heat transfer is usually a limiting factor in the load-side heat exchanger. A sink hot water tap may use 0.5 to 2.5 gal per minute, and a shower may use as much as 5 gal per minute. Cold water entering the $\frac{3}{4}$-in. copper coil mentioned previously would only be heated for 30 seconds. In this short time, 60°F service water would only be warmed to 110°F if the drainback tank temperature were 140°F. Even this figure is generous, since it is doubtful that 100 ft of $\frac{3}{4}$-in. copper coil could be squeezed into a 60- to 80-gal storage tank. A 60-ft coil is more typical.

The Separate Solar Storage Tank Drainback System

The design problems inherent in the load-side exchanger drainback system can be overcome with relative ease by separating solar storage from the load-side exchanger tank and either adding a small circulating pump or locating the load-side exchanger tank lower than solar storage, so that a thermosiphon loop is created between the storage tank and the exchanger tank.

Figure. 5.3 illustrates the piping for a thermosiphon loop. The greater the vertical distance between the top of the exchanger tank and the top of the storage tank, the faster thermosiphoning will take place. A floating/sinking ball check valve, similar to the one shown in Chapter 1 for the vertical tank thermosiphon system, is located on the riser next to the tank and will prevent any reverse thermosiphoning.

Figure 5.4 illustrates how forced heat transfer is accomplished through the repiping of the exchanger coil and the addition of a small circulating pump. In this configuration the tank on the left contains the load-side exchanger coil and the tank on the right is solar storage. No auxiliary water-heater tank is shown, but one can be added.

The cold service water from the bottom of the solar storage tank is pumped into the exchanger coil, passes through the coil, where it is warmed, and returns to the top of the solar storage tank. A differential controller that measures the difference in temperature between the load-side exchanger tank and the bottom of the solar storage tank turns the pump off and on.

This secondary loop could also be controlled by the same differential controller used to run the collector loop by wiring the pump in parallel with the collector pump.

System Conversion

In our opinion the compressed air drainback system does not represent a valid freeze-protected solar water-heater design because it is too prone to catastrophic freezing failure. The system should be drained and taken

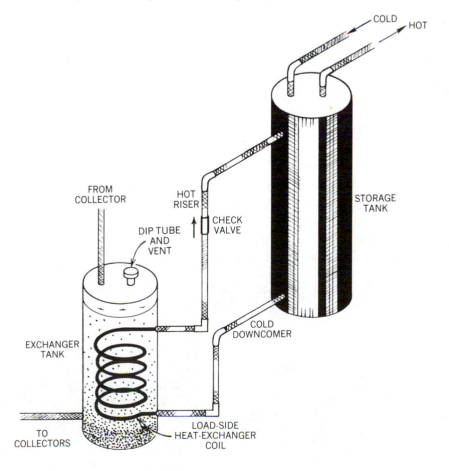

Figure 5.3 A load-side heat-exchanger coil system that uses a thermosiphon-connected separate solar storage tank. The auxiliary heater is not shown.

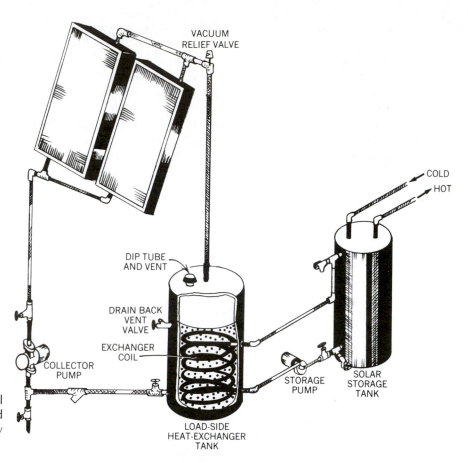

VACUUM
RELIEF VALVE

COLD
HOT

DIP TUBE
AND VENT

DRAIN BACK
VENT
VALVE

EXCHANGER
COIL

COLLECTOR
PUMP

STORAGE
PUMP

SOLAR
STORAGE
TANK

LOAD-SIDE
HEAT-EXCHANGER
TANK

Figure 5.4 A load-side heat-exchanger coil system that uses a pumped loop-connected separate solar storage tank. The auxiliary heater is not shown.

out of operation at the onset of freezing weather, or the system should be converted to a more reliable design.

The load-side exchanger system is usually an inefficient collector of solar energy unless solar storage is separated from the drainback tank and a thermosiphon or secondary pumped loop is placed in the system. We recommend that the solar serviceperson seriously consider the advantages of such a system conversion.

There are several drainback tanks on the market that could be added to such a system to provide efficient, year-round solar collection.

The State Industries Drainback System

State Industries, a leading manufacturer of hot water heaters, has been marketing drainback systems for several years. Figures 5.5, 5.6, and 5.7 show one of their earlier systems.

Looking at Figure 5.5, the solar storage tank contains four internal heat-exchanger tubes. These tubes are open at the top. The bottoms of the tubes are welded to the top of a water reservoir that forms a double-bottomed tank. At the top of the tank is another auxiliary reservoir. This top reservoir is normally filled with air. When the system drains down, this tank provides storage for the water from the collector array.

Water passes down from the top reservoir to a distribution fitting located at the top of the heater. Four pipes lead from the distribution fitting to the heat-exchanger tubes.

A small volume of water located in the heat-exchanger tubes and solar piping is constantly transferring heat to the storage tank. Such a configuration is called a *supply-side heat exchanger*.

The Operating Cycle

The tank is filled with cold service water. An outlet from the top of the tank leads to the auxiliary water heater, so that it is supplied with warm solar water.

The heat-exchanger tubes are filled with distilled, conditioned water. When the system is connected to an auxiliary heater, as shown, these heat-exchanger tubes are filled to the top, and the air void is located in the auxiliary reservoir.

The system can also be used alone, with no auxiliary heater, because the unit contains an electrical element halfway up the tank wall that is not shown in Figure 5.5.

When the electric element is used in place of an auxiliary heater, the auxiliary reservoir is eliminated from the system, and the heat-exchanger tubes are only filled to the height of the electric element when the pump is operating. The top halves of the exchanger tubes provide the drainback reservoir.

Distilled, conditioned water passes out from the heat-exchanger tubes to the pump, where it is pumped up to the bottom manifold of the collectors. It passes through the collectors, where it is warmed, returns to the auxiliary tank, and flows down into the heat-exchanger flues, where it gives up its heat to the water in the storage tank. When water is drawn by the user, the warm solar storage water feeds into the auxiliary heater tank or to the water tap.

The system is designed with a glass-lined, magnesium anode-pro-

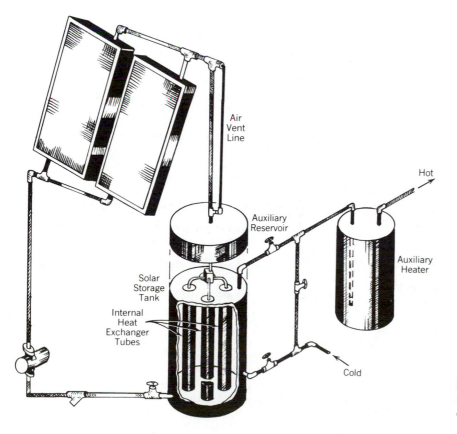

Figure 5.5 One State Industries' design uses four internal flues as the heat exchanger.

tected tank and a standard differential thermostat that uses a collector sensor, a storage tank sensor, and a high-limit shutoff device.

Figure 5.6, taken from State Industries' installation and maintenance manual, shows a typical installation without an auxiliary heater. Figure 5.7 shows the installation of the air tank for use with an auxiliary heater. Figure 5.8 is an exploded view, showing all the parts of the system.

In 1982, State Industries changed to a different drainback design that eliminated the internal flues and replaced them with a heat-exchanger jacket welded around the outside of the bottom half of the tank. A cross

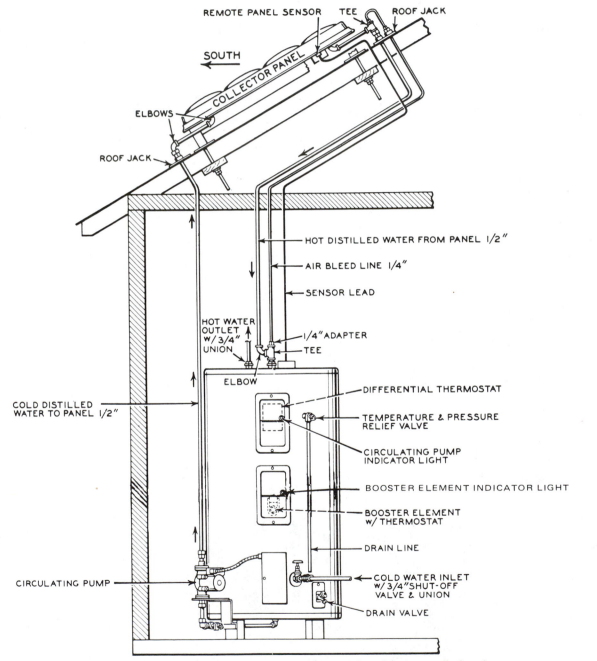

Figure 5.6 Details of the State Industries' internal-flue design installed without the auxiliary draindown tank. (Reprinted by permission from State Industries, Ashland City, Tennessee.)

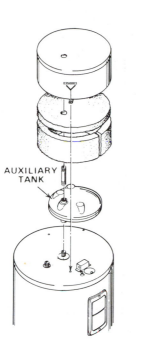

AUXILIARY TANK

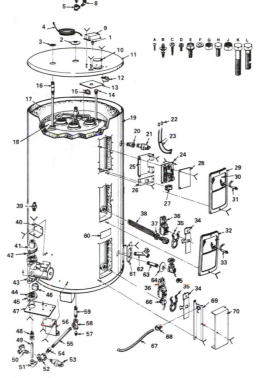

Figure 5.7 Details of the State Industries' auxiliary draindown tank installation. (Reprinted by permission from State Industries, Ashland City, Tennessee.)

Figure 5.8 An exploded view of the State Industries' internal-flue water heater. (Reprinted by permission from State Industries, Ashland City, Tennessee.)

section of this tank is shown in Figure 5.9, and a sketch of the tank is shown in Figure 5.10. Note that both illustrations contain the numbers 1 to 8. These will be used to guide you through the system's operation.

When the system pump (3) turns off, the fluid in the collectors drains back and fills the auxiliary drainback reservoir (2).

When the controller indicates that the collectors are sufficiently hotter than the water in the tank, the pump turns on and the collector fluid flows into the bottom of the heat exchanger (4) and around the baffles, finally exiting at the top of the exchanger (5), where it is returned to the collectors, warmed by the sun, and returned to the top of the auxiliary drainback tank (1). Operation continues until the controller turns the pump off.

An air bleed line runs from a tee fitting located on the top of the drainback tank (8) to the top of the collector array, as shown in Figure 5.11.

The air bleed or air transfer line replaces the vacuum relief and air vent valves seen in the previous drainback systems. It provides a path for air to enter the collectors when the water is draining out of them. It also routes the air in the collectors back to the reservoir when the pump begins to fill the collectors. By using an air bleed line, the manufacturer stops the introduction of fresh oxygen into the system during drainback. This helps eliminate any corrosion problems and prevents the hot water in the system from evaporating.

The use of air transfer line is an excellent design approach to drainback systems.

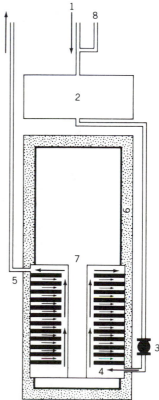

Figure 5.9 A cross-sectional view of State Industries' heat-jacketed design.

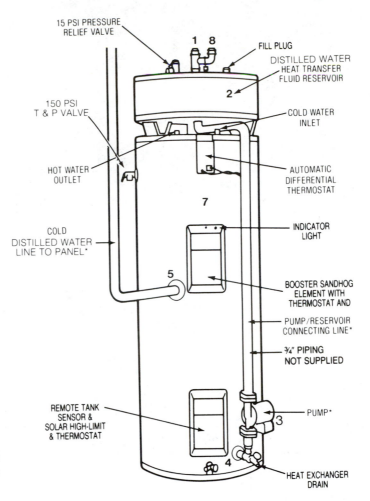

Figure 5.10 The details of the heat-jacketed design. (Reprinted by permission from State Industries, Ashland City, Tennessee.)

Controlling the Drainback System

The drainback system is controlled with a simple on/off differential thermostat that runs the pump and contains a high-limit switch. An ambient air sensor that prevents operation in extremely cold weather is optional equipment that can be added.

The flow rate is fine tuned using a ball or balancing valve as a throttle valve on the discharge side of the pump. If the system is noisy, the throttle valve can be moved to the downcomer line above the drainback tank.

A controller should be selected that will turn the pump on when the collectors are about 20°F hotter than the tank and turn the pump off when the differential drops to about 5°F. This high turn-on differential is used in drainback systems to prevent excess pump cycling. The high-limit switch should activate below 180°F maximum, and the cold air sensor, if included, should prevent pump operation at a temperature in the high teens to low twenties. The controller should contain one or more controlled outlets that operate at 120 V for running the pump. It should be connected to a 120-V, 15-A fused circuit.

CAUTION   If the pump is $\frac{1}{3}$ hp or larger, check if the amperage of the controller is large enough to handle both start-up and continuous running.

When the load-side exchanger is separated from solar storage and contains a pump, it should have a similar controller to control the

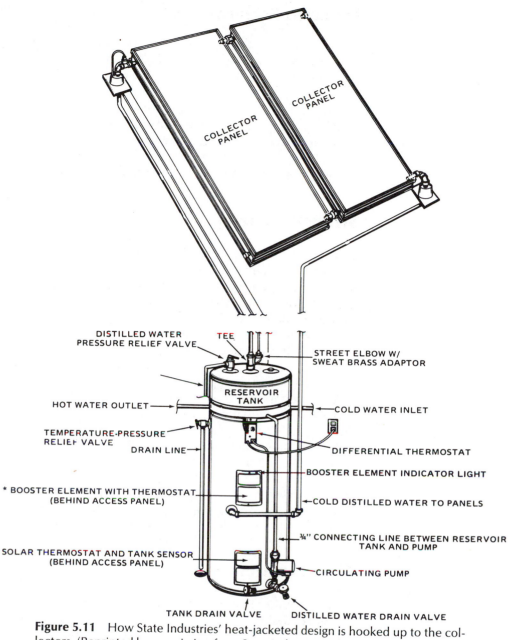

Figure 5.11 How State Industries' heat-jacketed design is hooked up to the collectors. (Reprinted by permission from State Industries, Ashland City, Tennessee.)

exchanger coil pump between the storage tank and the auxiliary heater. The pump on this side of the system will be extremely small, around $\frac{1}{35}$ hp or less. The hot sensor (collector) is placed on the load-side tank wall or on the load-side exchanger hot outlet pipe, and the cold sensor (storage) is placed on the pipe to the pump from the bottom of the solar storage tank. A check valve in this loop may be necessary to prevent reverse thermosiphoning.

It is possible to use one controller to run both pumps. In this case the pump turn on and turn off would be controlled by a collector plate sensor and an exchanger or solar storage tank sensor. Both pumps would run when the collector was a few degrees hotter than the tank, and there would be no way to control the secondary circuit separately. Such a system will operate satisfactorily but often will not be as efficient as a system with two separate controllers.

Expected Performance

Once again, the performance of the system is both site dependent and system dependent. The same assumptions used previously apply to the performance comments.

A properly designed and sized drainback system is less efficient than a pumped service-water or draindown system because the presence of the heat exchanger lowers the overall efficiency of the system. The sizing of the heat exchanger is critical in obtaining good system efficiency.

The lower system efficiency is offset by the ability to run the drainback system throughout the year in all but the coldest climates.

Sizing Parameters

The rule of thumb of 20 to 25 ft² per occupant to obtain 40 to 60% of the hot water needed usually holds true for well-designed drainback systems. However, a higher flow rate of 0.02 to 0.03 gpm/ft² of collector is used. A storage tank that provides 1.2 to 2 gal/ft² of collector is ample for water storage.

When the system drains back, the load-side exchanger tank needs to hold 1 gal of water per collector plus the water that fills the piping in the collector array ($\frac{1}{2}$-in. pipe will contain 1 gal/100 ft, $\frac{3}{4}$-in. pipe will contain $2\frac{1}{2}$ gal/100 ft, and 1-in. pipe will contain about 4 gal/100 ft).

As an example, in a two-collector system with 100 ft of $\frac{1}{2}$-in. pipe, the tank must be sized with a 3-gal minimum air void to allow the system to drain back. Also, the tank must always contain enough water to prime the pump adequately at all times. This can require as much as 3 ft of water above the pump.

The pump must be sized to pump against both the vertical head and the friction losses of the system. You can ignore the friction losses of the downcomer, because a siphon or gravity return effect is created once the water starts to return to the storage tank.

Field Experiences

First, remember that there are two types of drainback systems: the **open drainback system** and the **closed drainback system.** Open drainback systems allow fresh oxygen into the system through a vacuum relief valve during drainback. Closed drainback systems do not allow the introduction of fresh oxygen and use an air vent line instead of a vacuum relief valve. We prefer the use of the air vent line and regard this approach as a better design.

Closed systems should not be made into open systems. They may have been designed with materials that will corrode rapidly if fresh oxygen is introduced continually.

You can consider making a closed system out of an open system. This can decrease corrosion rates and lower evaporative water losses.

Empty collectors stagnating at high temperatures do not transmit their temperature rapidly to collector control sensors placed on the manifold piping. This can create a problem of control that is best overcome by placing the sensor on the absorber fin inside the collector if the collector backing can be cut and patched. A sensor placed near a riser tube in the upper half of the collector fin is much more responsive than one placed on the outside manifold piping. Sensor placement is especially critical in drainback systems, because periods of pump cycling cause very inefficient heat collection.

The higher suggested flow rate shown for drainback systems is an experience-dictated flow rate. In drainback systems, where cold water may be pumped into extremely hot empty collector arrays, it has been found that steam pressure can be generated and that this pressure will prevent the loop from filling correctly. A faster flow rate seems to stop this from happening.

If the system is experiencing problems draining or filling,

- Consider adding an air transfer line. This line should connect the top of the drainback tank to the highest point in the solar array. Installation is critical. The air line must slope continuously upward from the tank to the system high point so that water cannot be trapped in it.
- Check the return piping from the collectors. The piping must slope continuously from the exit point on the collector array to the top of the tank. No traps can be built into the piping, because they will not drain.
- Do not add a vacuum relief valve to a closed system unless the system manufacturer recommends it. If added, it should be placed at the highest accessible point in the collector array. Usually, a vacuum relief valve at the collector array should be complemented by an air vent at the drainback tank, and vice versa.

Other Considerations

For solar storage, a glass-lined tank is satisfactory. The tank should contain a magnesium anode to protect against premature corrosion failure.

NOTE In doing service work remember that the average consumer may not know that a worn-out magnesium anode can be replaced with a new anode to extend the tank life. Considering the high cost of replacing a solar tank, the cost of replacing the rod is minimal.

Aluminum water passages in the collector should be avoided. The collector loop water can be conditioned, but the heat-exchanger coil must be copper because of the presence of unconditioned, fresh water in the solar storage and auxiliary tanks. The piping and fittings should be all brass and copper to minimize any corrosion problems.

The size of the piping will depend on the length of the runs, the size of the pump, and the number of collectors. Generally, $\frac{1}{2}$-in. pipe is satisfactory for one collector, $\frac{3}{4}$-in. pipe is satisfactory for two to four collectors, and 1-in. pipe should be used for more than four collectors. The manifolds should be one pipe size larger than the runs if there are three or more collectors in the system. The top manifold and the downcomer piping must be the same size or larger than the bottom manifold and the riser to promote good draining. Externally manifolded collectors are shown in the drawings, but internally manifolded collectors may also be used. Details are shown in Chapter 9.

The materials used in the air vent line back to the upper manifold are less critical. Certainly, $\frac{3}{8}$-in. copper tubing would be best. However, high-temperature plastic piping or tubing can be used and will give good service life unless it is exposed excessive temperatures and too much sunlight.

The tank and, if possible, the associated piping, should be insulated with R-10 or better insulation that is protected against the weather, and all equipment must be securely mounted on well-designed supports that will carry their weight and not be affected by the weather.

Advantages

The drainback system has these advantages.

- Positive freeze protection in all but the coldest weather.
- Ability to be operated year round.
- No vacuum relief or air vent valve to clog up and cause unexpected system failure (closed systems only).
- No electrically actuated valves to service and maintain.
- The water in the collector loop can be conditioned to retard corrosion .

Disadvantages

- The system is less efficient than a system that contains no heat exchanger.
- Pump energy consumption is greater than in a draindown or recirculation system.
- When dual pumps and dual controls are needed, costs are higher and chances of failure are greater.

How to Test Performance

Performance testing of the drainback system consists of:

- Testing drainback operation.
- Testing heat collection operation.
- Testing collector loop pump and controller operation.
- Testing heat transfer to the solar or auxiliary storage tank.

System Drainback Testing

The system should drain back to the load-side exchanger tank whenever the collector loop pump is not running. Proceed to test for drainback as follows (see Figure 5.2).

- Turn the pump off.
- Allow the system to drain back.
- Screw a Schraeder air fitting on the riser drain valve and open the valve.
- Open the drainback fill valve and fill the drainback tank to a level equal to the top of the valve.
- Close the drainback fill valve.
- Close the valve on the pump leg coming from the exchanger tank.
- Using a hand or power air pump, pressurize the riser. This should force any water remaining in the system back into the tank.
  NOTE: If the vacuum relief valve does not remain closed, you must either pump faster or close off the valve.
- Open the valve on the pump leg to refill the pump leg.

- Open the drainback fill valve and collect any water that comes out.
- If the system is draining back, you should not collect any water. If you do, water is hanging up in the collector array.

Another Approach to System Drainback Testing

The telltale clue that determines inadequate drainback is the spring service call to fix leaks caused by freezing that can be due to either inadequate system venting or poorly sloped piping.

Go up on the roof. Using a soldering torch, unsweat an elbow off the lower collector header at the riser or anywhere that your level indicates inadequate pitch. If the joint comes apart easily, it is almost surely empty. If it will not heat up enough to unsweat, there is most likely water inside. Cut the pipe. Drain it and use blocks or other means to create a proper slope.

System Heat Collection Testing

First, look at Figure 5.2. Performance testing of this system consists of measuring tank temperature rise with time. Proceed as follows.

- The testing is carried out between 10 A.M. and 2 P.M. on a clear day in direct sunlight.
- The secondary loop to the auxiliary water heater must be valved off so that no heat can be drawn from the system.
- The pump is turned off.
- A temperature gauge is placed on the cold riser where it exits the tank and as close to the tank as possible.
- A temperature gauge is placed on the tank wall just below the top of the water level when the system is running or suspended through the dip tube into the tank water at the top.
- The gauges are allowed to come to equilibrium, which should take 3 to 4 minutes.
- The gauges are read and the readings are recorded on a chart like the one in Table 3.1.
- The pump controller is turned to automatic.
- If the pump starts up within 2 to 3 minutes, proceed with the test.
- If the pump does not start up on automatic, turn the controller switch to manual.
- If the pump starts up on manual but will not run on automatic, four possibilities exist.

  1. The controller is faulty.
  2. One or both of the sensors is faulty.
  3. The temperature differential between the collectors and the tank is not high enough to activate the controller (8 to 20°F, 20°F preferred).
  4. The tank temperature is high enough to activate the high-limit sensor.

- If the pump will run on manual, proceed with this test and check for temperature rise with time. When the test is completed, go to the pump/controller procedure.
- If the pump will not run on manual, terminate the test and go to the pump/controller test procedure.

Proceeding with the Test Assuming that the pump will run, continue the temperature testing with the pump on manual, as the preceding procedure indicates. If the pump will not run, return to this point in the test after repairing the pump or the controller.

- Allow 30 minutes to 1 hour to pass with the pump operating.
- Record the readings of the temperature gauges in a chart similar to Table 3.1.
- Fill out the chart as discussed in Chapter 3.

The Meaning of the Readings and the Calculations The meaning of the readings and the calculations are identical to the meaning of the readings and calculations shown in Chapter 3.

Interpreting the Results The results are interpreted in the same manner as the results in Chapter 3.

The Pump/Controller Test The pump/controller test is carried out in the same way as the pump/controller test in Chapter 3.

Testing Heat Transfer to the Separate Solar Storage Tank

In a system like the one shown in Figure 5.3, the rate of heat transfer from the solar storage tank to the auxiliary tank is important. If heat cannot be readily taken out of the load-side exchanger tank, the collection process becomes less efficient, because the collectors will operate at higher temperatures and suffer greater losses to the outdoor air. Proceed to test the rate of heat transfer in the following manner.

NOTE In order to carry out this test the water in the load-side exchanger tank must be at least 25 to 30°F hotter than the water in the solar storage tank.

- Suspend a thermometer in the tank water through the dip tube in the top of the load-side exchanger tank. Call this temperature T1.
- Place a temperature gauge on the pipe exiting from the bottom of the load-side exchanger tank to the collector pump close to the tank wall. Call this temperature T2.
- Place a temperature gauge on the exchanger coil pipe near the bottom of the load-side exchanger tank where water flows to the heat-exchanger coil from the heat-exchanger pump. Call this temperature T3.
- Place a temperature gauge near the top of the load-side exchanger

Table 5.1

TEMPERATURE CHART—HEAT-EXCHANGER PERFORMANCE TEST

TIME READ	TEMPERATURE HOT SUPPLY T_1 (°F)	TEMPERATURE COLD SUPPLY T_2 (°F)	TEMPERATURE COLD STORAGE T_3 (°F)	TEMPERATURE HOT STORAGE T_4 (°F)
Initial	_____	_____	_____	_____
15 min	_____	_____	_____	_____
30 min	_____	_____	_____	_____
1 hr	_____	_____	_____	_____

tank where the water exits from the heat-exchanger coil. Call this temperature T4.

- Shut off the water to the house or the auxiliary heater.
- Record the temperatures of the four thermometers in Table 5.1.
- Turn on both the collector loop and the heat-exchanger coil loop pumps using the manual controller settings.
- Run the system for about 15 minutes. Record the temperature of all four gauges.
- Continue to run the system for a total of 30 minutes. Record the temperature of the four gauges at the end of 30 minutes.
- Conclude the test, remove the gauges, and return the system to the correct operating conditions.

The Meaning of the Readings

The test that is being run is a **heat-exchanger effectiveness test.** Heat exchangers are spoken of in terms of their effectiveness, not in terms of their efficiency. Heat-exchanger effectiveness is a measure of how much heat is actually transferred versus how much heat would be transferred if there were no exchanger.

Effectiveness is measured in percent. A fully effective heat exchanger (no exchanger) would be 100% effective. For a heat exchanger to be 100% effective, it would have to be infinitely large and have no time lag in transferring the heat across the metal exchanger wall. This is economically and technically impossible, so heat exchangers are always less than fully effective.

Look at Figure 5.12 to visualize effectiveness. Hot supply fluid from the collectors enters at the top of the tank. Label the temperature of this fluid as T1.

The cooled supply fluid, which is returned to the collectors, exits from the tank at the bottom left. Label this temperature T2.

Cold solar storage tank water, which is contained in the heat-exchanger coil, enters at the lower right. Label the temperature of the incoming fluid T3.

The heat in the tank is transferred by conduction across the wall of the coil to the cold solar storage tank water, which exits back to the solar storage tank at the top coil exit. Label the temperature of the exiting fluid T4.

A 100% effective heat exchanger would lower the temperature of the exiting collector fluid, T2, to the temperature of the entering solar storage fluid, T3 so that T2 = T3. All the heat entering the tank from the collectors would be removed.

It is not possible to build a heat exchanger that is 100% effective, so T2 will always be hotter than T3.

T1 − T3 equals the maximum amount of heat that could be removed from the tank at 100% effectiveness.

T1 − T2 equals the actual amount of heat that is removed from the tank in any given situation.

Percent effectiveness is: actual heat removed/heat removed at 100%E, or

$$(T1 - T2)/(T1 - T3) \times 100 = \%E$$

Table 5.2 shows an example.

Acceptable Heat-Exchanger Effectiveness The higher the heat-exchanger effectiveness, the higher the system's efficiency. Heat exchangers in drainback systems should be at least 50% effective under

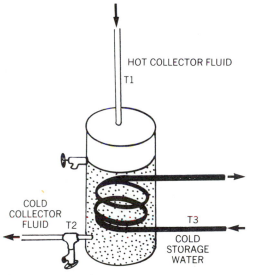

Figure 5.12 The temperature measurements for testing heat-exchanger effectiveness.

Table 5.2

SAMPLE TEMPERATURE CHART—HEAT-EXCHANGER PERFORMANCE TEST

TIME READ	TEMPERATURE HOT SUPPLY T_1 (°F)	TEMPERATURE COLD SUPPLY T_2 (°F)	TEMPERATURE COLD STORAGE T_3 (°F)	TEMPERATURE HOT STORAGE T_4 (°F)
Initial	155	110	85	105
15 min	150	120	95	115
30 min	150	125	100	120
1 hr	160	135	110	125

Initial: System is at rest; %E cannot be calculated.
15 min (150–120) × 100/(150–95)　= 55%E
30 min (150–125) × 100/(150–100) = 50%E
 1 hr　(160–135) × 100/(160–100) = 50%E

the conditions shown. Exchangers with less effectiveness will not transfer heat fast enough to give good system efficiency.[1]

Temperature Across the Coil　The size of the temperature differential across the heat-exchanger coil is of only minor importance compared with the difference in temperature between the storage tank and the coil exit. This temperature differential across the heat-exchanger coil is controlled by the rate at which the water flows through the coil. Faster flows will lower the differential, and slower flows will raise the differential, because the water will reside in the coil longer and get hotter before returning to the auxiliary tank.

If the temperature differential across the coil is less than 5 to 10°F, the flow could be lowered. If the temperature differential across the coil is in excess of 20 to 30°F, the flow may want to be increased. The throttle valve on the pump line is used to control this flow.

Determining the Total Heat Transferred

If the total amount of water in the solar storage tank is known, the total heat transferred can also be determined. It takes 1 Btu to raise the temperature of 1 lb of water by 1°F. Water weights 8.33 lb/gal, so the formula

(Difference in average tank temperature in 30 min) (8.33 lb/gal)(gal of water in the tank)(2) = Btu/hr

would give the number of British thermal units that were transferred in 1 hour. The average tank temperature is (T3 + T4)/2.

When a system shows a heat-exchanger effectiveness of less than 50%, the owner should be notified that the efficiency of the system is less than satisfactory. The only remedy for this low efficiency is to increase the size of the heat exchanger coil.

[1]The 50%E figure was chosen so that the system would have an overall efficiency greater than 25% on a clear day between 10 A.M. and 2 P.M. where:

Solar insolation = 300 Btu/ft²/hr
Collector efficiency = 50%
Solar collection rate = 150 Btu/ft²/hr
Exchanger effectiveness = 50% system heat-exchange rate = 75 Btu/ft²/hr
(300 × 0.50 × 0.50 = 75 Btu/ft²/hr)

WHAT CAN GO WRONG

Assuming that the system has operated successfully prior to the service call, nine things can go wrong.

1. The owner can change the valve settings.
2. The system can fail to drain back.
3. The vacuum relief valve can malfunction in open systems, or the air vent line can clog in closed systems.
4. The level of the water in the solar storage tank can be incorrect.
5. The collector pump can fail or refuse to pump against the vertical head.
6. The exchanger coil pump can fail.
7. Either or both controllers can fail.
8. The system can suffer physical damage or develop a leak.
9. The exchanger coil can become clogged with scale and/or sediment.

The serviceperson can determine which of these defects exists with a troubleshooting approach.

TROUBLESHOOTING CHART

1. Look for physical damage.

Item	Yes	No
Crushed pipes	_____	_____
Leaky valves or joints	_____	_____
Broken collector cover	_____	_____
Freeze damage	_____	_____
Structural movement	_____	_____
Broken sensor wire	_____	_____
Damaged or missing sensor	_____	_____
Electric service interruption	_____	_____
Broken or clogged air vent line	_____	_____
Malfunctioning vacuum relief valve	_____	_____
Malfunctioning air vent	_____	_____

2. Check the valve settings.

Valve Name	Solar System Mode		
	Operative	Drainback	Start-up
Drainback fill vent valve	Closed	Closed	Closed
Throttle valves	Varies	Varies	Varies
Drain valves	Closed	Closed	Closed
Temperature/pressure relief valves	Closed	Closed	Closed
Collector vacuum relief valve	Open	Open	Closed
Drainback tank vacuum relief valve	Closed	Closed	Open
Collector air vent valve	Closed	Closed	Open
Drainback tank air vent valve	Open	Open	Closed
Pump drain leg valve	Closed	Closed	Closed
Cold water feed valve	Open	Open	Open
Hot water delivery valve	Open	Open	Open

3. Drain the scale and sediment.

    * Turn the controller off.
    * Open the drain valves and run until clear.
    * On the load-side exchanger tank, replace any conditioned collector loop water.

4. Clean the strainer and check the flow.

    * Close the pump leg shutoff valve.
    * Remove and clean the strainer with a stiff brush.
    * Replace the strainer screen and plug.
    * Flush the drain leg.

5. Run the drainback test.

    * Check that the controller is off.
    * Let the system drain back.
    * Bring the exchanger water to the level of the exchanger vent valve.
    * Attach the air fitting to the riser drain valve.
    * Close the valve to lower the tank outlet to the pump.
    * Pressurize the system and force the water from the collectors.
    * Open the valve to lower the tank outlet.
    * Open the drainback tank vent valve. If the water drains, the system is not draining back.
    * Remove the air pressure and close the vent valve.

6. Run the collector loop performance test.

    * Perform the test on a clear day between 10 A.M. and 2 P.M.
    * Check that the controller is off.
    * Close down the secondary loop to the auxiliary tank.
    * Put a thermometer in the dip tube.
    * Put the temperature gauge on the riser.
    * Allow the gauges to come to equilibrium.
    * Read and record the temperatures in Table 5.1.

- Turn the pump controller to automatic. If the pump starts, proceed with the test.
- If the pump does not start, switch the pump to manual and conduct the test.
- If the pump will not run on manual, skip to the pump/control test procedure and do the repair. Then return to this point and run the performance test.
- Operate the pump for 30 minutes to 1 hour.
- Record the temperature gauge readings in Table 5.1.
- Evaluate the results.

  1. Does the pump run on automatic?
  2. Did temperatures rise at least 9 to 12°F in 1 hour?
  3. Does a 10 to 20°F temperature difference exist between the gauges?

  If the answers to these three questions are yes, the system is performing satisfactorily. However, the pump shutoff action has not been been checked. See Chapter 17 to perform this test.

Skip steps 7 and 8 if the pump is functioning correctly on the automatic controller setting.

7. If the pump will not run on manual;

   - Check for a blown fuse or faulty circuit breaker.
   - Check the pump on a separate circuit if it is the plug-in type.
   - Check the circuit with a trouble lamp if it is hard wired.
   - Isolate whether the pump or controller is faulty.
   - Troubleshoot the pump as shown in Chapter 16.

8. If the pump will not run on automatic;

   - Recheck the controller sensors and sensor wiring for damage.
   - Proceed to troubleshoot the control system as outlined in Chapter 17. Complete the collector performance test after repairing the pump and controller.

9. Test the heat transfer to the solar storage tank.

   - Make certain that the exchanger tank is at least 25 to 30°F hotter than the solar storage tank.
   - Place a thermometer in the dip tube of the exchanger tank (T1).
   - Place the temperature gauge at the bottom of the solar storage tank on the pump line (T2).
   - Place the temperature gauge on the pump return line to the solar storage tank (T3).
   - Shut off the water to the house or auxiliary heater.
   - Turn the collector loop and storage loop pumps on.
   - Record the three temperatures in the chart shown in Table 5.1.
   - Run the pump for 15 minutes; then read and record the three temperatures in the chart.
   - Continue to run the pump for 30 minutes; then read and record the three temperatures in the chart.
   - Conclude the test and return the system to operating condition.
   - Calculate the exchanger effectiveness.

- Calculate the total heat transferred. The system is operating satisfactorily if

   1. The exchanger effectiveness is over 50%.
   2. The pump works on automatic.
   3. The system is transferring 75 Btu/ft^2/hr to the auxiliary heater.

10. Visually check over the system.

   - Return operating valves to their proper positions.
   - Are the controllers placed in the automatic mode?
   - Are any structural repairs needed?
   - Is any painting needed?
   - Is there any loose or missing insulation?
   - Are there any leaks?

11. Finish up the job.

   - Clean up the area.
   - Remove all tools.
   - Secure the area.
   - Complete a service report for the owner.
   - Present a bill for services.

ON THE JOB

Your toolbox should include an assortment of boiler drain valves, air vents, vacuum relief valves, replacement control sensors, a voltmeter, a sensor substitution box, a hand or power air pump, a level, a long extension cord, and a trouble lamp or circuit tester.

Just as in draindown systems, gear your troubleshooting strategy to the problems being encountered by the owners. Repair any known physical damage first. Check out draindown action. Check out the pump and controller. Run the performance tests in steps 5 and 8 if needed. While the performance tests are running, perform steps 1 and 2. If performance is unsatisfactory, perform steps 3 and 4.

Any problems should now be corrected. Proceed to finish the job. If the collectors will not fill, suspect pressure buildup from steam or from faulty draining before you start considering pump changes. There is a good chance that increasing the flow rate or adding an air bleed line or vacuum relief and air vent valves will eliminate the problem.

If the system is not draining fully, get up on the roof with a level and torch and check the vertical pitch of the piping.

WHAT YOU SHOULD HAVE LEARNED

1. Drainback systems offer positive freeze protection in virtually all climates.
2. Even though there are a number of drainback system designs, they all have one thing in common. There is room in the solar storage tank or in the drainback tank to hold the water from the collector array when the pump is off.
3. The collector loop pump in a drainback system usually needs to be more powerful than the pump in a draindown or pumped

service-water system to overcome the vertical head that is created every time the loop drains back.

4. Compressed air drainback systems are subject to catastrophic failure in the case of a leak plus a power failure. They should be modified or drained in freezing weather.

5. Exchanger coil drainback systems that do not have a separate solar storage tank with a thermosiphon loop or a pump on the exchanger coil loop generally do not show high solar collection efficiencies.

6. The pumped secondary loop drainback system with a solar storage tank offers good efficiency and reliability, as does the State Industries supply-side heat-exchanger system.

7. A collector vacuum relief valve and a drainback tank air vent, or an air transfer line from the upper collector manifold to the air space in the load-side exchanger tank, is recommended for positive drainback.

8. The collector loop of the drainback system is controlled with a differential controller, two sensors, and a high-limit switch. A second controller of the same type can be used for the exchanger coil loop, or the two pumps can be wired in parallel from the collector loop controller.

9. Drainback systems are sized at 20 to 25 ft^2 per occupant for a 40 to 60% solar load. A flow rate of 0.02 to 0.03 gpm/ft^2 of collector is used, and 1.5 to 2.0 gal of storage per square foot of collector is suitable. The exchanger tank must also contain room to hold the water contained in the collector array when the system has drained back.

10. Glass-lined tanks with magnesium anodes may be used. Aluminum water passages should be avoided. The collector loop water should be conditioned. All piping and fittings should be copper and brass except for the air vent line, where plastic piping may be used.

11. The major advantage of a drainback system is that it contains a limited number of devices that can fail in a catastrophic mode. Therefore, it is very reliable.

12. There are four separate tests that a serviceperson should perform on a drainback system.

    • Drainback operation.
    • Heat collection performance.
    • Pump and controller operation.
    • Heat-transfer performance between the load-side exchanger tank and solar storage.

13. Heat-exchanger effectiveness is most important. The exchanger coil must be at least 50% effective for the drainback system to be efficient.

CHAPTER

6

Liquid Heat-Exchanged Systems

Liquid Heat-Exchanged System   A solar water-heater system that uses a *heat exchanger* containing a *collector fluid* other than water in a *closed loop* to transfer the solar energy from the collectors to a solar tank containing potable water.

As previously explained, a heat exchanger is a device, usually made of metal, that transfers heat across an internal wall that separates two fluids, two gases, or a fluid and a gas. The use of a heat exchanger allows the heat to be transferred without mixing the fluids or gases together. Figure 6.1 illustrates such a loop in its simplest form. The pump circulates the collector fluid up to the bottom of the collectors, where it picks up heat generated by the sun. The warm liquid passes out the top of the collectors and down through the heat exchanger which, in this case, is a coil located in the bottom of the solar storage tank. The heated fluid warms the wall of the heat exchanger, and the heat passes into the storage tank.

The **collector fluid** is the fluid that circulates through the collectors and the heat exchanger. It picks up the solar heat from the collectors and brings it to the heat exchanger. The collector fluid is also commonly known as the **heat transfer fluid.**

A **closed loop** is a piping system that is both closed to the atmosphere and separated from the storage water. This closed loop connects the collectors and the heat exchanger in a single loop of piping that recirculates the collector fluid again and again between the collectors and heat exchanger with a pump. Closed systems are completely filled with fluid and sealed. They are only opened to perform maintenance and service on the loop or the collector fluid.

Note the difference between this system and the secondary pumped exchanger coil loop shown in Figure 5.4. Now the heat transfer fluid is located within the exchanger coil instead of in the tank, so the system can be sealed; only a small amount of fluid is needed to run the system, and the solar storage and heat-exchanger tanks can be combined.

The Collector Fluids

Water is not usually used as a collector fluid in liquid heat-exchanged systems. Although water is the best collector fluid in terms of its ability to transfer heat, it freezes at 32°F and boils at 212°F. It can promote corrosion and it can build up high temperatures and pressures in the collector loop if the system is **stagnating.**

Stagnation is a condition in which the system is at rest and the fluid is not circulating. Stagnation can occur on a hot summer day because the high-limit switch has turned off the pump or because the pump is switched off or has failed. During stagnation, very high temperatures can be reached in the collector loop. These temperatures have been known to reach well over 300°F. At these temperatures, very high pressures may be generated in a water-filled closed collector loop. Collector loops are not often designed to withstand such high pressures.

There are five common classes of heat transfer fluids used in closed-loop, liquid heat-exchanged systems.

- Ethylene glycol/water mixtures.
- Propylene glycol/water mixtures.
- Hydrocarbon oils.
- Synthetic olefinic oils.
- Silicone oils.

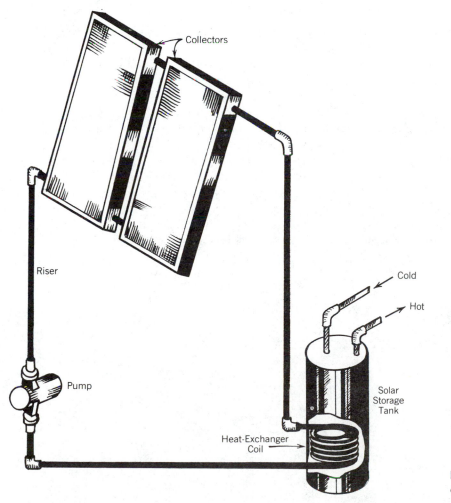

Figure 6.1 A simple heat-exchanged, closed-loop system.

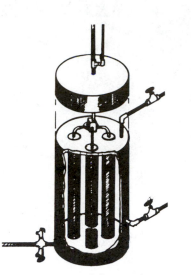

Figure 6.2 An internal-flue exchanger.

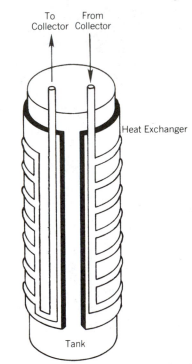

Figure 6.3 A plate-around-tank exchanger.

Each of these fluids is very adequate for the task, but each has its advantages and disadvantages.

All of these fluids differ from water in that they remain liquid well below 0°F and well above 300°F. That is, they are both freeze resistant and relatively insensitive to stagnation temperatures. However, their stability at high temperatures may create a fire hazard if combustible materials are located in or directly under the collector box.

The choice of collector fluid strongly influences the system design. Fluids cannot be substituted for each other unless the system has been designed for the replacement fluid. These fluids will be covered in depth in Chapters 10 and 15.

The Heat Exchanger

Many different types of heat exchangers are used in solar systems. Some commmon examples are

- The internal-flued tank.
- The plate around tank.
- The tank around tank.
- The cascade over tank.
- The coil in tank.
- The external separate exchanger.

Some of these exchangers have very high effectiveness. Others do not. Although many different styles of exchangers were used in early solar systems, only a few styles are still marketed today. Because the service-person will be called on to service all six styles of exchangers, he or she must be able to recognize all of them.

Figure 6.2 shows an internal-flued tank from State Industries. This is the same tank shown in Figure 6.5. State Industries recommends this tank for their drainback system only.

Figure 6.3 shows a plate-around-tank system. This exchanger has low effectiveness and is no longer marketed. Figure 6.4, which illustrates the exchanger plate as a flat surface, shows why this exchanger has low efficiency. Only the heat transfer fluid circulates through the exchanger, and the exchange is accomplished across the two walls created by wrapping the exchanger around the tank. This type of heat exchange is very slow. If the exchanger was an internal part of the tank, as in Figure 5.9, the

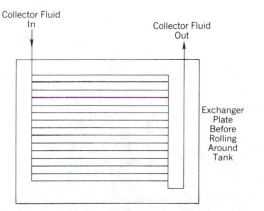

Figure 6.4 In a single-circuit, plate-around-tank exchanger, the heat exchange is very slow.

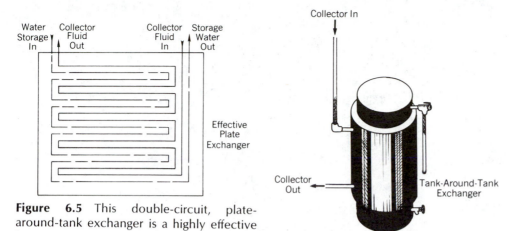

Figure 6.5 This double-circuit, plate-around-tank exchanger is a highly effective heat exchanger.

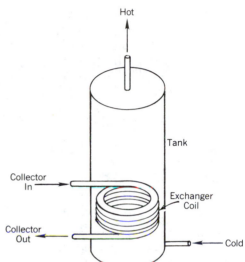

Figure 6.6 A concentric tank-around-tank exchanger.

exchange rate would be far better, because a single wall exists between the two liquids.

Figure 6.5 shows a more effective plate-around-tank system that has two liquid circuits flowing side by side within it. One circuit flows between the collector and the heat exchanger; the other circuit pumps the storage water through the panel. The heat transfer takes place within the plate across the fin between the channels as well as through the two walls. This exchanger has a high effectiveness.

Figure 6.6 shows a concentric tank-around-tank system. Again, two walls separate the fluids. The system had limited effectiveness and is no longer marketed.

Figure 6.7 shows a coil-in-tank system. These tanks, marketed by Ford Industries, A. O. Smith, and the Vaughn Corporation, have become one of the standard designs of the industry.

Figures 6.8 to 6.10 show separate shell-and-tube exchangers. Figure 6.8 shows an exchanger marketed by the Wolverine Division of UOP, Inc. Figure 6.9 shows an exchanger marketed by Doucette Industries,

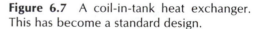

Figure 6.7 A coil-in-tank heat exchanger. This has become a standard design.

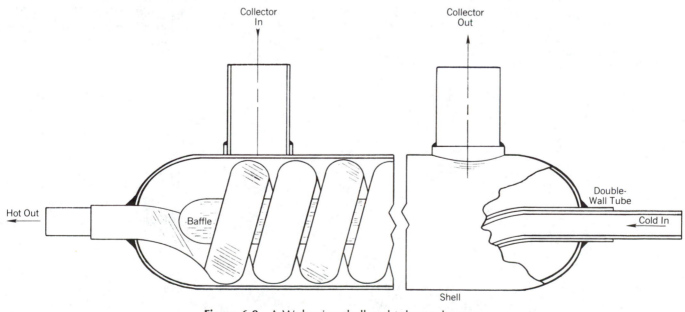

Figure 6.8 A Wolverine shell-and-tube exchanger.

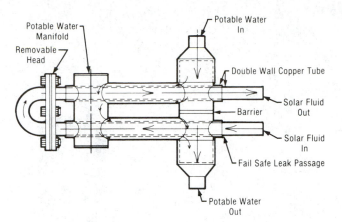

Figure 6.9 A Doucette Industry shell-and-tube exchanger.

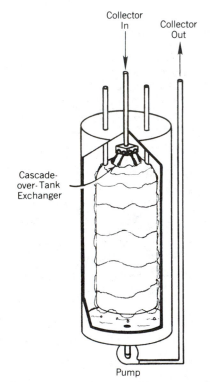

Figure 6.11 A cascade-over-tank exchanger.

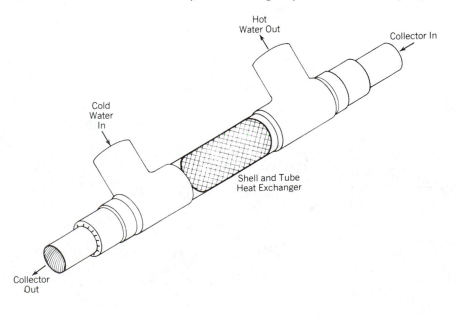

Figure 6.10 A Noranda Metal Industries shell-and-tube exchanger.

and Figure 6.10 shows one marketed by Noranda Metal Industries Inc. These shell-and-tube exchangers have also become industry standards.

Figure 6.11 shows a cascade-over-tank exchanger from Rudd Industries. This design was ineffective and is no longer available.

There are other variations of these heat exchangers, but they all fall into these general classes. Some municipalities require double-walled heat exchangers to prevent cross-contamination of domestic water with heat transfer fluid in the event of a heat exchanger leak. Heat exchangers will be explored in more depth in Chapters 10 and 15.

In this chapter you will study the coil-in-tank system, the separate heat-exchanger system, and how to convert the other systems to one of these two types.

The Coil-In-Tank System

Figure 6.12 shows a typical coil-in-tank system. The system is set up with an auxiliary water heater. It could be used without an auxiliary heater if desired.

Starting at the discharge end of the pump, the collector loop consists of the pump, a flow control valve, the manifolds, the collectors, the coil-in-tank heat exchanger, and a group of valves and gauges that are used to control and monitor the system. This group of valves and gauges,

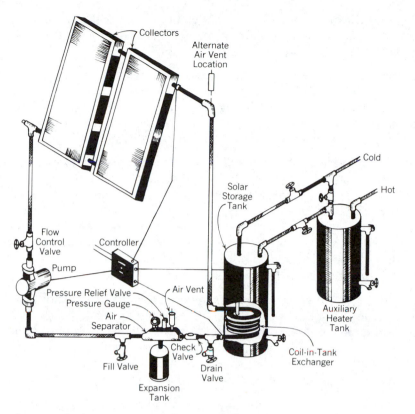

Figure 6.12   A typical coil-in-tank system.

along with the pump and the controller, is known as a **solar module.** A solar module is a preplumbed, panel-mounted system mechanical package that contains the pump, the controller, and the associated piping and valves. It may also contain a heat exchanger. The solar module usually comes mounted either on a separate panel or attached to the side of the tank. Some system designers also include an automatic air vent at the system high point.

The storage loop consists of the water preheater tank and its associated valves and parts. The collector liquid is separated from the service water by the heat exchanger.

The Solar Module

The function of the solar module is to monitor the collector loop, insure that it operates safely, allow the fluid to expand when it is heated, and eliminate any air from the system. Let us look at each of the parts of the module and their functions, using Figure 6.12 as an example.

Fill/Drain Valves   The fill/drain valves are common boiler drain valves that are used to fill the system at start-up and drain the system for maintenance.

Check Valve   The check valve is used to prevent the collector liquid from reverse thermosiphoning. It may be located between the drain and fill valves, where it also acts as a check valve during filling, or it may be located on the hot downcomer from the collectors. When the check valve is located in the downcomer, a gate valve should be placed in the pump leg between the drain and fill valves.

Air Separator or Air Eliminator   The air separator is used to remove any air in the collector loop. The air flows across a series of baffles, which helps the air bubble out of the liquid stream to the automatic air vent.

Many times the air eliminator is a handy connector for the other small items in the solar module, because it generally has a number of threaded openings that can be used to mount the various parts.

Automatic Air Vent or Float Vent   The automatic air vent will pass air but not liquid. Air bubbling out of the collector fluid in the air separator is eliminated from the system through the automatic air vent. This valve is similar to the valves used in the other system types shown in Chapters 2 to 5, but it is designed to operate at lower pressures.

Pressure Relief Valve   Most liquid heat-exchanged collector loops operate under low pressures, around 10 to 30 psi. This is in contrast to thermosiphon, pumped service-water, and draindown systems, which operate at service-water pressure. The pressure relief valve, usually rated between 30 and 60 psi, protects the collector loop against high pressures that would damage the loop. In some systems utilizing a mixture of water and antifreeze the designer may have utilized a 75- to 150-psi valve. Check the rating carefully before replacing the valve.

Pressure Gauge   This gauge monitors the pressure on the collector loop. Typically, the loop is set up at 10 to 20 psi when it is filled. The loop pressure rises as the fluid warms and drops as the fluid cools due to the expansion and contraction of the fluid. The gauge offers a quick check on the condition of the collector loop, because low pressure generally means that a leak has developed and fluid has been lost.

Expansion Tank   The expansion tank leaves room in the system for the fluid to expand with temperature. All fluids become less dense and occupy more space as their temperature rises. If there were no room in a solar system for the fluid to expand, the system could rupture from extremely high pressure. In the thermosiphon, pumped service-water, and draindown systems, fluid expansion was handled through the cold water connection. Any expansion merely was taken up by having the liquid expand back into the city water system. In the drainback system the fluid expands into the air void in the solar storage tank.

NOTE   The local code may require a *backflow preventer* on the house water line or on the solar system. Backflow preventers are valves that prevent water from the house from returning to the city lines. When a backflow preventer is placed in the lines, there may be no room for water expansion in the solar storage tank. If you are encountering high-pressure buildup in the solar storage tank, check for the existence of a backflow preventer. The addition of an expansion tank to the system will solve the problem.

Extrol diaphragm-containing tanks are almost universally used in closed-loop residential solar systems. The tank contains a rubber bladder with an air space on one side and the fluid on the other side. Air is compressible, so when the fluid expands, it compresses the air.

Flow Control Valve or Throttle Valve   This valve performs the same function as the throttle valve in the pumped service-water, draindown, and drainback systems.

The External Heat-Exchanger System

Figure 6.13 shows a typical separate heat-exchanger system. Again, the system is set up with an auxiliary water heater, but it may also be used alone. The collector loop is similar to the coil-in-tank system except that

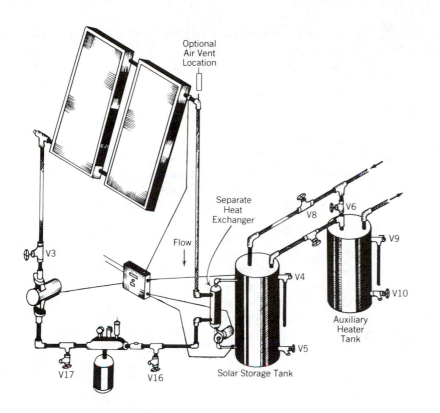

Optional
Air Vent
Location

Separate
Heat
Exchanger

Flow

V3

V8   V6

V9

V4

V10

Auxiliary
Heater
Tank

V5

V17   V16   Solar Storage Tank

Figure 6.13 A typical separate heat-exchanger system.

the heat exchanger is now external to the tank. The same parts are needed in the solar module, and the parts serve the same functions.

This system contains a second pump in a flow loop known as the exchanger-to-storage loop. In this loop the cold water from near the bottom of the solar storage tank flows to the pump, through the exchanger, and back to the upper part of the storage tank. The energy from the collector loop is transferred across the wall of the heat exchanger to the water flowing through the secondary loop and is then passed into the storage tank.

Note that the liquids in the exchanger flow in opposite directions. This configuration is known as **counterflow.**

The counterflow configuration is important in a system of this type because it removes heat more efficiently than a configuration where the two fluids flow in the same direction. In the counterflow mode the water exiting from the secondary side of the exchanger sees the hottest collector fluid as it exits. Therefore, it can be heated to a higher temperature than if the two fluids were flowing in the same direction.

Figure 6.13 shows the collector fluid passing through the shell and the storage water passing through the internal exchanger coil. In some exchangers this piping is reversed. Follow the heat-exchanger manufacturer's directions when piping the exchanger. In all cases use the counterflow configuration.

Controlling the Liquid Heat-Exchanged System

Both systems use the same type of control system. However, the controller for the separate heat-exchanger system must have two outlets for wiring in two pumps: a pump for the collector loop and a pump for the exchanger-to-storage loop.

A differential temperature control system is used to turn the pumps off and on. The system contains three sensors.

- A collector sensor, mounted at the top collector manifold.
- A tank sensor, mounted where the riser to the collectors leaves the tank.
- A high-limit sensor, mounted at the top of the tank.

Typically, a controller should be selected that will turn the pump on when the collectors are 8 to 20°F hotter than the storage tank and turn the pump off when the temperature differential between the collectors and the tank drops to 2.5 to 5°F. The high-limit switch should activate below 180°F and override the pump turn-on circuit so that the tank cannot overheat.

Some systems are designed without the high-limit sensor, because disabling the pump at high temperatures may cause water-containing heat transfer fluids to boil when the fluid is stagnating in the collectors. Whether or not this condition is satisfactory is a question of the system designer's goals. Boiling can result in the formation of deposits within the collector, pump damage, and system rupture.

In liquid heat-exchanged systems either the solar storage tank temperature and pressure relief valve opens when the tank overheats or the pressure relief valve on the collector loop opens when the fluid in the collector exceeds the system pressure limit. In the first case scalding hot water at 160 to 180°F is discharged. In the second case the loss of fluid from the collector loop can create a vacuum within the system that prevents the circulation of the fluid. No further solar heat can be collected until collector fluid is added.

If the discharge of scalding hot water can be tolerated, eliminating the high-limit sensor is acceptable. A more satisfactory solution is the redesign of the system to accommodate the high solar input that is causing the problem or by adding a heat rejection exchanger to the system.

Proportional Controllers

A second type of controller that can make the pump run at varying speeds can also be chosen for these systems. This controller is known as a **proportional controller.** A proportional controller delivers electricity to the pump in varying amounts proportional to the temperature difference between the collectors and the tank. This causes the pump to run faster or slower in proportion to the temperature differential. This controller will turn on when the collectors are only 3 to 5°F hotter than the tank and run the pump at a low speed. The low speed causes a low flow rate through the collectors that results in the collectors staying hotter than the tank when the day is hazy or overcast. A fast flow rate at this low temperature differential could cause the collector temperature to drop low enough to shut down the pump.

As the temperature differential between the collector and the tank rises, the controller delivers more and more electricity to the pump until, at 8 to 16°F temperature difference, the pump is running at full speed. In a proportional controller the pump turn-off temperature may also be lowered to a range of 2 to 3°F instead of 5°F so that the pump can run longer.

The advantage of a proportional controller is that the system runs longer and thus picks up more energy than a system controlled with a simple on/off controller. Its use is especially indicated in an area that has cloudy or quickly varying amounts of sunshine through the day.

The exchanger-to-storage loop pump turns on and off at the same time as the collector loop, so only one controller is needed. The controller should contain one or two controlled outlets for the pump or pumps with a rating of 10 A, and the controller should operate from a 120-V alternating current circuit fused at 15 A.

Proportional controllers are also used in other types of systems but to a lesser extent than in heat-exchanged systems. They cannot be used on drainback systems.

CAUTION If the pump is $\frac{1}{3}$ hp or larger, check as to whether a 10-A controller is large enough to handle both start-up and continuous running of the pumps.

Expected Performance

The performance of a liquid heat-exchanged system can vary tremendously, based on how well the heat exchanger is matched to the collectors. Heat-exchanger effectiveness is very important. Let us see why.

Figure 6.14 shows a graph of typical collector performance on a 65°F day with a clear sky between 10 A.M. and 2 P.M. The vertical axis shows the efficiency of collection, and the horizontal axis shows the temperature of the collector fluid entering the collector. As the collector fluid temperature rises above the temperature of the outdoor air, the efficiency of the collector drops and larger losses of heat to the air occur from the collector.

- At 95°F, the collector is 68% efficient.
- At 185°F, the collector is 48% efficient.

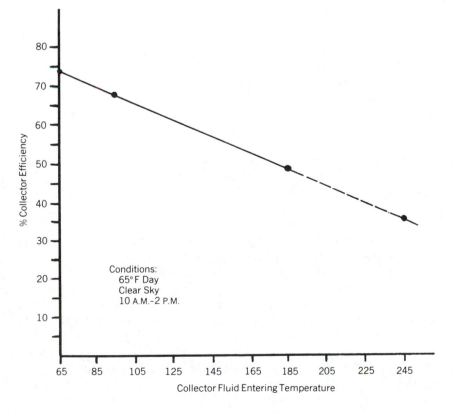

Figure 6.14 A graph of typical collector performance.

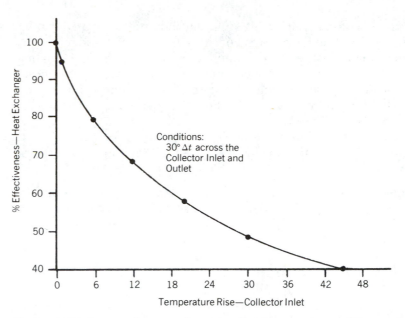

Figure 6.15 A heat-exchanger effectiveness curve.

Figure 6.15 shows a heat-exchanger effectiveness curve. The vertical axis represents the effectiveness of the heat exchanger under the conditions cited in Figure 6.14 on a 65°F day with a clear sky between 10 A.M. and 2 P.M.

This vertical axis represents the effect of the heat exchanger on the collector fluid temperature at the inlet when a 30°F temperature differential exists across the collector. If the exchanger were 100% effective, all the heat supplied by the collector to the exchanger would be removed and the collector fluid would be cooled to the temperature of the water entering from the storage tank before the fluid returned to the collector. This condition would be the same condition that would be experienced if there were no heat exchanger in the system and the storage water were circulating in the collector.

However, with heat exchangers always being less than 100% effective, not all the heat is removed from the collector fluid. Thus, the collector fluid returns to the collector at a higher temperature. The amount of temperature rise caused by the presence of the heat exchanger is a function of the exchanger's effectiveness. Figure 6.15 shows that at 80% effectiveness, the fluid would be 6°F hotter; that at 50% effectiveness, the fluid would be 30°F hotter; and that at 40% effectiveness, the collector fluid would be 45°F hotter.

This hotter collector fluid inlet temperature has an adverse effect on the performance of the collector under a given set of conditions. The collector runs hotter and its efficiency drops off. Figure 6.16 shows how much drop-off in collector performance is seen due to the heat exchanger's level of effectiveness. The vertical axis again represents the efficiency of the collector. Three horizontal axes are shown: one at 100%E for the exchanger; one at 70%E; and one at 40%E.

A collector running at 165°F with no heat exchanger (100%E) would run at 177°F with a 70%E exchanger and at 210°F with a 40%E exchanger. The respective collector efficiencies of these systems would be 49%, 47%, and 41%. Therefore, a 40% effective heat exchanger would result in an 8% system efficiency loss with the system running under the conditions shown.

Remember that the effectiveness of the exchanger varies up and down with the amount of heat being sent to it. Under hazy skies and earlier or

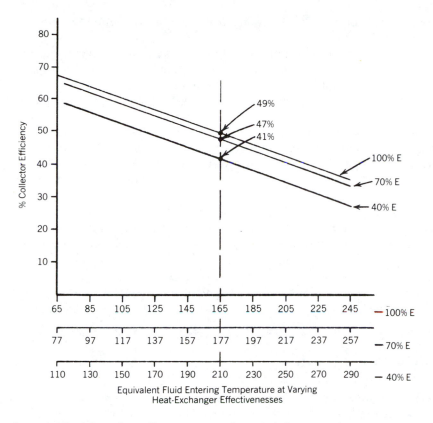

Figure 6.16 How exchanger effectiveness lowers collector performance.

later in the day, the effectiveness of any of these exchangers would be higher, because less heat is being supplied to the exchanger and the exchanger is being called on to do less work.

Generally, systems containing heat exchangers with less than 50% effectiveness should be considered for modification. Systems containing a 50% effective heat exchanger will suffer about a 5% efficiency penalty because of the heat exchanger.

The lower system efficiency is offset by the ability of the liquid heat-exchanged system to run year round in the coldest climates when it is properly designed and installed.

Sizing Parameters

The rule of 20 to 25 ft² per occupant for obtaining 40 to 60% of the needed hot water holds true for good liquid heat-exchanged systems. Again, the storage tank should be sized at 1.2 to 2 gal/ft² of collector surface. However, the flow rates will depend on the heat transfer fluid that is chosen. Let us examine why.

The ability of a fluid to transfer heat depends on the **density** of the fluid and the **specific heat** of the fluid. *Density* is a measure of weight per unit of volume. For instance, 1 gal of water weighs 8.33 lb, but 1 gal of silicone fluid weighs only 8 lb. The silicone fluid is about 4% lighter.

Specific heat is a measure of how much energy it takes to heat a given mass of the fluid by 1°F. In the English system of measurement specific heat is reported as the number of British thermal units it takes to heat or cool 1 lb of fluid by 1°F. Water has a specific heat of 1. It takes 1 Btu to raise or lower the temperature of 1 lb of water by 1°F.

The silicone fluid most commonly used in a solar collection loop has a specific heat of 0.40. It takes 0.4 Btu to raise or lower the temperature of 1 lb of this fluid by 1°F.

If the pump moved the same weight or mass of fluid through the system in a given period of time and the fluid gained the same in temperature, the water would pick up $2\frac{1}{2}$ times as much heat. That was determined by dividing the specific heat of water by the specific heat of the silicone.

It follows that to move the same amount of heat with each fluid, assuming the same temperature rise, it would be necessary to pump the silicone fluid $2\frac{1}{2}$ times as fast as the water. Table 6.1 gives the different flow rates that should be used with the different classes of heat transfer fluids.

The collector pump needs to be sized to pump only against the friction losses that are incurred in the piping. There is no vertical head encountered in a closed liquid heat-exchanged system. Again, pump size will vary, depending on the collector fluid chosen and the rate at which this fluid must be pumped. Generally, one pump size will work for water, ethylene glycol/water, and propylene glycol/water systems and another, larger pump size will work for hydrocarbon oils, olefinic oils, and silicone oils.

Other Considerations

The storage tank constructions for the two systems can differ. The coil-in-tank system has a tank with an internal heat exchanger that is about equivalent in cost to the tank itself. This tank should be chosen for long life, because to lose the tank is to lose the heat exchanger. Typically, the coil-in-tank heat exchanger is a large coil of finned copper piping. When a large mass of copper is placed in a steel tank or a glass-lined steel tank, the tank life tends to be shortened through corrosion of the steel.

The industry has been using a spun concrete-lined tank in most coil-in-tank systems. This tank is called a **stone-lined tank.** Stone-lined tanks seem to have very long lives. Although they are more expensive, heavy, and cumbersome to install, they represent the best tank for the coil-in-tank system.

The external heat-exchanger system keeps the tank and the exchanger separate, so the loss of the tank does not affect the heat exchanger. In these systems the traditional glass-lined tank with its 7- to 15-year life represents a good choice. The glass-lined tank should contain a magnesium anode rod to prevent premature tank failure.

When the oil collector fluids are used, aluminum liquid passageways may be used in the collectors. Aluminum passageways should be avoided with any collector liquids that contain water. The piping and fit-

Table 6.1

FLOW RATES FOR DIFFERENT COLLECTOR FLUIDS

COLLECTOR FLUID	FLOW RATE (GPM/FT² OF COLLECTOR)
Water	0.01–0.03
Ethylene glycol/water (50:50 ratio)	0.013–0.025
Propylene glycol/water (50:50 ratio)	0.013–0.025
Typical hydrocarbon oil	0.030–0.050
Typical olefinic oil	0.030–0.050
Typical silicone oil	0.030–0.050

tings should all be brass and copper in high-performance systems. Some of the systems constructed in the early 1970s used low-efficiency collectors and plastic piping. These should be maintained as presently constructed.

The manner in which the piping joints are constructed is very important in collector loops that use organic liquids as collector heat transfer fluids. These fluids are more prone to leakage, particularly through threaded joints. Generally, threaded joints should be avoided wherever possible in favor of tin/antimony soldered joints. When threaded joints are used, they must be sealed with a compound recommended by the fluid manufacturer.

Some manufacturers are using flare fittings and compression fittings with these fluids. These are not as reliable as soldered joints. In all cases, unless there are joint problems, you should keep using the type of joints used in the original system.

The size of the piping will depend on *the length of the runs, the size of the pump, the number of collectors, the flow rate, and the collector fluid used.* Choice becomes more critical in liquid heat-exchanged systems than it is in pumped service-water, draindown, or drainback systems.

These guidelines should be followed when the total piping run is less than 100 ft.

- In water/glycol systems $\frac{1}{2}$-in. pipe can be used for up to 50 ft^2 of collector area, $\frac{3}{4}$-in. pipe should be used for up to 120 ft^2 of collector area, and 1-in. pipe should be used for larger collector areas.
- In oil systems, where higher-viscosity liquids and faster flow rates exist, larger pipe sizes should be used; $\frac{1}{2}$-in. pipe generally will work well with one collector; $\frac{3}{4}$-in. pipe should be used for two to three collectors, and 1-in. pipe is used for four or more collectors. The use of this larger piping allows pumping horsepower to be kept to a minimum.

At any rate, through calculations the designer of the system should have established an effective pipe size, and the serviceperson should not consider making a change unless low flow rates are hurting the performance of the system. Never increase the flow rate over the manufacturer's maximum flow rate recommendation.

In choosing replacement parts for the mechanical systems, the serviceperson must be certain that the composition materials, pump gaskets, valve O-rings, expansion tank bladder, and other rubber and plastic parts of the system are compatible with the collector heat transfer fluid chosen. Many rubbers and plastics have poor life when exposed to various oils and glycols.

The tank, the heat exchanger and, if possible, the associated piping should all be insulated to R-10 or better with insulation that is appropriate for the particular location of the part being insulated.

All equipment must be securely fastened or mounted on adequate panels or foundations.

Retrofitting Systems to Overcome Low Exchanger Effectiveness

There are many liquid heat-exchanged solar water-heater systems with low exchanger effectiveness installed around the United States. These systems can be modified with an exchanger retrofit.

CAUTION Some of these systems contain aluminum liquid passages in the collectors, aluminum liquid passages in the present heat exchanger, plastic piping, and a water-containing glycol collector fluid in the collector loop. Such systems are not good candidates for retrofit because the exchangers available for retrofit are copper or copper and steel. Placing a copper or a copper/steel exchanger in the system could set up a corrosion condition that could rapidly degrade the aluminum in the system. This retrofit procedure is not recommended for such systems.

The retrofit consists of installing an external heat exchanger to boost the present heat exchanger's effectiveness. The job will require

- An external heat exchanger.
- A small circulating pump.
- A deep-probe temperature and pressure relief valve.
- Miscellaneous piping and fittings.

Figure 6.17 shows a tank-around-tank system that has been retrofitted with a separate exchanger while keeping the tank-around-tank exchanger in the loop.

Hot collector fluid enters the top of the shell-and-tube exchanger, exits at the bottom, goes to the top of the tank-around-tank exchanger, and exits at the bottom of the tank-around-tank exchanger to the collector pump. The two exchangers are now connected in series.

A small brass or stainless steel circulator pump circulates the storage water up through the shell-and-tube exchanger and out the top and returns it heated to the tank. Heat exchange takes place through both

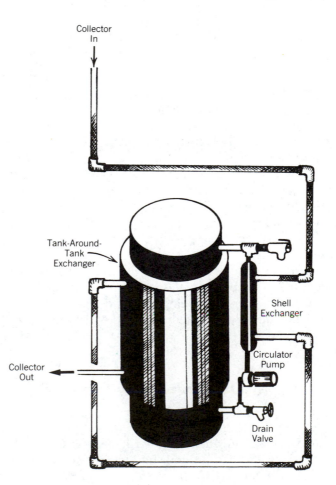

Figure 6.17 A tank-around-tank system retrofitted with a separate exchanger running in series with the old exchanger tank.

exchangers. Follow the exchanger manufacturer's recommendations for proper liquid flow configuration.

Like all water heaters, the inner tank, which contains the service water, has a safety valve near the top and a drain valve near the bottom.

If the proper tank outlets are not available, the temperature and pressure safety valve and the drain valve may have to be moved to accommodate the exchanger loop. They can be placed on the end of tee fittings used to pipe in the pump and exchanger. There may be a problem with the safety valve. Plumbing codes generally call for a valve with a probe long enough to reach into the tank. If the retrofit is such that this is not possible, the safety valve will have to be relocated to the hot water service pipe exiting at the top of the tank. A special close-coupled tee fitting is manufactured for this purpose.

This same retrofit can be used with a plate-around-tank system or with a coil-in-tank system that was installed with too small an exchanger coil.

Advantages

The liquid heat-exchanged system has these advantages.

- Positive freeze protection.
- Closed-loop operation, which allows the use of a nonfreezing collector fluid.
- No motorized valves or vacuum breaker valves to fail.
- The potential for longer system life than water-containing collector loop systems.

Disadvantages

The liquid heat-exchanged system has these disadvantages.

- The system is more complex than thermosiphon, pumped service-water, or drainback systems.
- The system needs careful design to insure that all the components match up.
- *Some collector liquids* will need periodic changeout on a biannual basis.
- There is a loss in collector efficiency as the system runs at a higher temperature.

How to Test Performance

Performance testing of the liquid heat-exchanged system consists of

- Testing solar heat collection operation.
- Testing pump/controller operation.
- Testing collector liquid condition.

Testing Solar Heat Collection Operation

In the drainback tests, you tested heat collection and heat transfer to storage separately. In closed-loop systems the tests are carried out as one test. Both energy transfers are occurring at the same time. Refer to Fig-

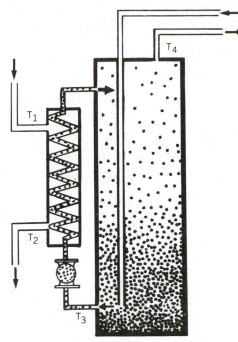

Figure 6.18 Temperature measurement locations in a separate exchanger system.

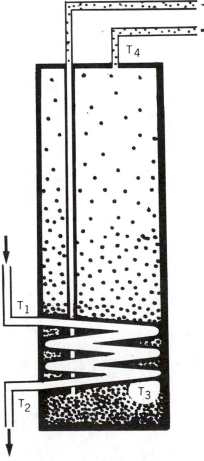

Figure 6.19 Temperature measurement locations in a coil-in-tank exchanger system.

ures 6.18 and 6.19 to see where the temperatures are taken. Testing proceeds as follows.

- The tests should be carried out between 10 A.M. and 2 P.M. on a clear day in direct sunlight.
- The hot water service must be shut off so that no hot water can be drawn from the system.
- The controller is turned off until the temperature gauges are placed and read.
- A temperature gauge is placed on the hot downcomer where it enters the heat exchanger (T1).
- A temperature gauge is placed on the cold riser where it exits the heat exchanger (T2).
- For tanks like the tank shown in Figure 6.19, a temperature gauge is placed on the wall of the tank about 1 ft up from the bottom (T3). **NOTE:** There may be some problems in placing this gauge on an insulated tank. If necessary, place it on the drain valve as close to the tank as possible.
- For tanks like the tank in Figure 6.18, place the gauge on the pipe leading to the heat exchanger (lower left).
- A temperature gauge is placed on the tank wall about 1 ft from the top or on the pipe returning to the tank from the heat exchanger (T4). **NOTE:** There may be some problems in placing this gauge on an insulated tank. If necessary, place it on the exit of the hot water pipe from the tank as close to the tank as possible.
- The gauges are allowed to come to equilibrium, which should take 3 to 4 minutes.
- The gauges are read and the readings are recorded on a chart like the one in Table 6.2.
- The pump controller is turned to automatic.
- If the pump starts within 2 to 3 minutes, proceed with the test. (**NOTE:** In the case of the external heat-exchanger system both pumps should start and stop at the same time.)
- If the pump does not start within 2 to 3 minutes, turn the controller switch to manual.
- If the pump starts on manual but will not run on automatic, four problems are possible.

  1. The controller is faulty.
  2. One or both of the sensors is faulty.
  3. The temperature differential between the collectors and the tank is not high enough to activate the controller (usually 8 to 20°F for off/on controllers and 3 to 5°F for proportional controllers).
  4. The tank temperature is high enough to activate the high-limit sensor, if one is present.

- If the pump will run on manual, proceed with the test and check temperature rise with time. When the test is completed, go to the pump/controller test procedure.
- If the pump does not run on manual, terminate the test and go to the pump/controller test procedure.

Proceeding with the Test   Assuming that the pump will run, continue the temperature testing with the pump on manual, as the preceding procedure indicates. If the pump will not run, you will return to this point, after you have corrected the pump or controller problem.

- Allow 30 minutes to 1 hr to pass with the pump operating.
- Record the readings in a chart like the one in Table 6.2.

Filling Out the Chart

1. Subtract the riser temperature from the downcomer temperature and place the result under T1 − T2. Do this for both readings.
2. Add the lower tank temperature to the upper tank temperature, divide the sum by 2, and place the result under (T3 + T4)/2.
3. Subtract the first riser temperature from the second riser temperature and record the difference under temperature rise.
4. Subtract the first downcomer temperature from the second downcomer temperature and record the difference under temperature rise.
5. Subtract the second T1 − T2 calculation from the first T1 − T2 calculation and place the result under temperature rise.
6. Subtract the second (T3 + T4)/2 calculation from the first (T3 + T4)/2 calculation and place the result under temperature rise.

The Meanings of the Readings and the Calculations

The test that you have just carried out has provided all the temperature readings needed to test the heat collection in the collector loop and the heat transfer from the exchanger coil to the storage tank. The configuration of both the coil-in-tank and the external heat-exchanger systems are such that the tests cannot be separated. The heat exchange from collector-to-exchanger and exchanger-to-storage systems are going on simultaneously.

Step 1 provides the temperature drop across both the collectors and the heat-exchanger coil. When the collector loop is at rest before the pump is started, T1 and T2 will read close to the same. These temperatures will also be close to the average tank temperature.

After the system is run for a brief period, a temperature differential will

Table 6.2

TEMPERATURE CHART—LIQUID HEAT-EXCHANGED SYSTEMS

TIME READ	TEMPERATURE DOWNCOMER T1 (°F)	TEMPERATURE RISER T2 (°F)	TEMPERATURE COIL ENTRANCE (LOWER TANK) T3 (°F)	TEMPERATURE COIL EXIT (UPPER TANK) T4 (°F)	T1 − T2	TANK AVERAGE TEMPERATURE (T3 + T4)/2 (°F)
°F Rise						

develop across the collectors and the exchanger coil. Under these test conditions, that difference should be around 10 to 20°F. Expect to see this 10 to 20°F difference in the second reading after you have run the system for 30 minutes to 1 hour.

Step 2 provides the average tank temperature when T3 is added to T4 and the sum is divided by 2.

Step 3 provides the rise in temperature at the exit of the exchanger. This temperature is identical, barring any line losses, to the collector entrance temperature.

Step 4 provides the rise in temperature at the entrance to the exchanger. Again barring any line losses, this temperature is identical to the collector exit temperature.

Step 5 provides the difference in temperature rise across the exchanger coil and the collector after the system has been run for a specified period of time.

Step 6 provides the average rise in temperature of the storage tank during the period of the test.

Interpreting the Results

A properly operating system should show a difference between the riser and downcomer temperature of 10 to 20°F after the system has been operating for 30 minutes. This determines whether the flow rate is correct. One of four conditions can exist.

1. The temperature difference can be between 10 and 20°F. This would indicate that the flow rate of the system was within the correct range.
2. A less than 3 to 5°F difference can exist or the initial temperature readings do not change. These conditions would indicate either excessively high flow or no flow. The controller cannot keep the pump on with a less than 3 to 5°F temperature differential using an off/on controller or a less than 3°F differential using a proportional controller.
3. The temperature difference can lie between 5 and 10°F. This would indicate too fast a flow rate.
4. The temperature difference can be higher than 20°F. This would indicate too slow a flow rate.

A no-flow condition would result in no energy being collected. Too slow a flow rate would result in decreased collector efficiency. Too fast a flow rate would cause the pump to cycle on and off too often, which will prematurely wear out the pump and can result in inefficient energy collection.

The throttle valve is used to control the flow rate. If the flow rate is too fast, close down the valve slightly to adjust the flow rate so that the temperature differential falls between 10 and 20°F. In making the adjustments give the system about 10 minutes to come to equilibrium between adjustments. If the flow rate is too slow, check the throttle valve to see if it can be opened to increase the flow. Check for any other valves in the collector loop that might be cutting down on the flow of the collector liquid.

This method of checking flow rate ignores the type of heat transfer liquid that may be used in the system. Obviously, higher flow rates are needed for oil-containing systems. Whatever the collector liquid, the system should be designed to keep the temperature differential in the 10 to 20°F range while operating between 10 A.M. and 2 P.M. on a clear day.

If there is a flow meter in the collector loop that can be used to check the flow rate directly, the flow rates should be in the ranges shown in Table 6.1. This is a second check that could easily be performed in that instance. Some systems contain a gate valve between the fill/drain valves, as shown in Figure 6.20. When the system is built in this manner, it is also possible to obtain an approximate flow rate by attaching a flow gauge between the fill/drain valves and closing the gate valve. However, the results of this test can be misleading, because flow gauges need careful installation, the boiler drain valves will add more head for the system to pump against, and many flow meters are calibrated for water and give erroneous readings when used with other fluids. This reading should only be used as a quick check.

The system should experience a minimum temperature rise of 6 to 9°F per hour. The higher the temperature rise, the better the system is operating. If the system does not experience at least a 6 to 9°F per hour temperature rise, it is not operating at a satisfactory efficiency and needs adjusting or repairs.[1]

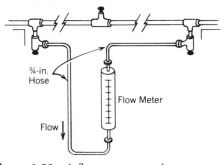

Figure 6.20 A flow meter can be temporarily placed between the drain and fill valves to check the approximate flow rate.

The Pump/Controller Test

For an off/on control system, the pump/controller test is run in the same manner as the pump/controller test in Chapter 3. If the controller is a proportional controller, the test must be run keeping in mind that the controller should turn the pump on at a slow speed at a 3 to 5°F temperature differential.

Testing Heat Transfer Fluid Condition   Each heat transfer fluid has its own set of test parameters, which are fully discussed in Chapter 10. In the case of glycol/water mixtures it is necessary to check the acidity/alkalinity or pH of the solution and to check its specific gravity or its refractive index to determine the freezing point.

In the case of hydrocarbon oils or synthetic oils, the oils must be checked for sludging and viscosity changes. Some oils need to be checked for pH changes.

In the case of silicone fluids no testing is required.

Ethylene glycol and proplyene glycol liquids are essentially pure liquids that vary little from manufacturer to manufacturer except in the inhibitors used in the liquid to prevent degradation. These inhibitors are used up with time. When the inhibitors are used up, the liquid must be changed to prevent system corrosion. Generally, most ethylene glycol or propylene glycol mixtures behave the same with the exception of Union Carbide's Prestone 2, which seems to have a longer life than other ethylene glycol liquids because of a unique inhibitor system.

Hydrocarbon oils vary tremendously from product to product. Each product can behave differently. One hydrocarbon oil should never be substituted for another, because sludging can result if the oils are from different base stocks. Additionally, the solvency effect of the hydrocarbon oils on rubber and plastic parts will vary greatly. Certain hydrocarbon oils being sold for solar systems should be carefully avoided. Others perform extremely well if they are changed out when they sludge.

The serviceperson working with a system containing a hydrocarbon oil

---

[1]The range of 6 to 9°F per hour temperature rise was chosen assuming that a system operating under the test conditions chosen would have a combined heat transfer effectiveness of 25% and would transfer a minimum of 75 Btu/ft^2 per hour to the tank when the incoming solar energy was 300 Btu/ft^2 per hour.

should always be certain that fluid makeup or changeout is done only according to the manufacturer's instructions and with a product recommended by the manufacturer.

The synthetic oils, such as polyolefins, will also vary from manufacturer to manufacturer. Again, they should not be diluted or mixed with other polyolefinic oils or other liquids. Generally, they cause fewer problems than the hydorcarbon oils and have a longer life cycle. Again, the manufacturer's instructions should be carefully followed.

The silicone oils are characterized as a 20 cSt dimethyl polysiloxane fluid. Examples of such fluids are Syltherm 444, Dow Corning 200 Fluid, 20 cSt, and General Electric SF-96, 20 cSt. These fluids are all compatible with each other and may be satisfactorily mixed together when necessary. None of these fluids should require replacement in a residential flat plate collector system over the system's lifetime.

The serviceperson suspecting that the collector liquid should be replaced should contact the manufacturer of the fluid and request that the fluid be tested. A sample of the fluid should be drawn from the collector loop according to the manufacturer's instructions and sent to the manufacturer prior to servicing the loop. The exception to this procedure would be when the manufacturer has provided specific guidelines in the operating manual for the serviceperson to follow.

Some General Guidelines

Some general guidelines for fluid replacement follow.

- Ethylene glycol/water mixtures are expected to last without full inhibitor loss for 2 to 3 years.
- Propylene glycol/water mixtures are expected to last without full inhibitor loss for 1 to 2 years.
- Hydrocarbon oils vary too much to provide a general guideline.
- Synthetic polyolefins should have a minimum 5-year life.
- Silicone fluids should last for the life of the system.

Testing Exchanger-to-Storage Loop Heat Transfer

Because the collector and storage loops are operating simultaneously, the performance of one versus the other cannot be separated. However, the effectiveness of the exchanger and the efficiency of the system can be calculated from the temperatures that were observed and placed in Table 6.2

You will recall from Chapter 5 that the heat-exchanger effectiveness formula is

$$(T1 - T2)/(T1 - T3) \times 100 = \%E$$

The exchanger should be at least 50% effective if the system is to have satisfactory efficiency. Assuming an average 50% collector efficiency and an incoming solar energy of 300 Btu/ft^2 per hour, the system will then place 75 Btu/ft^2 per hour into storage.

Determining How Much Heat Was Transferred If the gallons of water in the solar storage tank are known, the overall heat transfer from collector to storage can also be calculated by using the average tank temperatures.

Referring to Table 6.3, the average tank temperature at any given point in time is (T3 + T4)/2. At the end of $\frac{1}{2}$ hour, the average tank temperature had risen 4°F.

Table 6.3

TEMPERATURE CHART—LIQUID HEAT-EXCHANGED SYSTEMS

TIME READ	TEMPERATURE DOWNCOMER T1 (°F)	TEMPERATURE RISER T2 (°F)	TEMPERATURE COIL ENTRANCE (LOWER TANK) T3 (°F)	TEMPERATURE COIL EXIT (UPPER TANK) T4 (°F)	T1 − T2	TANK AVERAGE TEMPERATURE (T3 + T4)/2 °F
30 min	130	110	97	121	20	109
1 hr	140	120	102	124	20	113
°F Rise	10	10	5	3		4

30 min (130 − 110)/(130 − 97) = 61%E
1 hr (140 − 120)/(140 − 102) = 53%E

The formula is:

Gal of storage × wt/gal × specific heat
$$\times \text{ temperature rise} = \text{heat gain, Btu}$$

For example, let

Storage water = 60 gal
Water's weight = 8.33 lb/gal
Specific heat water = 1
Length of test = 30 min
Heat gain = 4°F

Then

$$60 \times 8.33 \times 1 \times 4 = 1999 \text{ Btu/30 min} \quad \text{or} \quad 3998 \text{ Btu/hr}$$

Knowing the square feet of collector will also allow the calculation of how many British thermal units per square foot per hour are being placed in the storage tank. Merely divide the total heat gained by the square feet of collector in the system.

Using the result from above and 50 ft² of collector,

$$3998 \text{ Btu/50 ft}^2 = 80 \text{ Btu/ft}^2\text{/hr}$$

These calculations would include any standby losses occurring during the period of testing and yields the overall system efficiency per unit of operating time. It does not include any standby losses occurring when the collector loop is at rest.

WHAT CAN GO WRONG

Assuming that the system has operated successfully prior to the service call, five things can go wrong.

1. The owner can change the valve settings.
2. The system can suffer physical damage or develop a water leak.
3. The collector fluid can leak out.
4. One or both pumps can fail.
5. The control system can fail.

The serviceperson can determine which of these defects exists with a troubleshooting approach.

TROUBLESHOOTING CHART

1. Look for physical damage.

Item	Yes	No
Crushed pipes	_____	_____
Leaky valves or joints	_____	_____
Broken collector cover	_____	_____
Freeze damage	_____	_____
Structural movement	_____	_____
Broken sensor wire	_____	_____
Damaged or missing sensor	_____	_____
Electric service interruption	_____	_____
Sticking check valve	_____	_____
Damaged heat exchanger	_____	_____
Sticking air vent valve	_____	_____
Leaky expansion tank	_____	_____
Leaky pressure relief valve	_____	_____
Loss of heat transfer fluid (no pressure on gauge)	_____	_____
Leaky boiler drain valves	_____	_____

2. Check the valve settings.

Valve	Solar System Mode		
	Operating	At Rest	At Start-up
Throttle valve	Varies	Varies	Varies
Temperature and pressure relief valves	Closed	Closed	Closed
Tank drain valves	Closed	Closed	Closed
V6	Closed	Closed	Closed
V7	Open	Open	Open
V8	Open	Open	Open
V16	Closed	Closed	Closed
V17	Closed	Closed	Closed

3. Drain the scale and sediment.

- Turn the controller off.
- Open the storage tank drain valve and run until clear.
- Open the auxiliary tank drain valve and run until clear.

4. Check the heat transfer fluid liquid level.

    - Read the pressure gauge. The system should have at least 5 lb of gauge pressure.
    - Crack open the top vent in the upper collector manifold and bleed off any entrapped air. If only fluid emerges, the system is full.
    - Recheck the gauge pressure for a minimum 5-lb level.

If the gauge reads less than 5 lb of pressure after this procedure is carried out, proceed as follows.

    - Check the automatic air vents for spitting or leakage.
    - Check the safety valve for leakage.
    - Examine the joints in the system for leaks.
    - Examine collector absorber plate assembly for leaks.
    - Replace valves or repair joints as required.
    - Check the pressure in the bottom of the expansion tank. This should read about 10 to 15 lb.
    - Recharge to 12 lb if pressure is low.
    - Clean and replace the pump strainer screen.
    - Add heat transfer fluid to the system, following the manufacturer's directions carefully. Refer to Chapter 10 if no manufacturer's information is available.

5. Check the condition of the heat transfer fluid.

    - Water/glycol systems 18 to 24 months old or older should have their fluid pH and inhibitor levels checked.
    - Oil-containing systems 2 years old or older should be checked for sludging.

6. Heat collection test.

    - Choose a time between 10 A.M. and 2 P.M. on a clear day.
    - Check that the controller is off.
    - Close V7.
    - Place a gauge on the hot downcomer (T1).
    - Place a gauge on the cold riser (T2).
    - Place a gauge about 1 ft up from the bottom on the tank wall (T3).
    - Place a gauge about 1 ft down from the top of the tank on the tank wall (T4).
    - Allow the gauges to come to equilibrium. Read and record the readings in Table 6.2.
    - Turn the pump controller to automatic.
    - If the pump starts in 2 to 3 minutes, proceed with the test.
    - If the pump does not start in 2 to 3 minutes, turn the controller to manual.
    - If the pump runs on manual, proceed with the test. Otherwise go to procedure 7, the pump/controller test procedure. Repair the pump/controller; then return to this point.
    - Let 30 minutes to 1 hour pass with the pump operating. Read and record the temperatures of all four gauges in Table 6.2.
    - Fill out the temperature chart.
    - Perform the calculations.
    - Evaluate the results.

1. Was a 10 to 20°F temperature differential developed between T1 and T2?
2. Did the tank temperature rise at least to 6 to 9°F per hour?
3. Did the controller turn on and turn off the pump properly?

If the answers to these questions are all yes, the system is operating satisfactorily. Skip steps 7 and 8 if the pump is functioning correctly on the automatic controller setting.

7. Run the pump/controller test.

   - Determine whether the system has one (coil-in-tank system) or two (external heat-exchanger system) pumps.
   - Identify which pump is for the collector loop.
   - Identify which pump is for the storage loop.
   - Identify which pump is malfunctioning.

8. If the pumps will not run on manual,

   - Check for a blown fuse or faulty circuit breaker.
   - Check the pump on a separate circuit if it is the plug-in type.
   - Check for electricity at the pump power cord with a trouble lamp or voltmeter if the pump is hard wired.
   - Isolate whether the pump or control system is faulty.
   - Proceed to troubleshoot the pump.

9. If the pumps will not run on automatic,

   - Recheck the sensors and sensor wiring for damage.
   - Proceed to troubleshoot the controller.

Complete the system performance test after repairing the pump and controller.

10. Run a storage loop heat transfer test.

    - Perform the heat transfer calculations.
    - Determine the heat-exchanger effectiveness. If the effectiveness is over 50%, the system is operating satisfactorily; skip to step 11.

If the effectiveness is less than 50%,

    - Check out the heat-exchanger design.
    - Look for pipe constrictions, partially closed valves, or similar problems that would prevent proper flow.
    - Drain and open the loop as required to clean out the liquid passages. This procedure will vary, depending on the design of the heat exchanger. Follow the manufacturer's instructions. If no instructions are available, see Chapter 10.
    - Close up the loop, recharge, and retest the system.

11. Visually check over the system.

    - Return the operating valves to their proper positions.
    - Is the controller placed in the automatic mode?
    - Are any structural repairs needed?
    - Is any painting needed?
    - Is there any loose or missing insulation?
    - Are there any leaks?

12. Finish up the job.

- Clean up the area.
- Remove all tools.
- Secure the area.
- Complete a service report for the owner.
- Present a bill for services.

ON THE JOB

In addition to the tools and supplies carried in the toolbox for the drain-down and drainback systems, carry the instruments needed to check out the heat transfer fluids. These are specified in Chapter 19. Which instruments you elect to carry will depend on what types of systems you normally service.

The general order of troubleshooting is similiar to the order shown in Chapters 2 to 5. Quiz the homeowner carefully to define just what type of problem is being encountered.

Once any physical damage has been repaired, check out the sealed collector loop and verify that the loop is filled with heat transfer fluid, as shown in step 4.

Then check the pump/controller operation, as shown in steps 7 to 9. Repair any control system/pump problems.

Start the heat collection performance test shown in step 6.

While the performance test is underway, perform steps 2, 3, and 5. Finally, if the performance test is not going well, perform step 1.

Any problems should have been isolated and repaired by now. Proceed to finish up the job.

In this type of system you must be concerned with collector and insulation damage caused by excess system stagnation. When the system stagnates throughout a hot summer while people are vacationing, overheating can degrade many heat transfer fluids, damage collector arrays, or ruin insulation. In extreme cases a fire hazard may even be present. In step 1 you will be looking more for damage from overheating than for damage from freezing.

Fluid toxicity can be a problem when the homeowner or a less-than-knowledgeable serviceperson or installer has mistakenly put the wrong fluid in the system. You must be certain that toxic heat transfer fluids are only used under the proper conditions. Keep this in mind when you are working with the fluids.

WHAT YOU SHOULD HAVE LEARNED

1. A liquid heat-exchanged system is a system that uses a collector fluid other than water in a closed loop that contains a heat exchanger to prevent the collector fluid and the service water from mixing.
2. Liquid heat-exchanged systems eliminate the problems associated with the use of water because of its high freeze point, low boiling point, and tendency to cause corrosion.
3. There are five classes of heat transfer fluids in general use. All are adequate, but each has its advantages and disadvantages.
4. Different classes of fluids cannot be substituted for each other because system designs change with heat transfer fluid class changes.

5. There are six common types of heat exchangers in use. The industry has settled on the coil-in-tank and external heat-exchanger classes as the most acceptable.

6. The use of a liquid heat-exchanged system requires the addition of a mechanical package to the collector loop.

7. The efficiency of a liquid heat-exchanged system is highly dependent on the effectiveness of the heat exchanger. Exchangers should be at least 50% effective to allow the collection of reasonable amounts of energy.

8. Older systems with poor heat-exchanger effectiveness can be improved by adding an external heat exchanger.

9. The liquid heat-exchanged system is sized at 20 to 25 ft^2 per occupant to obtain 40 to 60% of the total hot water requirements. The storage tank should be sized at 1.2 to 2.0 gal/ft^2 of collector area.

10. Pump sizes will vary according to system size and by the class of heat transfer fluid chosen, because the density of the fluid and the fluid's specific heat determine the required flow rate.

11. Coil-in-tank systems generally use a spun concrete storage tank, which has a very long life because the heat exchanger is part of the tank. The external heat-exchanged system can use glass-lined tanks that may have shorter lives.

12. When the system uses a water-containing heat transfer fluid, aluminum liquid passages should be avoided.

13. The oil classes of heat transfer fluids will require higher pumping rates and larger pipe runs than the water/glycol classes of fluids.

14. The liquid heat-exchanged system offers positive freeze protection, low corrosion rates, and freedom in piping layout. It has no motorized valves or vacuum relief valves to malfunction. The system has the potential for longer life than water-containing collector loop systems.

15. Some collector liquids will need periodic maintenance. The system's components must be carefully matched, and the design is more complex than thermosiphon, pumped water, or drainback systems.

16. The performance testing of a liquid heat-exchanged system consists of testing heat collection, testing pump/controller operation, and testing heat transfer fluid condition.

CHAPTER 7

Air Heat-Exchanged Systems

Air Heat-Exchanged System A solar water-heater system that uses air for the collector fluid and an air-to-water heat exchanger to transfer the air's heat to the water in the solar storage tank. Such solar water-heater systems are generally part of a space-heating system that circulates the air from the building through the solar collectors.

Figure 7.1 shows the water-heater portion of such a system. Hot air from the solar collectors is passed through an air-to-water heat exchanger prior to the air being circulated to storage or to the house. The cooled air from storage or the house is returned to the collectors through an **air handler.** An air handler is a mechanical package containing an air blower or fan, appropriate directional dampers, and controls that circulate, direct, and control the flow of the air used as the heat-exchange fluid.

In the summer months, when space heating is not required, bypass ductwork is used to bypass the house and storage bed heating ducts.

The Air-to-Water Heat Exchanger

Figures 7.2 and 7.3 show an air-to-water heat exchanger, commonly called a **finned tube heat exchanger.** In this exchanger a liquid such as water circulates through banks of tubes arranged in a serpentine fashion (see Figure 7.2). A series of closely spaced metal fins, as shown in Figure 7.3, are mechanically crimped to the tubes. The hot air passes through the fins and heats them. The water circulating through the tubes carries the heat away. An automobile radiator and the condenser on the back of a room air conditioner are typical examples of finned tube heat exchangers.

The Air-Exchanged System

Looking again at Figure 7.1, cold water from the bottom of the solar pre-heater tank passes through a strainer to a small centrifugal circulating pump and is pumped through a throttling valve to the heat exchanger. It gathers the heat from the coils and passes down through a check valve to the top of the solar preheater tank.

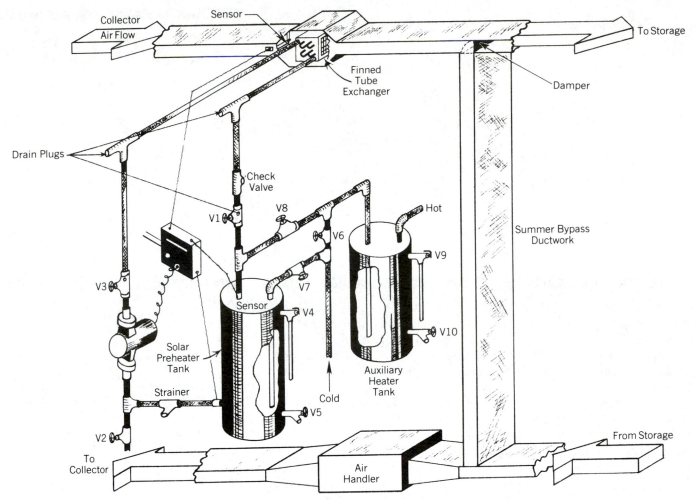

Figure 7.1   An air heat-exchanged system. (The full system is not shown.)

When water is drawn in the home, the water from the top of the solar preheater tank is fed to the bottom of the auxiliary heater. If the solar-heated water is hot, the auxiliary heater does not turn on. If more heat is needed, the auxiliary heater turns on and provides it.

The Preheater Package

The purpose of the solar preheater package is to circulate the cold service water in the solar storage tank through the heat exchanger, remove and store the collected solar energy, and pass the preheated water to the auxiliary water heater. Let us look at the various parts of the package (Figure 7.1) and review the function of each part.

The Solar Storage Tank   The solar storage tank is a standard glass- or stone-lined, insulated water storage tank containing a drain valve, a temperature and pressure relief valve, a magnesium anode, a tapped fitting for the pump leg, and hot and cold water connections. The cold water inlet connects to a dip tube that extends most of the way down the inside of the tank.

In some tanks the tapped fitting for the pump leg may not exist. In that case the drain valve is removed; a tee fitting is inserted, and both the pump leg and the drain valve are attached to the tee.

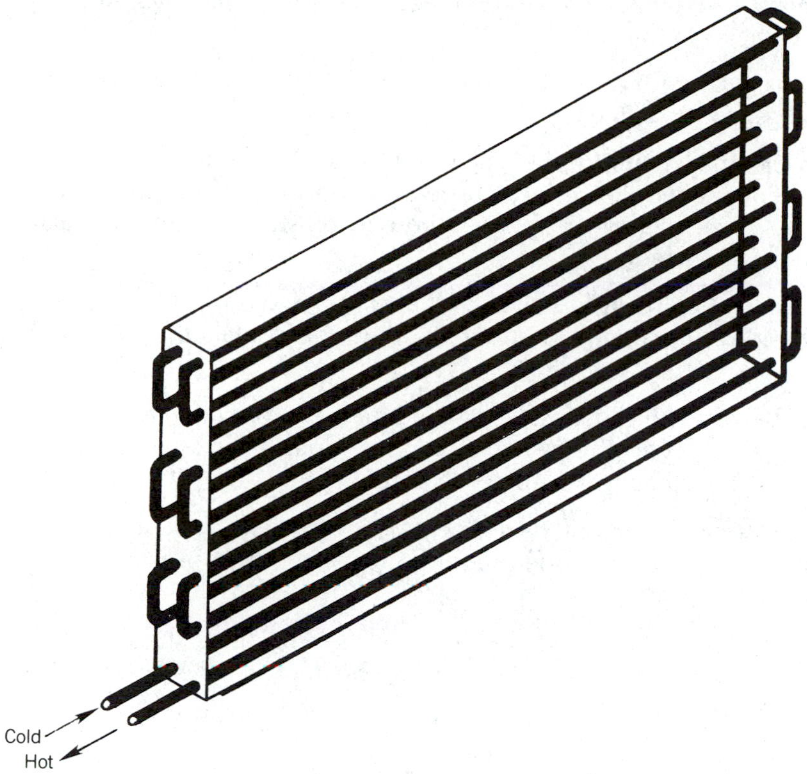

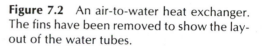

Cold
Hot

Figure 7.2 An air-to-water heat exchanger. The fins have been removed to show the layout of the water tubes.

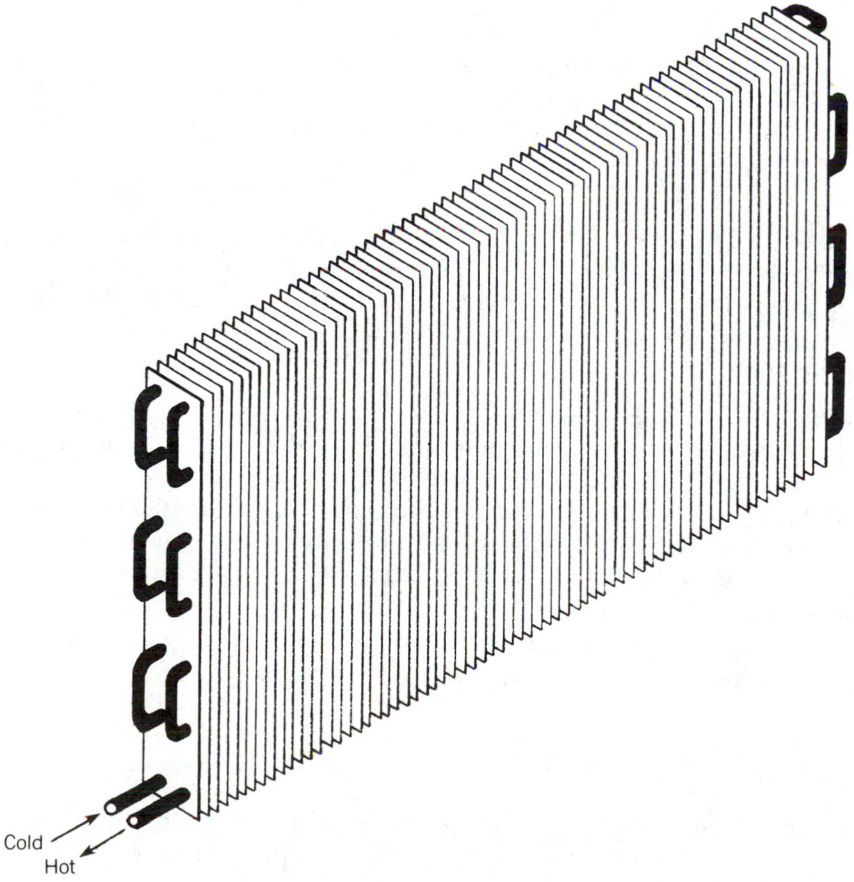

Cold
Hot

Figure 7.3 An air-to-water heat exchanger. The frame has been removed to show the fins clearly.

Strainer   The strainer in the pump leg protects the pump against foreign solid particles in the water.

Pump Leg Drain Valve   The pump leg drain valve allows the heat exchanger to be drained.

Pump   A small, centrifugal circulator pump that is turned off and on by a differential controller is located below the top level of the preheater tank. Any reliable, properly sized bronze or stainless steel circulator pump may be used.

Flow Control or Throttle Valve   A throttling valve is located on the discharge side of the pump. This valve is adjusted to obtain the correct flow rate.

Check Valve   The check valve, located on the return pipe from the heat exchanger, prevents the heated tank water from reverse thermosiphoning when the exchanger is colder than the tank.

Shutoff Valve   A shutoff valve, located below the check valve, allows the heat exchanger and the pump leg to be removed from service for maintenance.

Supply Valves   Three shutoff valves are used to direct the flow of hot and cold water. They also allow the preheater package to be isolated from the auxiliary heater.

Controlling the System

A differential temperature controller is used to control the system. The controller uses three temperature sensors.

- A collector air temperature sensor located on the face of the heat exchanger.
- A tank water temperature sensor located on the tank near the pump leg fitting.
- A high-limit sensor located near the top of the solar preheater tank.

Usually, a controller is selected that turns the pump *on* when the face of the heat exchanger fins is 8 to 20°F hotter than the bottom of the tank and turns the pump *off* when the temperature drops to 2.5 to 5°F hotter than the lower tank temperature as measured by the tank sensor located at the pump leg.

NOTE   One reviewer suggested a different control strategy that is more closely related to a combined control system where pump turnon takes place at 20 to 40°F and pump turnoff takes place at 10 to 20°F. We are not familiar with this control strategy. It would seem that the strategy is closely related to running both air space heating and water heating simultaneously from the same controller. This type of control strategy should work well.

The high-limit switch activates at or below 180°F and overrides the pump turnon circuit so that the tank cannot overheat.

Expected Performance

Like any system containing a heat exchanger, the system's performance will vary depending on heat-exchanger effectiveness, which is a function of exchanger design and size.

However, the heat-exchanger selection parameters for an air-exchanged system are not as important as they are in liquid systems, because the heating of water is only one of the uses for the collected solar heat.

During the heating season, the system is removing additional heat from the airstream either in the heat storage bed or in the home. In the non-heating season the collectors are oversized for the water-heating load, and maximum efficiency is generally not required. Any exchanger operating at over 20% effectiveness should be considered adequate.

If the System Is Only for Water Heating

Occasionally, air-exchanged systems are encountered that are used only for water heating. These systems should have an exchanger with at least 50% effectiveness as measured at solar noon on a clear day with a luke-warm tank.

Why 20% Exchanger Effectiveness Is Adequate

Some heat-exchanger effectiveness calculations show why 20% effectiveness is adequate in a combined water-space-heating system.

Exchanger effectiveness in this exchanger configuration (see Figure 7.4) is calculated by the following formula.

$$(T1 - T2)/(T1 - T3) \times 100 = \%E$$

In Case 1, assume

$$T1 = 140°F$$
$$T2 = X$$
$$T3 = 90°F$$
$$E = 0.20 \ (20\%)$$

In Case 2, assume

$$T1 = 140°F$$
$$T2 = X$$
$$T3 = 90°F$$
$$E = 0.60 \ (60\%)$$

Then

$$(140 - X)/(140 - 90) = 0.20 \qquad (140 - X)/(140 - 90) = 0.60$$

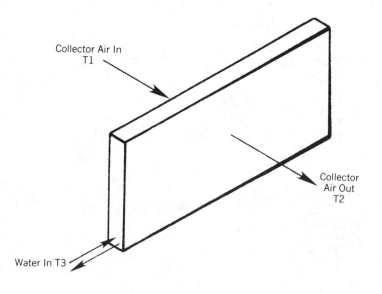

Collector Air In
T1

Collector
Air Out
T2

Water In T3

Figure 7.4 How exchanger effectiveness is calculated.

Solving for X,

$$140 - X = (50)(0.20) \qquad 140 - X = (50)(0.60)$$
$$X = 130°F \qquad\qquad X = 110°F$$

In Case 1 the air temperature drop across the exchanger is 10°F, and the 130°F exiting air can be used effectively to heat the house or the storage bed.

In Case 2 the exiting air has dropped 30°F to 110°F. This temperature is not as effective for heating the storage bed or the house.

Sizing Parameters

The sizing parameters for the collector loop are generally determined by the spaceheating load plus the water-heating load imposed on the system. Because the designers of combined systems use different design strategies, the number of collectors or gallons of storage water will not be recommended in this chapter.

The heat exchanger should be sized (see Figure 7.5) to provide 1 ft² of face area per 400 ft³ per minute of air flow unless the exchanger manufacturer specifies a different rate of flow.

The rate of water flow through the exchanger tubes is set by the exchanger manufacturer to provide optimum heat transfer at the specified air flow rate. Both flow rates are generally stamped on the manufacturer's nameplate for your reference.

There are many variations in tube configurations, so no rule of thumb can be given for liquid flow rate.

In some installations the ductwork from the collectors may be split so that only part of the incoming airstream is routed through the heat-exchanger coil while the balance of the air goes directly to storage or to the house. In this case high heat-exchanger effectiveness is appropriate.

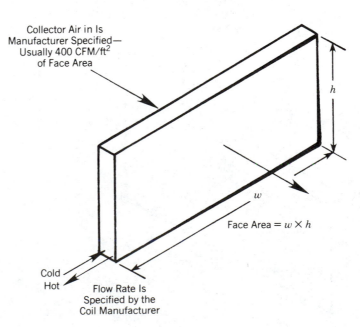

Figure 7.5 Exchanger operating conditions are set by the exchanger manufacturer.

Other Considerations

The storage tank should be lined to prevent corrosion. Glass- or stone-lined tanks are recommended. All piping and fittings should be copper. Usually $\frac{1}{2}$-in. pipe is satisfactory for short runs; $\frac{3}{4}$-in. pipe may be necessary for long runs and large heat exchangers.

A bronze or stainless steel pump should be used, and the controller sensors must be carefully installed per the controller manufacturer's instructions.

Generally, the water-heating system is operated as an auxiliary system to the space-heating system and the design cannot be optimized.

Advantages

The air-exchanged system has four major advantages.

- Positive freeze protection.
- Noncorrosive collector fluid.
- Minimum liquid piping.
- No electrically actuated or vacuum relief valves.

Disadvantages

The air-exchanged system has two major disadvantages.

- Less efficient solar collection.
- Nonoptimized sizing.

How to Test Performance

Performance testing of the air-exchanged system consists of

- Testing the heat collection and the exchanger loop heat transfer.
- Testing pump/controller operation.

Testing Heat Collection/Heat Transfer Operation

Testing proceeds as follows.

- The tests should be carried out between 10 A.M. and 2 P.M. on a clear day in direct sunlight.
- The hot water service must be shut off so that no water can be drawn from the system.
- The collector loop controller is turned off. This controller is part of the air space-heating system and has not been previously mentioned.
- The storage loop controller is turned off.
- The summer bypass ductwork dampers are set so that no air goes to storage or to the house.
- A temperature gauge is placed in the collector air duct ahead of the heat exchanger to measure T1 (see Figure 7.6).
- A temperature gauge is placed in the collector ductwork slightly downstream from the heat exchanger to measure T2.

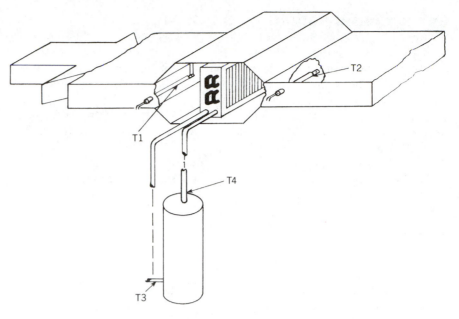

Figure 7.6 Where the temperature gauges are placed during the performance tests.

- A temperature gauge is placed on the hot downcomer from the heat exchanger as close to the storage tank as possible to measure T4.
- A temperature gauge is placed on the pump leg close to where it exits from the preheater tank to measure T3.
- The temperature gauges are allowed to reach equilibrium, which will take about 3 to 4 minutes.
- Read the temperature on the four gauges. Record the temperatures on a chart such as Table 6.2.
- Turn on the collector loop controller and allow the collector loop to run for 4 to 5 minutes.
- Now turn the storage loop differential controller to automatic.
- If the pump starts within 2 to 3 minutes, proceed with the test.
- If the pump does not start on automatic, turn the controller switch to manual.
- If the pump starts on manual but will not run on automatic, four faults are possible.

  1. The controller is faulty.
  2. One or more of the sensors is faulty.
  3. The temperature differential between the bottom of the tank and the face of the heat exchanger is not high enough to activate the controller (usually 8 to 12°F or 20 to 40°F, depending on the system).
  4. The tank temperature is high enough to activate the high-limit sensor.

- If the pump will run on manual, proceed with the test and check the temperature rise with time. When the test is completed, go to the pump/controller test procedure.
- If the pump does not run on manual, terminate the test and go to the pump/controller test procedure.

Proceeding with the Test Assuming that the pump will run, continue temperature testing with the pump on manual as the preceding procedure indicates. If the pump will not run, you will return to this point after you have corrected the pump or controller problem.

- Allow 30 minutes to 1 hour to pass with the pump operating.
- Record the readings in a table like Table 6.2.

Filling Out the Chart

Step 1   Subtract T2 from T1 and place the answer in the column headed T2 − T1. Do this for both gauge readings.

Step 2   Add T3 to T4 and divide the sum by 2. Place the result in the column headed T3 + T4/2. Do this for both gauge readings.

Step 3   Subtract the initial T3 + T4/2 result from the final T3 + T4/2 result. Place the result on the tank temperature rise line.

The Meaning of the Readings and Observations

T1 measures the heat-exchange inlet air temperature. Initially, with the collector loop fan off, this reading should be close to the temperature of the ambient air surrounding the ductwork. When the collector loop fan is running, the temperature will rise to the temperature of the air exiting the collector less any ductwork losses.

T2 measures the heat-exchanger outlet air temperature. With the collector loop off, this temperature should be about the same as T1. When *only* the collector loop is running, T1 and T2 should read the same. When the collector loop *and* the storage loop are *both* running, T2 should read lower than T1. The difference between the two temperatures is a measure of how much energy is being transferred to the storage tank. The larger the temperature difference, the greater the amount of energy being transferred.

T3 measures the temperature at the bottom of the storage tank, and T4 measures the temperature at the top of the storage tank. When these two temperatures are averaged (T4 + T3/2), the average temperature of the tank is obtained.

The amount of energy added to the tank during the test is obtained by subtracting the average tank temperature at the beginning of the test from the average tank temperature at the end of the test and multiplying by the amount of water in the tank. Use this formula.

(Tank temperature rise)(gal water)(8.33) = Btu of energy added

Interpreting the Results

In a stand-alone system, where the size of the collector loop is matched to the storage loop, the difference in temperature between the exchanger entrance air and the exchanger exit air should be 15 to 30°F when the system is operating properly.

In the case of an unmatched system, where the collector loop is sized to handle both the water-heating load and the space-heating load, the difference between the exchanger entrance air and exchanger exit air will be much smaller. It may be only 1 to 2°F when the water heater is a minor part of the total load.

In both stand-alone and combined systems, the difference between the top and bottom tank temperatures (T3 and T4) should be at least 5°F when the system is operating. The difference may be as high as 25 to 35°F.

T3 and T4 are also measures of the exchange liquid entrance and exit temperatures. A differential must exist if heat is being transferred to the tank.

In the case of a stand-alone system follow these guidelines for determining proper liquid flow rate.

1. The flow rate is correct if the temperature difference lies between 10 and 20°F.
2. The flow rate is slow if the temperature difference is over 20°F. Open the throttle valve.
3. The flow rate is fast if the temperature difference is below 10°F. Close down the throttle valve.

In the case of combined systems follow these guidelines.

1. If the temperature differential is less than 10°F, the throttle valve should be closed down. Otherwise, the pump will cycle off and on too frequently in the early morning or late afternoon and on cloudy days. Attempt to set the throttle valve to give a 10 to 20°F temperature differential.
2. If the temperature differential is more than 20°F, the throttle valve should be opened to bring the system back into the 10 to 20°F range. The collector system may be too large to allow this. If so, leave the throttle valve wide open.

A properly operating system should raise the average tank temperature at least 6 to 9°F per hour. The higher the hourly temperature rise, the better the system is operating.

The Pump/Controller Test

The pump/controller test is run in the same way as the pump/controller test in Chapter 3, for an off/on controller. Proportional controllers are generally not used in these systems.

It is also possible to find a single controller system that is operating both the water-heating and the space-heating loops. In this case you should consult the controller manufacturer's troubleshooting information to check out the controllers and the sensors.

WHAT CAN GO WRONG

Assuming that the system has operated successfully prior to the service call, five things can go wrong.

1. The owner can change the valve settings.
2. The system can suffer physical damage or develop a leak.
3. The collector system can fail.
4. The pump can fail.
5. The control system can fail.

TROUBLESHOOTING CHART

1. Look for physical damage.

Item	Yes	No
Crushed pipes	_____	_____
Leaky valves or joints	_____	_____
Broken collector cover	_____	_____

Structural damage	_____	_____
Damaged or missing insulation	_____	_____
Broken sensor wire	_____	_____
Damaged or missing sensor	_____	_____
Electric service interruption	_____	_____
Sticking check valve	_____	_____
Damaged heat exchanger	_____	_____
Impeded collector air flow	_____	_____

2. Check the valve settings.

	Solar System Mode		
Valve Number	**Operating**	**At Rest**	**At Start-up**
V1	Open	Open	Open
V2	Closed	Closed	Closed
V3	Varies	Varies	Varies
V4	Closed	Closed	Closed
V5	Closed	Closed	Closed
V6	Closed	Closed	Closed
V7	Open	Open	Open
V8	Open	Open	Open
V9	Closed	Closed	Closed
V10	Closed	Closed	Closed

3. Drain the scale and sediment.

- Turn the storage loop controller off.
- Open V10, run until clear, close.
- Open V5, run until clear, close.
- Open V2, run until clear, close.

4. Flush the heat-exchanger circuit.

- Close V1 and V3.
- Remove drain plug A.
- Open V3 and flush through drain plug A until it is clear.
- If the system will not flush, the heat-exchanger tubes are clogged.
- Clean and replace the pump strainer screen.

5. If the heat-exchanger is plugged,

- Close V1.
- Close V3.
- Remove the exchanger clean out plugs A and B.
- Blow compressed air through the exchanger.
- Replace the plugs.

6. Refill the heat exchanger/pump leg.

- Close V1, V2, and V7.
- Loosen the drain plug above V1 so air can escape.
- Open V7.
- Run water until no water comes out of the drain plug above V1.
- Close the drain plug on V1.
- Switch the controller to manual and start the pump.
- Crack drain plug A to remove any air in the system; then close A.

7. Perform a heat collection test.

- Perform the test between 10 A.M. and 2 P.M. on a clear day.
- Close V8.
- Turn off the collector loop controller.
- Check that the storage loop controller is off.
- Open the summer bypass ductwork.
- Place a temperature gauge in the duct near the heat-exchanger face.
- Place a temperature gauge in the duct downstream from the heat exchanger.
- Place a temperature gauge on the hot downcomer close to the tank.
- Place a temperature gauge on the cold pump leg close to the tank.
- Allow the gauges to come to equilibrium.
- Read and record the gauge temperatures in Table 6.2.
- Turn on the collector loop controller.
- Run the collector loop for 4 to 5 minutes.
- Turn the storage loop controller to automatic.
- If the pump starts within 2 to 3 minutes, proceed with the test.
- If the pump does not start, turn the storage loop controller to manual.
- If the pump runs on manual, proceed with the test; otherwise skip to step 8 and perform the pump/controller test procedure and repair the pump or controller. Then return to this point to continue the test.
- Let 30 minutes to 1 hour pass with the pump operating. Read and record the temperatures of all four gauges in Table 6.2.
- Fill out the temperature chart.
- Perform the calculations.
- Evaluate the results.

  1. Did the collectors provide heat to the heat exchanger?
  2. Was a temperature differential developed between T1 and T2?
  3. Was a 5°F or greater temperature differential developed across the liquid side of the heat exchanger?
  4. Did the average tank temperature rise at least 6 to 9°F per hour?
  5. Did the controller operate the pump properly?

  If the answers to these questions are yes, the system is operating satisfactorily.

Skip steps 8 and 9 if the pump is functioning correctly on the automatic controller setting.

8. If the pump will not run on manual,

   • Check for a blown fuse or faulty circuit breaker.
   • Check the pump operation on a separate circuit if it is the plug-in type.
   • Check for electricity at the pump power cord with a trouble lamp or voltmeter if the pump is hard wired.
   • Isolate whether the pump or controller is faulty.
   • Proceed to troubleshoot the pump as outlined in Chapter 16.

9. If pump will not run on automatic,

   • Recheck the sensors and sensor wiring for damage.
   • Proceed to troubleshoot the controller as outlined in Chapter 17.

Complete the system performance test (step 7) after repairing the pump and controller.

10. Visually check over the system.

   • Return the operating valves to their proper positions.
   • Are the controllers placed in the automatic mode?
   • Are any structural repairs needed?
   • Is any painting needed?
   • Is there any loose or missing insulation?
   • Are there any leaks?

11. Finish up the job.

   • Clean up the area.
   • Remove all tools.
   • Secure the area.
   • Complete a service report for the owner.
   • Present a bill for services.

WHAT YOU SHOULD HAVE LEARNED

1. An air heat-exchanged system is a system that uses air as the heat-exchange fluid.
2. Air heat-exchanged water-heating systems are usually part of a combined space- and water-heating system.
3. An air-to-liquid heat exchanger (finned tube exchanger) located in the ductwork ahead of solar storage is used to transfer energy to the water-heating system.
4. The inclusion of bypass ductwork in the system allows the heating of water year round without operating the space-heating system when space heating is not required.
5. Because most air heat-exchanged systems have a dual load, the collectors are generally not sized correctly for the water storage loop.
6. This size mismatching problem results in high collector temperatures in the summer months.
7. Heat-exchanger effectiveness can be as low as 20% in a combined air-exchanged system as additional heat is removed by the space-heating load on the system.
8. The circulator pump type and size are not critical as long as a bronze or stainless steel pump is chosen and the flow rate rec-

ommended by the heat-exchanger manufacturer can be obtained.

9. An air flow rate of 400 ft^3 per minute per square foot of exchanger face area is normal for a typical finned tube exchanger unless the manufacturer specifies differently.

10. Either glass- or stone-line tanks are satisfactory.

11. The water-containing side of the system should use all copper pipe and fittings. The valves should be brass.

12. The performance testing consists of testing heat collection, heat transfer to storage, and pump/controller operation.

CHAPTER
8

Phase-Change Heat-Exchanged Systems

Phase-change heat-exchanged system: A solar water heater that uses a boiling liquid as the heat transfer fluid.

The solar water-heating systems you studied in Chapters 2 through 7 transferred heat from the solar collector to the solar storage tank using **sensible heat.** Sensible heat is heat accompanied by a change in the fluid's temperature.

Phase-change heat-exchanged systems transfer heat using the **latent heat of vaporization.** The latent heat of vaporization is heat accompanied by a phase change from the liquid state to the gaseous state. There is no change in the fluid's temperature during phase change.[1]

> **WARNING** We recommend that servicepeople do not attempt to repair phase-change liquid collector loops unless they have been fully trained in the maintenance and repair of refrigerant-type solar systems. The liquids used in these loops have low boiling points and build up high pressures even at relatively low temperatures. Do not use a torch on these systems, attempt to disassemble them, or cut into the fluid passages of the system. Serious injury may result.

Physical States of Matter

All substances can exist in three different physical states: solid, liquid, or gas. The physical state of a substance at any time is a function of the nature of the substance, the pressure exerted on it, and its temperature. It takes heat energy to make a substance change state, but that energy does not change the temperature of the substance.

---

[1]We have limited personal experience with phase-change heat-exchanged systems, as does much of the solar industry. The American Society of Heating and Air Conditioning Engineers (ASHRAE) has been funding protracted studies of such systems. We recommend you obtain their latest information before purchasing or repairing these systems.

Water is a convenient substance to examine as an example. At standard atmospheric pressure, water is solid (ice) below 32°F, liquid from 32°F to 212°F, and gaseous (steam) above 212°F.

It takes 1 Btu of heat energy to raise the temperature of 1 lb of water by 1°F. This is the substance's specific heat capacity. To raise the temperature of 1 lb of ice from 0°F to 32°F takes 32 Btu.

$$1 \text{ lb} \times 32°F \times 1.0 \text{ specific heat} = 32 \text{ Btu}$$

However, to change the 32°F ice to 32°F water requires additional heat. This is called the **heat of fusion.** The heat of fusion for water is 144 Btu/lb—far more heat than the heat required to change the temperature from 0°F to 32°F.

Once this 1 lb of ice has melted to water, the addition of further heat energy will raise the water's temperature. To raise the temperature from 32°F to 212°F requires another 180 Btu of heat energy.

$$(212 - 32) \times 1 \text{ lb} \times 1.0 = 180 \text{ Btu}$$

To change the 212°F water to 212°F steam also requires additional heat. This is called the **heat of vaporization.** The heat of vaporization for water is 970 Btu.

So, as shown in Figure 8.1, it requires a total of 1326 Btu of heat energy to make 1 lb of 212°F steam from 1 lb of 0°F ice—212 Btu of sensible heat and 1114 Btu of latent heat.

Every substance has its own unique specific heat, heat of fusion, and heat of vaporization. Every substance also has its own unique **melting point** and **boiling point.** The melting point is the temperature at which the substance changes from a solid to a liquid. The boiling point is the point at which the substance changes from a liquid to a gas.

Effect of Pressure Change

When the pressure on a substance is increased, the melting point and the boiling point are raised. At standard atmospheric pressure (14.7 psi), water boils at 212°F; at 10 psi, it boils at 193°F; at 50 psi, it boils at 281°F; and at 100 psi, it boils at 328°F.

When the liquid is held in a closed system and its temperature is raised above the boiling point, the increase in temperature is accompanied by an increase in pressure, which changes the boiling point.

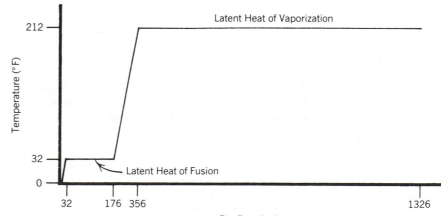

Figure 8.1 It takes 1326 Btu to turn 1 lb of ice into 1 lb of steam.

Collector Temperature and Solar System Efficiency

The efficiency of a solar collector decreases as its temperature increases above the temperature of the surrounding air because of increasing heat losses out the front, back, and sides of the collector. In a solar system that uses sensible heat to transfer the energy, the collectors operate at much higher temperatures than the storage water. This lowers their efficiency.

Phase-change systems operate at lower temperatures—at the boiling point of the liquid used to transfer the heat. They transfer energy using the latent heat of vaporization with less increase in collector temperature—making the collectors more efficient.

One manufacturer of solar water-heater systems compared its phase-change system with its hydronic system in side-by-side tests using identical collectors and storage tanks. The results during a typical day are shown in Figure 8.2.

When these two identical systems were compared, the phase-change system collector's temperature was lower than the hydronic system collector's temperature, and the tank temperature was hotter at the end of the day.

Even though this test would seem to indicate that phase-change systems are vastly superior to hydronic systems, this is not the case.

Throughout this book, we have stressed that liquid systems should be designed so that the collectors operate at a maximum temperature only 10°F higher than the bottom storage tank temperature to be efficient.

The hydronic collectors in this system are operating as much as 60°F hotter than the tank, indicating that the hydronic system has insufficient water flow in the storage loop and thus does not meet our design criteria.

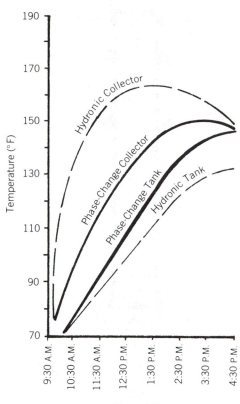

Figure 8.2 This "racing" test compares one manufacturer's phase-change and hydronic systems.

Boiling-Heat Transfer Fluids

Water is not the fluid of choice for phase-change solar collectors. It has the wrong melting and boiling points. Instead, **halocarbon fluids** are commonly chosen. Halocarbon fluids are fluorocarbon chemicals that are commonly used throughout industry as refrigerants and aerosol propellents. A common trade name is Freon.

To work, a boiling heat transfer fluid must be used above its boiling point and must have a freezing point lower than the ambient conditions to which it is exposed. Additionally, it must have a low order of toxicity, be nonflammable, and be nonhazardous to human life.

To be useful in solar collectors, the boiling point—at the operating pressure—must be in a range that will allow efficient solar collector operation. For solar water heaters, this collector temperature would be in the range of 70 to 150°F.

FLUID NAME	BOILING POINT AT ATMOSPHERIC PRESSURE (°F)	FREEZING POINT AT ATMOSPHERIC PRESSURE (°F)	PRESSURE, 150°F[a] (psia)
R-11	75	−168	52
R-113	118	− 31	26
R-114	39	−137	96

[a]psia = lb/in.² absolute.

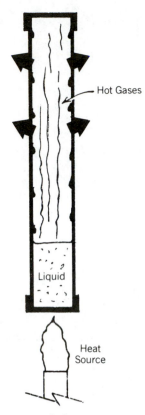

Figure 8.3 A simple heat pipe.

In phase-change heat-exchanged solar water heaters, one of three halocarbon fluids is usually used as the heat transfer fluid: R-11 (trichlorofluromethane), R-113 (trichlorodifluoroethane), or R-114 (dichlorotetrafluoroethane). Here are the boiling points, the freezing points, and the vapor pressure at 150°F of these three fluids.

All three of these fluids are classified as nonflammable, as having a low order of toxicity (UL Laboratories), and as the least hazardous (ANSI Standard) of all available low-temperature boiling heat transfer fluids. However, they should not be discharged into the atmosphere. The Environmental Protection Agency claims they are harmful to the environment.

The Common Types of Phase-Change Systems

There are three common types of phase-change solar water heaters: heat pipe, gravity return, and pumped return. The industry calls heat pipe and gravity-return systems passive systems; pump-return systems are called active systems.

Heat Pipes

Figure 8.3 shows a simple heat pipe. A metal tube containing a small amount of a boiling liquid is sealed closed. When heat is applied to the bottom of the tube, the liquid boils and vapor, being lighter, rises rapidly and heats the entire pipe.

The heat is conducted across the wall of the tube and heats the air surrounding it. This loss of heat causes the vapor to cool and condense on the tube's walls, whereupon it trickles back to the bottom of the tube by gravity to be reheated.

The Finned Tube Heat Pipe

In Figure 8.4 a fin has been placed on the lower two-thirds of the heat pipe, and the upper third of the pipe has been inserted into a water bath. Now, when heat from the sun falls on the finned tube, the liquid boils and rises into the tank section of the tube, where it gives up its energy to the water, condenses, and returns by gravity to the bottom of the heat pipe.

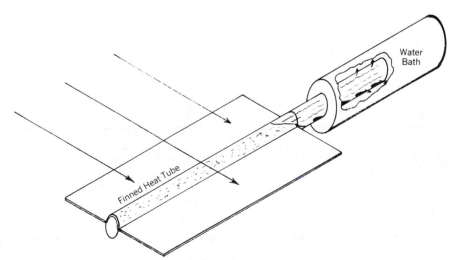

Figure 8.4 A finned heat pipe inserted into a water bath.

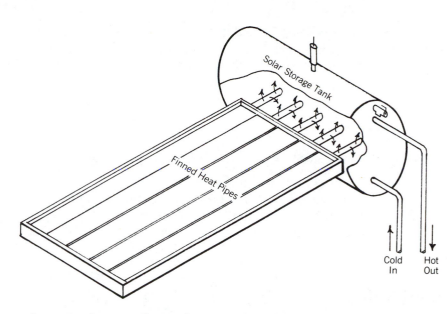

Figure 8.5 A horizontal tank gravity-return system.

Cold
In

Hot
Out

The Horizontal Tank Gravity-Return System

This heat pipe principle can be used to build a horizontal tank gravity-return solar water heater. Figure 8.5 shows a typical configuration. A series of heat pipes is built into a solar collector that is attached to a storage tank so that the heat pipes protrude into the tank. Each heat pipe operates independently.

A pressure relief valve, a drain valve, and hot and cold water connections complete the hookup. The water storage system operates in the same manner as the horizontal tank thermosiphon system shown in Chapter 2.

These systems are built and charged with liquid at the factory and installed as a complete operating unit. A typical system would contain approximately 24 ft² of collector area and 40 gal of water storage.

Such a unit weighs about 160 lb empty and about 500 lb when the storage tank is filled with water. Extreme care must be taken to locate the system where this heavy load can be properly supported.

Other Considerations

The storage tank should be stainless steel and insulated to R-24. No magnesium anode is required.

The heat pipes must be rated for 300 psi pressure.They should be all copper when using R-113 and R-114, but they should be steel when using R-11. To meet some local codes, the heat-exchanger portion of the heat pipes located in the storage tank must have a double wall that is vented to the atmosphere.

All brass valves and copper piping should be used throughout the system, and provision must be made for draining the system. The outdoor piping must be insulated to R-5 or better.

The system must be mounted at an angle of at least 15° from the horizontal for the heat transfer fluid to drain back properly to the bottom of the heat pipes.

The system should be sized at 20 to 25 ft² per occupant to supply 40 to 60% of the hot water needed by one person. Multiple units may be piped together in parallel to increase capacity.

Although the collector will not freeze, the tank can. However, the volume of water is extremely large, and it would take a long time to freeze the tank solid. Provision should be made to take the system out of service during extremely cold periods in northern climates or at high altitudes.

Advantages

The horizontal tank gravity-return system has these advantages.

- May be ground or roof mounted.
- A compact arrangement that requires minimal structural bracing.
- A factory prefabricated system that installs readily.
- A simple design that requires minimum maintenance.
- A nonfreezing collector.

Disadvantages

- Many roofs are not engineered to withstand the weight.
- The system must be installed outdoors, with the storage tank higher than the collectors.
- The tank can freeze in extremely cold weather. It does not have positive freeze protection.

Gravity-Return Evaporator/Condenser Collectors

In many instances it is not practical to place the water storage tank higher than the collectors. **Evaporator/condenser collectors** can be used in these instances. An evaporator/condenser collector is a collector that

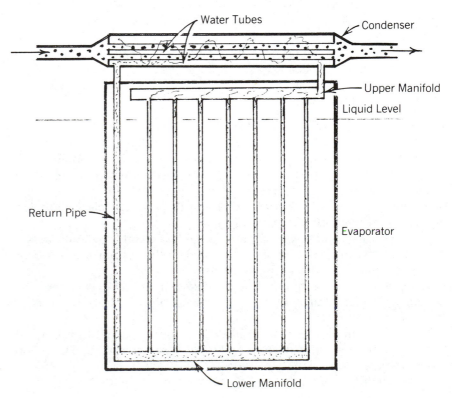

Figure 8.6 An evaporator/condenser collector.

contains a heat exchanger to boil the liquid (the evaporator) and a second heat exchanger to condense the vapor (the condenser). The solar collector plate is the evaporator, and the collector-to-storage heat exchanger is the condenser. Figure 8.6 shows a sketch of such a collector.

When the system is at rest, the liquid heat transfer fluid fills the collector's lower manifold and its risers. As the sun heats up the collector plate, the liquid boils and passes first into the upper manifold and then into the shell of the heat exchanger.

Water pumped from the bottom of the storage tank enters the tubes of the heat exchanger, is warmed by the hot vapors, and returns to the top of the tank. This cools the vapor, which then condenses and trickles down the return line to the lower manifold to be reheated.

Storage Loops for Evaporator/Condenser Collectors

The ability of the evaporator/condenser collector to operate efficiently and place more energy in storage is dependent on the ability of the storage loop to carry heat away from the collector.

If no heat could be removed from the system, the collector would heat up until the heat losses out the front, back, and sides equaled the incoming heat energy. This would be the stagnation point of the collector. At the stagnation point, the collector would contain a mixture of liquid and vapor at high pressure. No further energy could be collected.

Recirculating Water Storage Loops

Some evaporator/condenser system storage loops are being operated with a small, continuously operating circulator pump and no controller. We do not recommend such systems. Figure 8.7 shows such a system.

The circulator pump takes water from the bottom of the solar storage tank, pumps it up to the condenser, where it picks up heat, and returns the heated water to the top of the solar storage tank. No check valve is required. A temperature and pressure relief valve is used to protect the system, which can be drained by the valve at the bottom of the pump leg.

No provision is made in this configuration to protect the storage tank from overheating. The tank or the storage loop is only protected by temperature and pressure relief valves, which can open suddenly and discharge scalding water.

Although the collector portion of the system will not freeze, the heat exchanger, which contains storage water, may. There will be some loss of collected heat from the tank due to water circulation in nonsunshine hours.

Figure 8.8 shows a similar system with a differential controller added to turn the pump off and on and a check valve placed in a downcomer to prevent any reverse thermosiphoning. Even though an evaporator/condenser solar collector does not reverse thermosiphon below its boiling point, heat losses can be experienced in the piping and the condenser.

When the condenser sensor is 8 to 20°F hotter than the tank sensor, the pump turns on. When it drops down to 2.5 to 5°F hotter than the tank, the pump turns off.

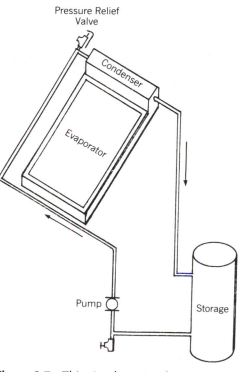

Figure 8.7 This simple recirculation system does not protect against freezing or overheating.

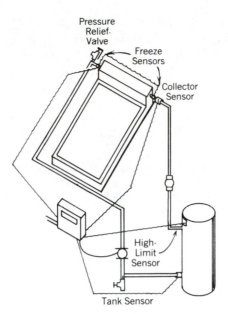

Figure 8.8 This recirculation system protects against overheating but offers inadequate freeze protection.

A high-limit sensor at the top of the tank overrides the differential sensors and turns the pump off if the tank has reached 160 to 180°F; freeze sensors located along the piping and at the condenser turn the pump on at 40 to 45°F to circulate hot water and prevent freeze-up.

This configuration is better, but we recommend its use only in very mild climates. Freezing of the water in the condenser can occur if power is lost.

For troubleshooting, testing, materials of construction, maintenance, and other considerations, refer to Chapter 3.

Draindown Storage Loops

A draindown storage loop may also be used with evaporator/condenser collector systems. Figure 8.9 shows such a system with two collectors connected in parallel.

When the controller is energized, if the temperature is above 40 to 45°F, the draindown valve closes and water from the storage tank starts to fill the piping and the condenser. Any air in the system is discharged through the air vent valve. When the system is full, the air vent valve closes and the system is ready to operate.

When the differential controller signals that the condenser is 8 to 18°F hotter than the bottom of the tank, water from the bottom of the solar storage tank is pumped up to the condenser and returns through the downcomer to the inlet near the top of the solar storage tank.

If the freeze sensors signal that the temperature of the piping or condenser has dropped below 40 to 45°F, the draindown valve opens and dumps the water in the storage loop to the drain. The vacuum relief valve allows air to enter the system to aid draining.

Precautions must be taken to make certain that the condenser and the piping is pitched to drain fully when the dump valve opens.

A more complete description of draindown loops and their testing, troubleshooting, and maintenance is located in Chapter 4.

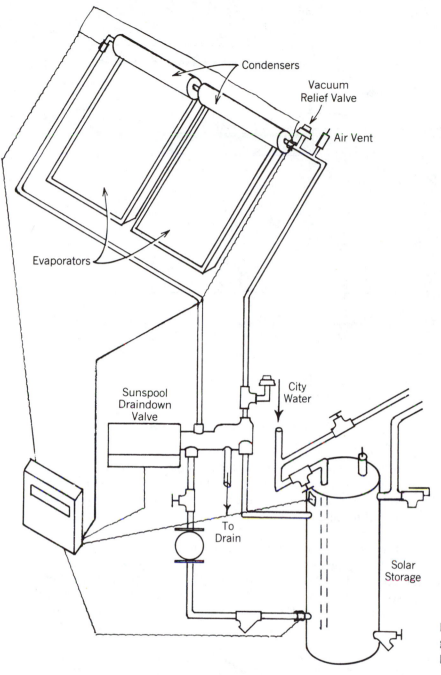

Condensers

Vacuum
Relief Valve

Air Vent

Evaporators

Sunspool
Draindown
Valve

City
Water

To
Drain

Solar
Storage

Figure 8.9 A draindown storage water loop gives good freeze and overheating protection.

Other Considerations

The condenser is usually insulated by the manufacturer. If not, use R-10 or better insulation.

The materials of construction of the condenser will also be determined by the manufacturer. The condenser should be rated for 300 psi pressure, and a temperature and pressure relief plug should be located in the wall of the shell. To meet some local codes, the condenser's water coil may need to have a double wall and be vented to the atmosphere.

All brass valves and copper piping should be used throughout the system. The outdoor piping must be insulated to R-5 or better.

Follow the manufacturer's instructions on mounting angles to assure

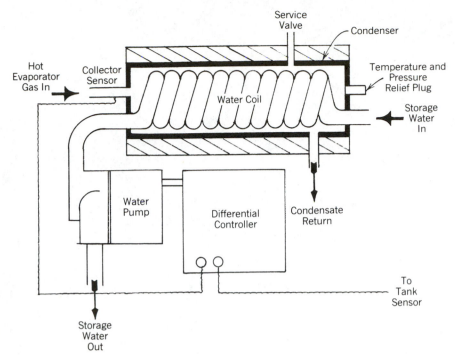

Figure 8.10 A cross-sectional view of a separate condenser and its storage water loop pump and controller.

that the condensed heat transfer fluid will drain back properly to the bottom of the evaporator.

Separate Condenser Gravity-Return System

As long as the condenser is higher than the evaporator and the lines connecting the two allow the condensed fluid to drain back to the evaporator by gravity, the condenser and the evaporator may be mounted separately. This allows the condenser to be placed indoors, where it is subjected to milder conditions.

Figure 8.10 shows a schematic of a commercial solar module containing a condenser, a differential controller, and a pump for the storage water loop. Figure 8.11 shows the module mounted above the collectors. When the condenser sensor is 8 to 20°F hotter than the bottom tank sensor, the pump turns on. When it drops to 2.5 to 5°F hotter than the bottom tank sensor, the pump turns off. A high-limit sensor set at 160 to

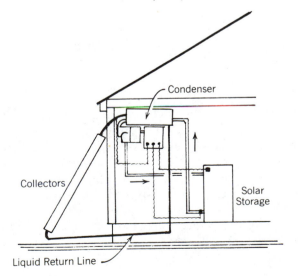

Figure 8.11 A simple gravity-return system installation schematic using a separate condenser.

180°F overrides the differential sensors and prevents pump turn-on when the tank reaches 160 to 180°F.

Hot vapor from the evaporator enters the condenser shell while water from the storage tank is circulated through the condenser coil by the pump. The vapor condenses on the cooler water coil and drains back to the bottom of the collectors.

This system has positive freeze protection. Only the refrigerant lines are outdoors. In all cases keep the condenser as close to the evaporator as possible to insure maximum system efficiency.

The operation, materials of construction, troubleshooting, maintenance, and other considerations are similiar to the evaporator/condenser system.

Pumped Phase-Change Systems

It is not always possible to mount the condenser higher than the collectors. In these cases the vapors will still flow to the condenser, but the condensed liquid must be pumped back to the evaporator (solar collector).

When the condenser is lower than the evaporator, a special pump—known as a liquid refrigerant pump—must be used to return the liquid to the evaporator. Liquid refrigerant pumps are *not designed to pump gases—only liquids,* so system design is critical.

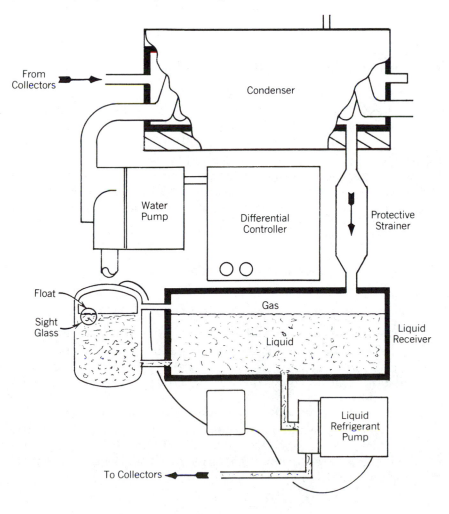

Figure 8.12 A cross-sectional view of a pumped-return condenser module.

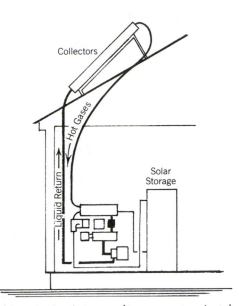

Figure 8.13 A pumped-return system installation schematic.

In addition to the pump the system must contain a receiver, a float switch assembly, and a protective strainer. The pump inlet must be located below the liquid receiver to keep the pump primed and to allow any gases formed in the pump to rise back into the receiver. A preassembled factory-built module, such as the one shown in Figure 8.12, should be used where the hot gas flows from the evaporator (collector) manifold down into the condenser and condenses on the surface of the storage water coil. The liquid then trickles through the protective strainer into the liquid receiver, which is attached to a float valve that actuates a switch to turn the refrigerant pump off and on. The pump returns the liquid to the bottom of the evaporator.

Figure 8.13 illustrates a typical pumped phase-change system. Hot gases from the upper manifold of the collector pass down to the condenser, condense on the condenser water coil, and flow by gravity through the filter into the liquid receiver.

When the liquid in the receiver has risen to the proper level, the float switch turns the refrigerant pump on and pumps the condensate back to the lower manifold of the collector.

The water loop operates independently. When the hot gases entering the condenser are 8 to 18°F hotter than the water in the bottom of the storage tank, the differential thermostat turns the water pump on and storage water flows through the water coil in the condenser, picks up the heat from the hot gases, and is returned to the upper part of the storage tank. When the tank reaches 160 to 180°F, the high-limit sensor, located on the storage tank, overrides the differential sensors and turns the water pump off.

Checking Out the Collector Loop

WARNING We recommend that servicepeople do not attempt to repair phase-change liquid collector loops unless they have been fully trained in the maintenance and repair of refrigerant-type solar systems. The liquids used in these loops have low boiling points and build up high pressures even at relatively low temperatures. Do not use a torch on these systems, attempt to disassemble them, or cut into the fluid passages of the system. Serious injury may result.

Gravity-Return Collector Loop Checkout

There are some simple observations and tests that you can perform to determine whether the collector loop is operational.

- Determine whether the collector array faces within 30° east or west of geographic south and is unshaded during sunlight hours.
- Measure the collector tilt angle and determine whether the manufacturer's recommendations for minimum tilt angle have been followed.
- If the evaporator and the condenser are separate, make certain that the condenser is mounted higher than the top of the evaporator (collector). Check the manufacturer's recommendations for minimum vertical separation. A good rule of thumb is an 18-in. vertical separation.

- Trace the return line from the condenser to the bottom manifold of the evaporator (collector) and make certain that liquid can drain freely from the condenser to the evaporator.
- Make certain that the line carrying the hot vapors from the upper manifold to the condenser slopes upward with no traps where vapors can condense and become trapped.
- Make certain that the condenser is installed so that the liquid return line is at the lowest point.
- Make certain that the controller's sensors are properly attached and that the wiring is intact.
- Look the collector system over carefully for any signs of physical damage, leaks, or ruptures.
- If the system contains sight glasses that allow the liquid to be observed, make a visual check of the fluid as per the manufacturer's instructions.

Pumped-Return Collector Loop Checkout

First, perform the checkout recommended for the gravity-return loop except for the location of the condenser and the lines to and from the evaporator. Then perform the following checks.

- The condenser in a pumped-return system is located below the top of the collector. Make certain that the lines from the evaporator (collector) upper manifold flow down to the condenser without any traps that would impede condensate flow.
- Make certain that the line from the evaporator to the condenser is fully insulated.
- Check out the operation of the condenser, filter, receiver, switch, and refrigerant pump by *carefully following the manufacturer's checkout procedure. If the procedure is not available and you do not have the proper training, do not attempt to check out the module.*

Measuring Condenser Exchanger Effectiveness In heat-exchange systems using sensible heat to effect the transfer you were able to measure heat-exchanger effectiveness by using the changes in temperatures for the collector loop fluid and the storage loop fluid. You cannot perform the same test in a boiling system, because the collector fluid is transferring heat by changing state, not temperature. You would have to know the flow rate and the latent heat of vaporization for the boiling fluid. Those figures are almost never available to a serviceperson.

Checking Out the Solar Storage Loop

As previously noted, the solar storage liquid loop characteristics are similar to those of the recirculation systems shown in Chapter 3 or the drain-down systems shown in Chapter 4. Follow the procedures shown in those chapters to check out the solar storage loop.

System Sizing

Phase-change systems should be sized at 20 to 25 ft² per occupant to provide 40 to 60% of the total hot water load. Solar storage should be sized at 1.25 to 2.0 gal/ft² of collector.

The flow rate in the collector is regulated automatically by the amount of incoming solar energy and—in *flooded evaporators*—is equal to the fluid's latent heat of vaporization divided by the incoming heat.

$$\frac{\text{Latent heat of vaporization, Btu/lb}}{\text{Incoming heat flux, Btu/hr}} = \text{flow rate, lb/hr}$$

A flooded evaporator is an evaporator (collector) fully filled with liquid at temperatures below the liquid's boiling point.

The flow rate of the storage water loop usually ranges from 0.04 to 0.06 gal per minute per square foot of collector, which is faster than in most liquid heat transfer fluid systems.

Testing Performance

The performance of the system is tested on a clear day between 9 A.M. and 3 P.M. by taking system temperatures and calculating the heat gain of the storage tank, and examining the operating conditions of the condenser. Proceed as follows.

- In measuring the temperatures of a phase-change system you will be measuring very small temperature differences, so you should perform a preliminary calibration of your temperature gauges by placing four gauges in one hot location to determine whether they all record the same temperature. If not, add or subtract the differences between them when recording your test readings.
- Place a temperature gauge on the line between the upper manifold and the condenser as close to the condenser as possible. Call this location T1.
- Place a temperature gauge on the line between the condenser and the lower manifold as close to the condenser as possible. Call this location T2.
- Place a temperature gauge on the cold water line running from the bottom of the solar storage tank to the condenser. Place the gauge as close to the condenser as possible. Call this location T3.
- Place a temperature gauge on the hot water line running from the top of the solar storage tank to the condenser. Place the gauge as close to the condenser as possible. Call this location T4.
- Wait 3 to 5 minutes for the gauges to reach equilibrium. Read the gauges and record the temperatures in a chart like the one in Table 8.1 on the line labeled inital.
- Wait 60 minutes. Read the gauges and record the temperatures in Table 8.1 on the line labeled 60 min.

Table 8.1

TEMPERATURE CHART—PHASE-CHANGE HEAT EXCHANGER PERFORMANCE TEST

TIME READ	TEMPERATURE EVAPORATOR IN T1 (°F)	TEMPERATURE EVAPORATOR OUT T2 (°F)	TEMPERATURE STORAGE IN T3 (°F)	TEMPERATURE STORAGE OUT T4 (°F)	TANK AVERAGE TEMPERATURE (T4 − T3)/2 (°F)
Inital	_____	_____	_____	_____	_____
60 min	_____	_____	_____	_____	_____
°F Rise	_____	_____	_____	_____	_____

Performing the Calculations

First, determine the temperature rise in the tank over the 1-hour test as follows.

- For the second set of readings (60 min), add T3 to T4 and divide the result by 2. This gives the average storage tank temperature at the end of the test.
- Perform the same calculation for the first set of readings (inital). This gives the average temperature of the tank at the start of the test.
- Subtract the average tank temperature at the start of the test from the average temperature of the tank at the end of the test. This gives the heat rise in the tank in 1 hour.

Second, look at the operating temperatures of the condensing heat exchanger. The relationships to examine are T1 — T4 and T4 — T3.

The correct values for these two relationships are established by the system manufacturer and will vary from one design to another. *You must have the manufacturer's engineering information to determine whether the relationships are meeting the design specifications.*

As an example of the proper relationships, one manufacturer specifies that T1 — T4, the difference between the hot collector gases and the water exiting back to the tank, be ideally 2°F, and that T4 — T3, the difference between the cold and the hot storage water, be 3°F.

There will be little or no difference between T1 and T2, the entering hot collector gases and the exiting condensed liquid.

The Meaning of the Results

A properly functioning system will give a storage tank heat rise of at least 6 to 9°F in 1 hour. If the heat rise is less, the system is not performing properly. When the piping and standby losses are ignored, the heat gain of the storage tank is a function of the incoming solar heat pickup, the effectiveness of the condenser heat transfer, and the ratio of collector area to water volume. The following formula can be used to calculate the maximum heat that could be placed in the tank.

$$\frac{\text{Solar heat collected, Btu/ft}^2\text{/hr} \times \text{collector area, ft}^2}{\text{Storage, gal} \times 8.33 \times \text{specific heat}} = \text{temperature rise}$$

For example, if

Solar heat collected = 150 Btu/ft^2/hr
Collector area = 40 ft^2
Storage = 60 gal

Then

$$\frac{150 \times 40}{60 \times 8.33 \times 1} = 12°F \text{ maximum temperature rise/hr}$$

and if

Solar heat collected = 150 Btu/ft^2/hr
Collector area = 40 ft^2
Storage = 80 gal

Then

$$\frac{150 \times 40}{80 \times 8.33 \times 1} = 9°F \text{ maximum temperature rise/hr}$$

If the difference in temperature between the incoming and exiting water, T4 — T3, is within the manufacturer's specifications, but the difference between the hot collector gases and the exiting water, T1 — T4, is larger than the specifications call for, there is a problem with the collector circuit. One common problem would be air in the system.

If the difference in temperature between the incoming and exiting water, T4 — T3, is larger than the manufacturer's specifications, but the difference between the hot collector gases and the exiting water, T1 — T4 is within the specifications, there is a problem with the storage water circuit. One common problem would be a low storage water flow rate.

WHAT CAN GO WRONG

Assuming that the system has operated successfully prior to the service call, nine things can go wrong.

1. The owner can change the valve settings.
2. The water loop can become airbound.
3. The collector loop can leak and lose its refrigerant.
4. The water loop can leak or become clogged with scale or sediment.
5. The system can suffer physical damage.
6. The water pump or liquid refrigerant pump can fail.
7. The controller or sensors can fail.
8. The float or switch in a pumped-return system can fail.
9. In a draindown system the drain valve, vacuum relief valve, or air vent valve may fail.

THE TROUBLESHOOTING CHART

1. Look for physical damage.

Item	Yes	No
Crushed pipes	_____	_____
Leaky valves, joints, or pipes	_____	_____
Collector cover or case damage	_____	_____
Freeze damage	_____	_____
Sensor wire damage	_____	_____
Damaged or missing sensor	_____	_____
Electric service interruption	_____	_____
Sticking air vent valve	_____	_____
Clogged vacuum relief valve	_____	_____
Inoperative check valve	_____	_____
Pump or controller damage	_____	_____

Storage tank damage   _____   _____

Structural movement   _____   _____

2. Check the valve settings. There is so much variation from system to system that a generalized chart cannot be given. Consult the troubleshooting charts in Chapters 2 to 4 for the storage loops. Use the manufacturer's information for the collector loop. Make no changes in the collector loop if the manufacturer's information is unavailable.

3. Drain scale and sediment from the storage water loop. The mechanics of accomplishing this are dependent on the type of system being worked on. See the troubleshooting charts in Chapters 2 to 4.

4. Clean the storage loop pump strainer and check the flow.

   - Isolate the strainer by closing the throttle valve or the pump isolation valve and the water supply valve or the valve between the strainer and the tank.
   - Remove and clean the strainer with a stiff brush.
   - Open the water supply valve or the valve between the strainer and the tank to flush the strainer case.
   - Stop the water flow and replace the strainer screen and plug.
   - Open the pump leg drain valve to flush the drain leg.
   - Restore the system's valves to their operating positions.

5. Flush the riser and check the flow. The specifics will depend on the valve configurations of the system. Follow the procedures outlined in the troubleshooting charts in Chapters 2 to 4.

6. Run the system draindown test (if it is a draindown storage water system). Follow the troubleshooting chart in Chapter 4.

7. Run the system performance test as outlined previously.

   - Install the temperature gauges.
   - Check for clear sky and time of day.
   - Run the test and record the temperatures in Table 8.1.
   - If the pump will not run on automatic or the draindown valves will not close, skip to step 8.
   - Evaluate the results.

     1. Does the pump operate on automatic?
     2. Was a 6 to 9°F tank temperature rise experienced?
     3. Do the temperatures for T1 to T4 agree with the manufacturer's specifications?

If the answers to these questions are yes, the system is performing satisfactorily. Except that you have not checked the pump shutoff action. See Chapter 17 to check this function.

Skip steps 8 and 9 if the pump is functioning satisfactorily on the automatic controller setting.

8. If the pump will not run on manual or a draindown system will not fill,

   - Check for a blown fuse or faulty circuit breaker.
   - Check the pump on a separate circuit if it is the plug-in type.

- Check for electricity, using a trouble lamp, if the pump is hard wired.
- Isolate whether the pump or the controller is faulty.
- Proceed to troubleshoot the pump if it is faulty.
- Check for electricity at the valve motors.
- Check for frozen or stuck valves.

9. If the pump will not run on automatic,

- Recheck the sensors and sensor wiring for damage.
- Troubleshoot the controller and sensors as outlined in Chapter 17.

10. Visually check over the system.

- Restore all valves to their operating positions.
- Place the controller on automatic operation.
- Check for any needed structural repairs.
- Is any painting needed?
- Is there any loose or missing insulation?
- Are there any leaks?

11. Finish up the job.

- Clean up the area.
- Remove all tools.
- Secure the area.
- Complete a service report for the owner.
- Present a bill for services.

ON THE JOB

Your toolbox should include an assortment of boiler drain, air vent, vacuum relief, and draindown valves. It should include replacement sensors, a voltmeter, a sensor substitution box, a level, a long extension cord, and a trouble lamp or circuit tester. If you are qualified to work on phase-change systems, you should also carry the parts and tools required.

Gear your troubleshooting strategy to the owner's problems.
Repair any known physical damage first.
Check out the pump and controller.
In draindown systems, check out the draindown action.
Run the performance tests outlined in step 7.
While the performance tests are running, perform steps 1 and 2.
If performance is unsatisfactory, perform steps 3 to 5.
If performance has not been restored, there may be a fault in the sealed collector system. Proceed to troubleshoot this system, if qualified, or call in a phase-change solar collector specialist.
Proceed to finish up the job.

WHAT YOU SHOULD HAVE LEARNED

1. Phase-change heat-exchanged systems use a boiling liquid to transfer heat from the collector to the storage water. These systems transfer heat using the latent heat of vaporization instead of using sensible heat.
2. The liquid of choice is usually a halocarbon fluid, several of

which have boiling and freezing points in the correct range, have a low order of toxicity, are nonflammable, and are nonhazardous to human life.

3. Phase-change systems can be built on the heat pipe principle. They must be mounted so that the condensed fluid drains back to the bottom of the heat pipe.

4. Phase-change systems can be built with either integrated or separate evaporators and condensers.

5. When the condenser is mounted above the evaporator, the condensate can return to the evaporator by gravity, and no pump is required.

6. When the condenser is located below the evaporator, a liquid refrigerant pump with a liquid receiver and a float switch is required. This operates independently of the storage water loop pump.

7. Both recirculating water systems and draindown systems are commonly used with both gravity-return and pumped-return phase-change systems when the storage tank is located below the condenser.

8. Phase-change collector loops operate under high pressures. They should never be heated, disassembled, or cut into by unqualified service people. *Serious injury can result.*

9. Heat-exchanger effectiveness cannot be determined with only temperature measurements, because the evaporator and condenser operate on latent heat transfer, which is not accompanied by a temperature change.

10. Performance testing is best accomplished by running a tank temperature rise test. The average system should increase tank temperature by at least 6 to 9°F per hour under clear skies between 9 A.M. and 3 P.M.

Evaluate the Installation

After the system type has been identified, the physical installation must be evaluated. The installation is evaluated by checking how the system is laid out, examining the system for structural integrity, rating the system's components, determining if the system is sized correctly for the load, and looking for any safety- or health-related problems.

Chapters 9 through 13 cover the evaluation of the major solar system subsystems. The tasks that need to be accomplished vary from one part of the installation to another.

At the end of these five chapters, readers should be able to:

- Pinpoint layout problems.
- Recognize component-related problems.
- Determine if the installation is structurally sound.
- Match the size to the load.
- Locate and correct any safety- or health-related problems.

The Collector Array

The collector array must be oriented in the right compass direction, tilted at the proper angle from the horizontal, and unshaded for at least 6 hours during midday. It must be securely fastened down, properly plumbed, and protected against adverse environmental conditions. The collectors must be of proven design, performance rated, and built from durable materials. The array must be matched to the load imposed on the system, and the installation must not create any safety or health hazards.

Solar Radiation

The sun's energy reaches the collector in three ways—as *direct, diffuse,* and *reflected* radiation. Direct radiation consists of parallel rays coming directly from the sun; diffuse radiation consists of scattered, nonparallel rays that can come from any point in the sky; and reflected radiation consists of radiation that is reflected onto the collectors from a nearby object.

All three types of energy contribute to heating the collector, but direct radiation makes up the greatest portion—about 70 to 80% of the total on a clear day.

Therefore, the solar system's collectors must be oriented to take maximum advantage of direct radiation.

Solar Movement

The sun rises in the east, moves across the southern sky, and sets in the west. However, its exact position depends on the latitude of the location and the day of the year.

Think of the sky as a dome, as shown in Figure 9.1, with the center of the dome being the collector array. The path of the sun for any day of the year can be drawn on this dome.

When the sun's path on the shortest day of the year, December 21, is drawn on the dome, the lowest path of the sun during the year is described. When the sun's path across the sky on the longest day of the year, June 21, is drawn on the dome, the highest part of the sun during the year is described.

On any other day of the year, the sun's daily path lies somewhere between these two lines. For example, on March 21, the path of the sun is through the middle of the "solar window."

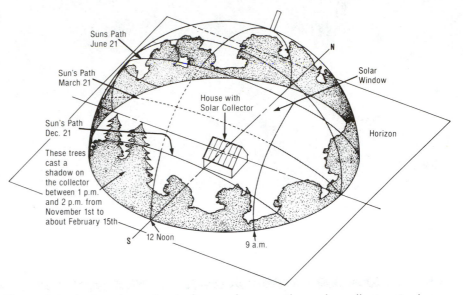

Figure 9.1 An imaginary dome above a house with a solar collector. Such a dome can be used for drawing a "solar window." (Reprinted by permission from *The Solar Decision Book*, R. Montgomery and J. Budnick, Wiley, New York, 1978.)

The best hours for collecting solar radiation are from 9 A.M. to 3 P.M.; so if we were to locate the position of the sun at 9 A.M. on December 21 and June 21 and draw a line on the dome between these points, we would describe the easterly side of the solar window through which most direct solar radiation passes. A similar line drawn between the 3 P.M. locations of the sun on December 21 and June 21 would describe the westerly side of the window. At 12 noon, the sun would be in the center of the window, which is located at geographic south.

Figure 9.2 shows a side view of this window. You can see from this side view that the sun's height above the horizon changes, moving through an arc of 47° from December to June.

The same solar window can be plotted on a flat surface as a *Mercator projection*. Figure 9.3 shows a typical Mercator projection. Now the height of the sun above the horizon is described by horizontal straight lines running from left to right, and the compass orientation is described by vertical lines running from the horizon to directly overhead.

Note the two trees sticking up into the solar window in Figure 9.3. Obviously, these will shade the collector during part of the afternoon from December to almost March but will not shade the collector from March to June.

Figure 9.4 shows this Mercator protection in a little more detail.

Figure 9.2 A side view of the dome with a 40°N latitude "solar window" shown. (Reprinted by permission from R. Montgomery and J. Budnick, *The Solar Decision Book*, Wiley, New York, 1978.)

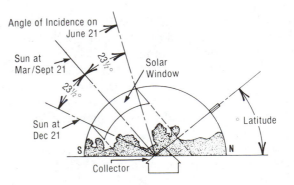

146 Evaluate the Installation

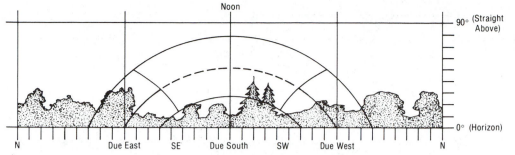

Figure 9.3 A Mercator projection of the sky dome. (Reprinted by permission from R. Montgomery and J. Budnick, *The Solar Decision Book,* Wiley, New York, 1978.)

December 21 is known as the **winter solstice,** and June 21 is known as the **summer solstice.** This particular projection is for 28°N latitude.

Figure 9.5 contains projections for 24°N, 28°N, 32°N, 36°N, 40°N, 44°N, 48°N, and 52°N latitudes. You can photocopy the proper solar window for your latitude or you can reproduce it in larger form on graph paper by plotting the lines. If the latitude of your location is not shown here, use the projection closest to your latitude. That will give you enough accuracy for your purposes in examining the installation site.

Orientation, Tilt, and Shading

Orientation is the compass direction that the collector faces. **Tilt** is the angle between the face of the collector and the ground. **Shading** is the existence of objects in the solar window that prevent the sun's rays from striking the collector.

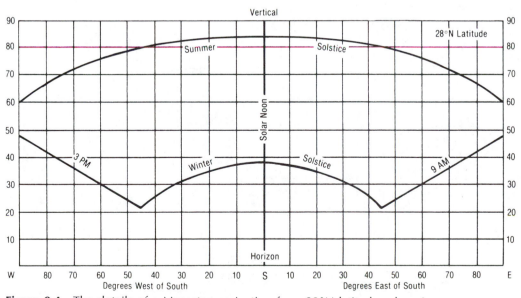

Figure 9.4 The details of a Mercator projection for a 28°N latitude solar window. (Reprinted by permission from R. Montgomery and J. Budnick, *The Solar Decision Book,* Wiley, New York, 1978.)

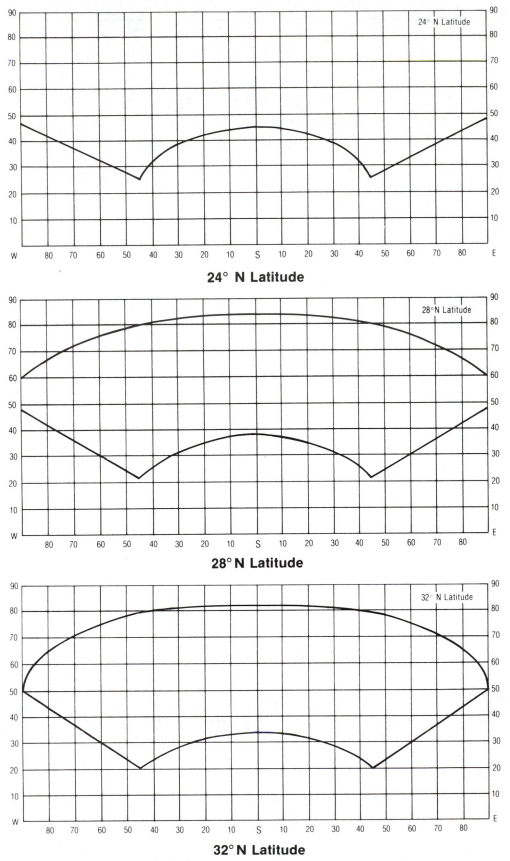

24° N Latitude

28° N Latitude

32° N Latitude

Figure 9.5 Mercator projections of solar windows for the United States. Covers from 24°N to 52°N latitude. (Reprinted by permission from R. Montgomery and J. Budnick, *The Solar Decision Book,* Wiley, New York, 1978.)

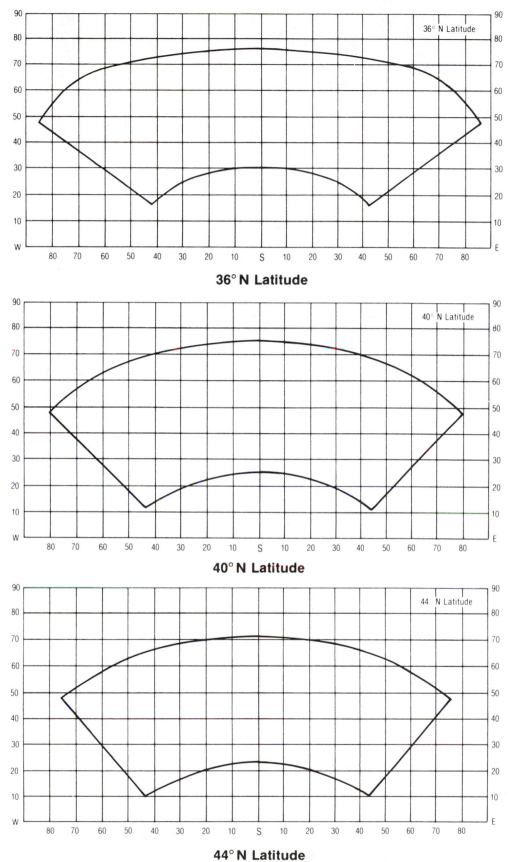

36° N Latitude

40° N Latitude

44° N Latitude

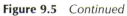

Figure 9.5 *Continued*

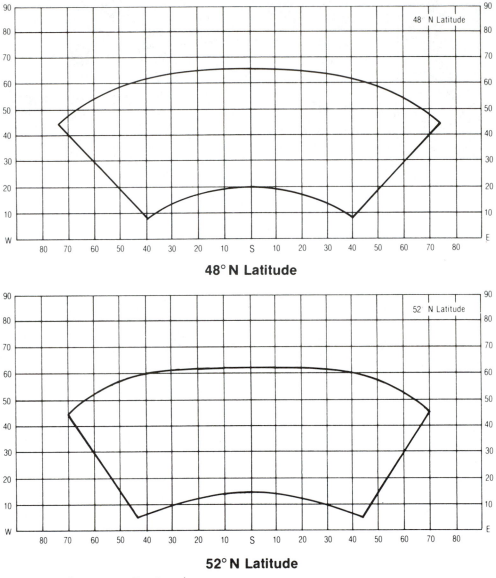

48° N Latitude

52° N Latitude

Figure 9.5 *Continued*

Orientation

In most locations a collector facing due south is ideally oriented. The exceptions would be when the local weather conditions result in more cloudy mornings or more cloudy afternoons or where the collectors are shaded in the morning or the afternoon.

Here is a simple rule to follow for the proper orientation of a solar collector array.

> **Rule for proper orientation**   The collector array should face true south ±30°. For all practical purposes, an orientation of south ± 15° causes no loss in energy collection. Orientations greater than 15° cause a loss that reaches 10% at ±30°.

Determining True South   There are four ways to determine true south: maps, compasses, stars, and solar time calculations. The use of a local street map or a compass are the most practical ways for a serviceperson.

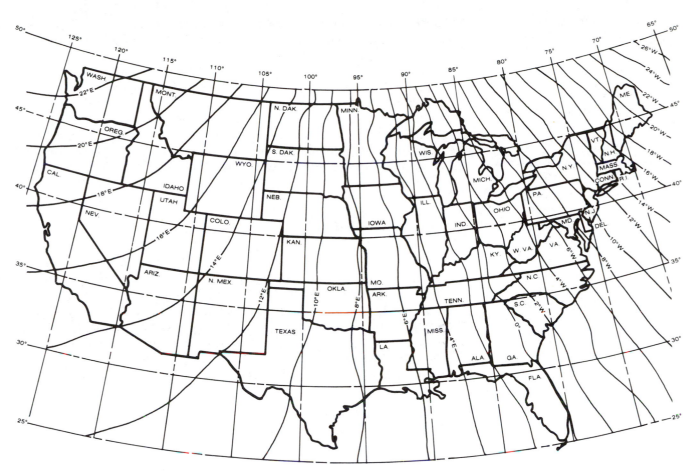

Figure 9.6 An isogonic chart of the United States. (Reprinted by permission from R. Montgomery and J. Budnick, *The Solar Decision Book,* Wiley, New York, 1978.)

When the stars are used, find the North Star. True geographic south is 180° from the North Star.

When a compass is used, the magnetic reading provided by the compass must be corrected for magnetic deviation to show true geographic north and south. Magnetic deviation changes with geographic location.

Figure 9.6 is an **isogonic chart** of the United States. An isogonic chart is a map that shows the magnetic lines of deviation from north for all geographic locations. For instance, if the magnetic deviation for your location is 10° east, true north will be 10° west of the compass-indicated magnetic north, and true south will be 10° east of the compass-indicated south. Once true north is determined, true south always lies 180° away.

Tilt

Figure 9.7 shows a collector tilted so that the face of the collector is perpendicular to the sun's rays. This is called a *right angle-collector tilt angle.* A solar collector collects the maximum amount of heat when its surface is tilted at a right angle to the sun. Thus the collector tilt angle is an important consideration when evaluating the collector layout.

As previously noted, the sun's altitude at solar noon changes during the year (see Figure 9.2). On March 21 and September 21 the sun's noon altitude is equal to 90° minus the latitude of the collector array's location. On December 21 the sun's altitude is $23\frac{1}{2}°$ lower, and on June 21, the sun's noon altitude is $23\frac{1}{2}°$ higher. For example, at latitude 40°N, the

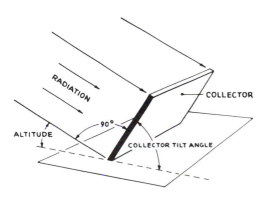

ALTITUDE + 90° + TILT ANGLE = 180°
TILT ANGLE = 180° - 90° - ALTITUDE
OR, TILT ANGLE = 90° - ALTITUDE

Figure 9.7 The right angle-collector tilt angle. (Reprinted by permission from R. Montgomery and W. Miles, *The Solar Decision Book of Homes,* Wiley, New York, 1982.)

The Collector Array **151**

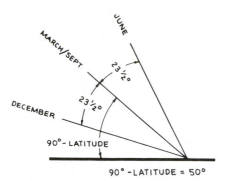

Figure 9.8 The sun's altitude at solar noon changes with the season. (Reprinted by permission from R. Montgomery and W. Miles, *The Solar Decision Book of Homes,* Wiley, New York, 1982.)

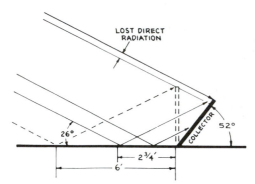

Figure 9.9 The collection angle can be increased for winter energy collection if the collectors are mounted on a highly reflective surface. (Reprinted by permission from R. Montgomery and W. Miles, *The Solar Decision Book of Homes,* Wiley, New York, 1982.)

sun's altitude at noon on March 21 and September 21 is 50° from the horizontal. On December 21 that angle is $26\frac{1}{2}°$. Finally, on June 21, the angle is $73\frac{1}{2}°$.

Figure 9.8 shows the best collection angles for any location. If the collection of the solar energy is to be maximized during the winter months, the collectors should be set at a tilt angle equal to the latitude plus 11 to 12°. If the collection of solar energy is to be year round, the collectors should be tilted at an angle equal to the latitude. If solar collection is to be maximized during the summer months, the collectors should be tilted at an angle equal to the latitude minus 11 to 12°.

There is an exception to these tilt angles. If the collection is to be maximized during the winter months and if the collector is mounted on a highly reflective horizontal surface, the collector tilt angle can be increased up to 90° to take advantage of the reflected energy (see Figure 9.9).

In actual practice it is rarely possible to tilt the collectors at the best collection angle, because the rooftops of most homes, the most popular place for collector mounting, are generally tilted at very low angles.

When the system is being operated year round, variations in tilt angle from latitude to ±10 to 15° will not affect the annual collection of solar energy to any great degree. It will merely change the months in which the energy collection is optimized. The tilt angle is not critical within these guidelines.

How to Determine Tilt Angle   The easiest way to determine tilt angle is with an inclinometer. When an inclinometer is not available, the tilt angle can be determined with a plumb bob and a protractor or with a level and a protractor. Figures 9.10 and 9.11 show how this is done. In Figure 9.10 measure angle *a* with a protractor. Subtract angle *a* from 90° to obtain angle *b*. Angle *b* is the tilt angle. In Figure 9.11 measure angle *b* directly by holding a carpenter's level at the level and measuring with a protractor.

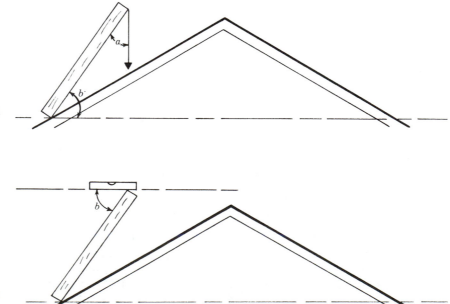

Figure 9.10 Determining tilt angle with a plumb bob and a protractor.

Figure 9.11 Determining tilt angle with a level and a protractor.

Shading

The shading rule is simple:

> **Rule for shading** It is *essential* that the collector not be shaded during the prime collection hours. The prime collection hours are 9 A.M. to 3 P.M. Thus, little or no shade should appear in the solar window.

For the serviceperson, shade determination for an existing system is an eyeball proposition. The serviceperson merely has to ascertain that gross shading problems do not exist.

Commercial instruments for determining shading are available and should be utilized in laying out new installations. Two of these instruments are discussed in Chapter 19.

Fastening, Piping, and Environmental Protection

Fastening is properly supporting and bolting or lagging the collectors to a solid framework. **Piping** is properly interconnecting the collectors and leading the connections to the rest of the system. **Environmental protection** is properly protecting the collectors and the associated piping from the weather.

Fastening

Four major types of collector mountings must be considered.

- The collectors can be an integral part of the roof.
- The collectors can be mounted against a building's roof or sidewall.
- The collectors can be mounted up off a roof or sidewall and parallel to the roof or sidewall.
- The collectors can be mounted on racks so that they are not parallel to the roof or sidewall.

In all four instances the collectors or racks must be bolted or lagged into solid timbers. They cannot be safely nailed or screwed to the side or the roof of a building structure.

Integral Roof or Wall Mounting When the solar collectors are an integral part of the roof or wall, these precautions must be taken (see Figure 9.12).

1. The outer glazing of the collector must be part of the roof's waterproof membrane. As such, it must be built as part of the roof. Metal collectors expand and contract as they heat and cool. Therefore, the outer glazing should not be part of the collector itself.
2. The lower edge of the glazing must lap over the roofing material and be waterproofed with the proper mastic and/or caulking.
3. The upper edge of the glazing must extend up under the roofing material and be waterproofed with mastic and/or caulking. The mastic or caulking chosen must not run down onto the glazing when the roof is hot, and the mastic must be capable of cycling from $-20°F$ to $+200°F$ without failure.
4. The sides of the glazing must be flashed to each shingle course if

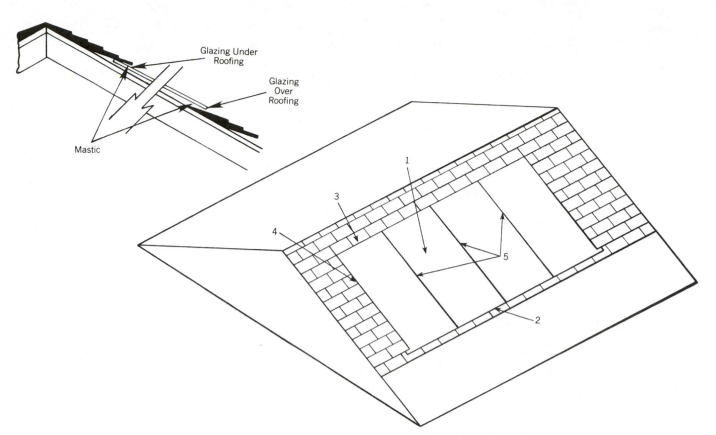

Figure 9.12 Precautions to take when the collectors are an integral part of the structure.

the glazing protrudes above the shingle. If it does not, the glazing must extend under the shingle and be waterproofed with mastic and/or caulking.

5. The outer glazing must be supported in a way that will ensure the glazing's ability to carry a proper snow loading for the geographical location. All seams must be carefully cemented with a mastic and/or caulking. Provision must be made for glazing expansion and contraction, and all glazing penetrations must be carefully sealed with mastic.

When the collectors are an integral part of the roof, piping is usually inside under the roof. No piping penetrations are necessary.

Rooftop or Sidewall Mountings When the collectors are mounted directly on and against the roof or the sidewall, these areas must be considered (see Figure 9.13).

1. The waterproof membrane of the roof or sidewall must extend fully under all collectors so that the roof or sidewall is watertight without the collectors.
2. The collectors must be firmly fastened to supports that have been bolted through the roof to a 2- × 4-in. or larger timber that spans the rafters, or they must be lag bolted securely to the rafters or sidewall studs (see insets to Figure 9.13). It is not satisfactory to fasten the collector mounts to just the roof sheathing. The mounts should be set in mastic, and the bolt heads should be mastic covered.

When fastening down the collectors, provisions must be made for expansion and contraction during heating and cooling. This is accomplished by placing a flexible spacer cushion between the collector box and the mounting screws or by using a hinged mount that can move freely. Failure to allow for this expansion and contraction will cause the roof mounts to "work." This will eventually lead to leaks in the roof around the mounting.

3. The upper edge of the collectors must be flashed so that water, ice, and snow cannot collect against the top of the collector and back up under the roofing or leak down under the collector. The flashing surface must be pitched downward. The edge must extend over the upper edge of the collectors, and the ends of the flashings must be boxed in.

4. If the collectors abut each other, the crack between them must be flashed and caulked so that no water, ice, or snow can get between them. On the other hand, if the collectors are spaced away from each other, the lower roof support brackets must be fashioned to allow free drainage. Otherwise, ice can build up and expand, forcing the sides of the collectors apart and causing leaks in the interconnecting piping.

5. The piping from the collectors must be carefully insulated with insulation suited for the environment. The penetration of the pipes through the roof must be carefully made through flexible roofing pipe collars designed to prevent water from penetrating the roof membrane. These penetration seals or collars must be properly

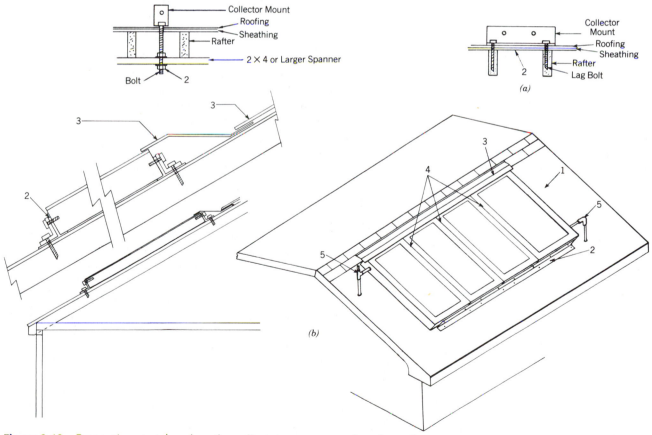

Figure 9.13 Precautions to take when the collectors are mounted on the surface of the structure.

mounted within the membrane. Seal the point where the pipes enter the collar with a high-quality, heat-resistant, flexible caulking compound.

Similiar considerations are needed for sidewall-mounted collector arrays. Additionally, the bottom collector brackets in sidewall mounts must be sturdy enough to carry the full weight of the loaded collectors, and the brackets must be bolted through or lagged into the wall studs, shoe, or joist band.

Rooftop Mounting Problems Collectors mounted directly on and against the roof have serious disadvantages. It is almost impossible to construct a flashing system that will provide a waterproof seal around and between collectors on either a shake or shingle roof. Any openings will provide a point of entrance for water to fill the space between the collector backs and the roof surface. It is almost impossible to promote the free drainage of this water. Algae growth and extended wetting of the roof by standing water will shorten the life of the roof. Flashing systems to help keep the roof dry are expensive and labor intensive. Therefore, collectors should never be installed directly against the roof. If you encounter a system where repairs are necessary, remount the collectors up off the roof.

Collectors Mounted on Standoffs[1] It is commonly accepted practice to mount the collectors on **standoff brackets.** A standoff bracket is a collector bracket that holds the collector up off the roof so that water, ice, and snow are free to collect and drain from under the collectors. The use of standoff brackets is recommended by most authorities as the best way to mount collectors on an existing roof or sidewall. These areas must be carefully considered when using standoff brackets.

1. The standoff brackets must be set in mastic and either lagged to the rafters or bolted through spanners running under the rafters. All lag screw and bolt heads should be mastic covered. Since the wind can now exert high lifting forces on the back of the collector array, the brackets must be engineered to withstand uplift. Expansion and contraction are less of a problem, but allowances must be made to prevent the bracket bolts or lags from "working."
2. Piping penetrations must be made through the proper flexible gasketing, and all piping must be carefully insulated against the environment.
3. The connection of the brackets to the collectors must be rugged enough both to support the collectors' weight and to withstand all the forces exerted on them. A minimum of four mounting points, two top and two bottom, is essential.
4. The space between the collector box bottom and the roofing should be at least $1\frac{1}{2}$ in. In areas where snow buildup occurs, a 3-in. offset spacing will insure good drainage under collectors.
5. Collector support members such as wood sleepers must run vertically on pitched roofs to promote draining and avoid damming water on the roof.
6. There is no need to caulk or flash between collectors, but they

[1]Each collector manufacturer has a unique standoff design that is engineered for its collectors. No single design is satisfactory for all systems. Consult with the collector manufacturer for details or follow a design that has proven satisfactory through previous use.

should be spaced far enough apart to allow for expansion and contraction.

Rack Mounts   Collectors can also be rack mounted. This procedure consists of mounting the collectors on racks built to support them. Racks are commonly used on low-pitched or flat-roof houses or where the collectors are ground mounted. The racks are usually fabricated from metal angle iron, round tubing, or square tubing. They can be arc welded or firmly bolted together. When rack mounting collectors, the following areas must be considered (see Figure 9.14).

1. All rack foot plates on buildings must be carefully set in mastic and firmly bolted or lag screwed to timbers or spanners.
2. Cross bracing must be used to prevent lateral movement and subsequent rack collapse under high wind forces.
3. The collectors must be firmly attached to the racks. Collector expansion and contraction must be allowed for. If the collector box, the brackets, and/or the racks are dissimilar metals, a nonconducting spacer should be used between them.
4. If the racks are mild steel, they must be thoroughly rustproofed and painted. Galvanized steel and aluminum may be left unpainted.
5. Ground-mounted racks should be set on concrete piers. The piers are best poured wedge shaped to prevent uplift from the winds. The mounting plates should be fastened to 6- to 8-in. J bolts set into the concrete before it hardens. If the piers do not extend below the frost line, piping should be flexible enough to allow for frost heaving.
6. All racks must be professionally engineered throughout to withstand high wind loads.

Standoff and Rack-Mounting Problems   Both the standoff and rack-mounting methods are superior to rooftop mounting but, again, there are common precautions to be observed.

Extreme care must be used in mounting metal brackets on asphalt composition roofing systems, since thermal expansion and wind loads on these brackets can cause the brackets to cut the roofing.

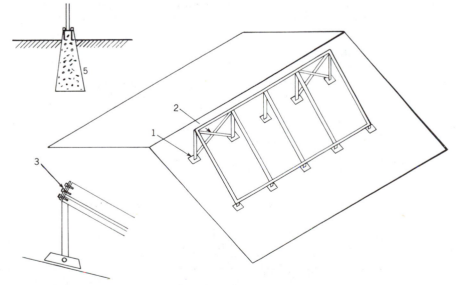

Figure 9.14   Precautions to take when the collectors are rack mounted.

Inspect the points where the brackets contact the roof for signs of this problem and check the mastic bedding around and under the brackets for deterioration. Applying fresh mastic and/or aluminum flashing may resolve the problems. However, a safer, long-term solution is to remove and remount the collectors on brackets separated from the roof by treated wood blocks that have no sharp edges and little thermal expansion and contraction.

In making new installations there are four critical issues to consider.

1. Use as few roof penetrations as possible and make certain that each penetration provides a strong connection to the rafters or trusses. Large lag bolts on 4-ft centers are better than small lag bolts every 2 ft.
2. Set all roof mounts that penetrate the roof membrane in a heavy bed of cold-process mastic.
3. Use *no nails,* only bolts and lag screws.
4. Use only galvanized or stainless steel screws or rivets to fasten aluminum collectors to racks. Insulate dissimilar metals from each other whenever possible.

Piping

Piping is interconnecting the collectors and leading *supply* and *return* lines to the solar storage tank or to the heat exchanger. The supply line is the pipe leading to the bottom of the collector array; the return line is the pipe leading to the top of the array.

Manifolds, Headers, and Risers   Figure 9.15 shows the parts of a typical flat plate solar collector. The absorber plate of the collector is an integral part of the piping system. Absorber plates come in many different shapes and configurations. Usually, the absorber plate contains top and bottom **headers** and finned **riser tubes.** The headers may or may not be utilized as the **manifold.** The finned risers are the small finned tubes running from the top to the bottom of the collector; the headers are the larger tubes running across the top and bottom of the collector. The risers are attached to, and end in, the headers. The manifold is the piping that feeds and removes the fluid from the collectors as it circulates.

Figure 9.16 shows an array of four collectors connected through the use of external manifolds. These manifolds are typically fabricated by the installer after the collectors have been fastened to the mounts. They are custom cut for the job. The four collectors are shown in a **reverse-return parallel connection.** Reverse-return parallel-connected collectors are collectors that are individually piped to a supply and return manifold in reverse order. That is, the first collector piped to the supply manifold is

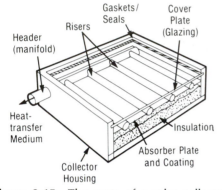

Figure 9.15   The parts of a solar collector. (Reprinted by permission from R. Montgomery and J. Budnick, *The Solar Decision Book,* Wiley, New York, 1978.)

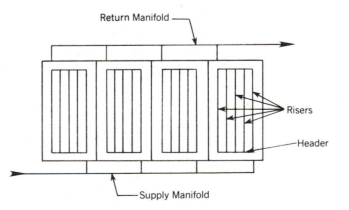
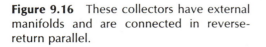

Figure 9.16   These collectors have external manifolds and are connected in reverse-return parallel.

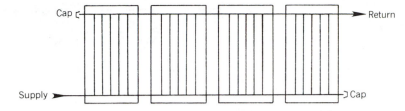

Figure 9.17 These collectors have internal manifolds and are connected in reverse-return parallel.

the last collector piped to the return manifold, and the last collector piped to the supply manifold is the first collector piped to the return manifold.

This order of connection is extremely important, because it equalizes the pressure drop across each collector; this assures that the same volume of liquid flows through each collector.

In reverse-return parallel-connected installations the flow of the fluid is divided among all the collectors in the array, and the same fluid only passes through one collector before being returned to storage or the heat exchanger. Figure 9.17 shows an array of four collectors interconnected in a reverse-return parallel configuration through the use of internal manifolds. These manifolds are part of the collector and are fabricated at the factory when the collector is manufactured.

In Figure 9.18 slanted external manifolds have been used. This is common in draindown and drainback systems. The slant in the manifolds allows complete drainage.

Figure 9.19 shows a different configuration. In this case the collectors contain no manifolds, headers, or risers. Instead, a single finned tube is snaked through the collector box in an s-shaped or serpentine fashion, and the two collectors have been connected in *series* instead of in parallel. A series-connected collector array is an array where all the fluid flows through all the collectors and the flow is not divided among the different collectors.[2]

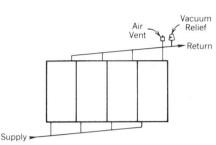

Figure 9.18 In draindown and drainback systems manifolds are usually slanted for drainage.

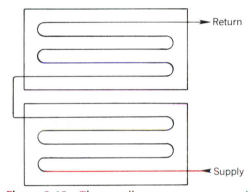

Figure 9.19 These collectors are connected in series.

Reverse-Return Parallel-Connected Versus Series-Connected Arrays

When two or more solar collectors are connected in reverse-return parallel, each collector operates at the same temperature, and all the collectors operate at the same efficiency.

When two or more solar collectors are piped in series, the collectors operate at different temperatures. The supply liquid to each collector in the series is at the liquid exit temperature of the previous collector. The first collector supplied operates at the lowest temperature. Each succeeding collector operates at a higher temperature. The last collector, which is located at the return end of the piping, operates the hottest. Thus, each collector operates at a different efficiency, with the coolest collector being the most efficient. Most state-of-the-art systems no longer use series-connected collectors.

It is not uncommon to see combined series-parallel flow configurations such as the configuration shown in Figure 9.20 in older systems. Here, two banks of series-connected serpentine collectors have been

---

[2]The use of serpentine collectors should be limited to either nonfreeze-protected systems or to closed heat-exchanged systems because serpentine collectors cannot be readily drained. Do not use serpentine collectors in draindown or drainback systems. When using serpentine collectors, install only in strict accordance with the manufacturer's recommendations.

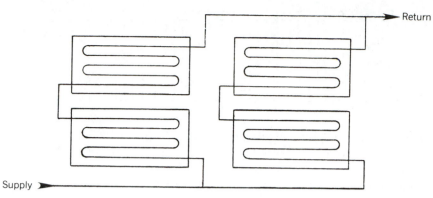

Figure 9.20 These collectors are series-parallel connected.

piped in a reverse-return parallel configuration to top and bottom manifolds.

Whether collectors should be connected in series, in reverse-return parallel, or in combined series-parallel configuration depends on how the collectors are engineered. If the collectors are designed with small risers, only reverse-return parallel configurations should be used. By dividing the flow across many risers, the reverse-return parallel connection lowers the velocity of the fluid in the risers and creates the lowest *resistance to flow*. Resistance to flow in pipes, which is also known as **pressure drop,** is caused by the friction created as the liquid flows through the pipe.

If a liquid were being supplied to the collector array in Figure 9.16 at a rate of 4 gpm, and each collectors contained 5 risers, the flow through each riser would be 4/20 or 0.2 gpm per riser. If the four collectors had been connected in series, the flow through each riser would have been 4/5 or 0.8 gpm per riser (four times as fast). This could create an excess pressure drop, because the pressure drop through a pipe varies as the square of the velocity of the liquid traveling through the pipe, making it necessary for the pump to overcome 64 times as much resistance to flow in this series configuration. Such a configuration would require much higher pumping power.

The collector connection configuration is specified by the collector manufacturer, who has engineered the diameter of the piping to obtain the lowest possible pressure drop and the highest possible heat transfer from fin to liquid under a recommended installation configuration.

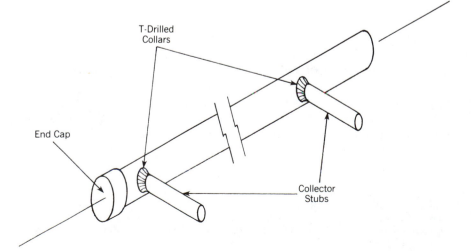

Figure 9.21 A T-drilled external manifold.

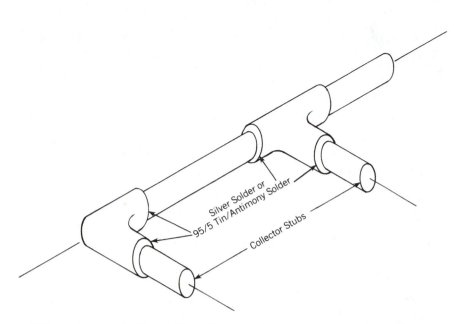

Figure 9.22 A component-assembled external manifold.

Building External Manifolds Two techniques are used for building external manifolds: *T drilling* and *component assembling*. T drilling is a shop technique; component assembling can be either a shop technique or an on-site technique. Very few small arrays use a T-drilled external manifold.

In T drilling, a raised seat is manufactured during the drilling of the hole in the pipe and a stub is brazed or silver soldered to the seat. It requires a special T-drill tool and jig. Figure 9.21 shows such a manifold.

Figure 9.22 shows a component-assembled manifold made from tees, piping, and elbows.

T-drilled manifold stubs should be brazed, temporarily plugged, and tested during manufacture. Component-assembled manifolds should be 95/5 tin/antimony soldered on the job. A 50/50 tin/lead solder should not be used, because it loses most of its strength when subjected to high collector stagnation temperatures.

External Manifold/Collector Connections

A high **differential thermal expansion** takes place in the collector array using external manifolds. Differential thermal expansion is expansion and contraction that take place at different rates or in different directions. This differential expansion must be considered when connecting arrays containing more than two collectors.

Figure 9.23 shows the problem. The absorber plates in the collector

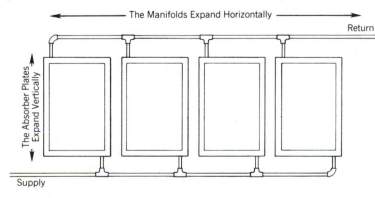

Figure 9.23 The collector array differential thermal expansion problem.

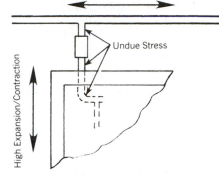

Figure 9.24 This array does not properly allow for differential expansion.

expand in the vertical direction, and the manifolds expand in the horizontal direction. As a result, high stress is placed on the manifold joints and the collector's internal piping if the manifold is close coupled and hard soldered, as shown in Figure 9.24.

Allowing for Expansion

Differential expansion is allowed for by using flexible connectors between the collectors and the manifolds. Figure 9.25 shows four different types of flexible connectors that are commonly used.

Bellows Connector Figure 9.26 shows a bellows connector. The bellows connector allows the vertical expansion of the absorber plate without stress and will take some horizontal expansion. However, when using the bellows connector, the manifold stubs must be carefully aligned both horizontally and vertically. Misalignment, as shown in Figure 9.27, causes bellows connector failures.

Rubber Hose Connector A high-temperature, ultraviolet-resistant rubber hose may be used as the connector. When rubber hose is used, the following areas must be considered (see Figure 9.28).

1. The hose must be capable of repeated thermal expansion and contraction from −20°F to +300°F.
2. The hose must be ultraviolet radiation resistant, or it must be located so that it is protected from the sun's rays.
3. Both the manifold and collector stubs should have an expansion ring built into them.
4. The hose clamps should be matched to the hose and should have

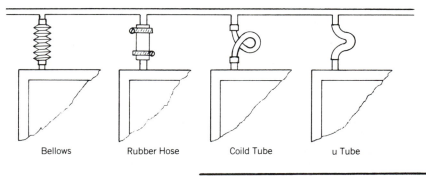

Figure 9.25 Four connectors that allow for differential thermal expansion.

Bellows   Rubber Hose   Coild Tube   u Tube

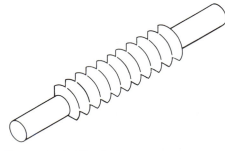

Figure 9.26 The bellows connector.

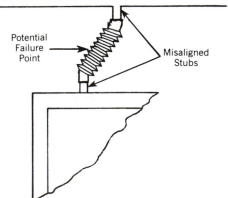

Figure 9.27 Misalignment of the bellows connector causes premature connector failure.

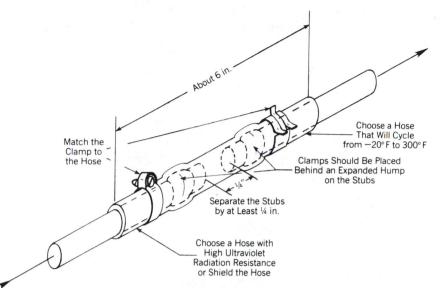

About 6 in.

Match the Clamp to the Hose

Choose a Hose That Will Cycle from −20° F to 300° F

Clamps Should Be Placed Behind an Expanded Hump on the Stubs

Separate the Stubs by at Least ¼ in.

Choose a Hose with High Ultraviolet Radiation Resistance or Shield the Hose

Figure 9.28 The rubber hose connector.

inside shoulders so as not to cut into the hose. The wrong clamps can quickly ruin a hose.

5. There should be ample expansion room between the stubs so that they cannot touch when the array is hot.

6. Hoses are apt to leak and may not be appropriate for some closed loop systems. Manufacturers' installation guidelines frequently require use of sealants and special tightening procedures. The guidelines should be followed carefully.

Elastomers vary even within polymer class. It is not safe to assume that any hose will give long-term satisfaction just because it is made from a certain type of rubber. Hose is a fabric-reinforced elastomeric product. As such, it must be compounded and fabricated for the application.

In the absence of such information the serviceperson can look at the elastomer class and the general end-use recommendations to rule out products that are not likely to work. Generally, hoses made for high-temperature and high-pressure use fabricated from Viton fluoroelastomers, Silastic silicone elastomers, Hyplon chlorosulfonyl polyethylene elastomers, and EPDM ethylenepropylenediene elastomers will be satisfactory; products made from Neoprene®, butyl, natural rubber, buna N, and buna S elastomers will be marginal or unsatisfactory.

When servicing a system 5 years old or older, check any rubber hose connectors for hardening, cracking, and leaking. You may want to advise the owner to replace them.

Coiled Pipe Connector A 360° coil of soft, type M or type L copper tubing serves as a good conductor that will take both vertical and horizontal expansion. However, the coil is hard to insulate and will not drain fully in draindown and drainback systems. Install the coil in a horizontal position, as shown in Figure 9.29. Both ends may be swaged for making solder joints, or swage-lock fittings may be used to make the connections. Swage-lock fittings are quick connect/disconnect fittings for piping.

The U-tube connector in Figure 9.30 is used in a similiar way and will allow both horizontal and vertical expansion. Problems can also be experienced in draining U tubes unless they are installed in a horizontal position.

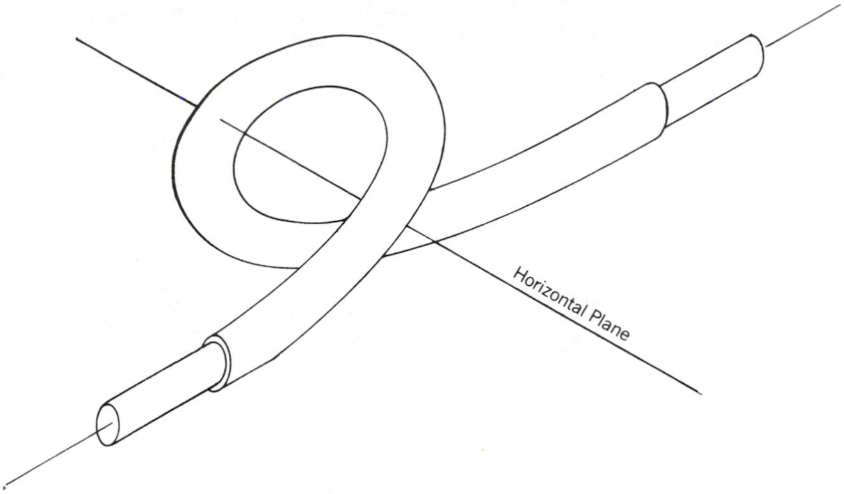

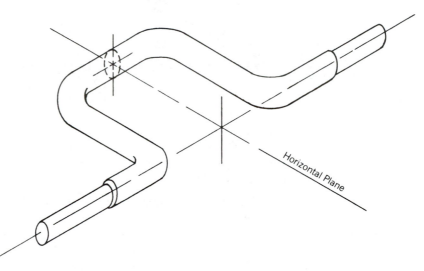

Figure 9.29 The 360° coil connector. It should be mounted with the coil in a horizontal position.

Figure 9.30 The U-tube connector. It should be mounted with the U in a horizontal position.

Soldered Connection   A soldered connection should be made with 95/5 tin/antimony solder. A 50/50 tin/lead solder loses its strength at high stagnation temperatures, and brazing can melt a joint within the collector, burn out the rubber grommet around the collector stub, or prevent the joint from being unsoldered for repairs at some future date. A swage-lock fitting, used at the collector nipple end, also eliminates all three problems.

Internally Manifolded Collector Connections

Differential expansion with internally manifolded collectors is less of a problem, but it still exists. It is best allowed for between collectors. Otherwise, the entire array can expand and work the mounting plates sideways or buckle an absorber plate if the mountings will not move. The bellows connector and the hose fitting are used in this application. There is usually not enough room between collectors to use coiled tube connectors.

Internally manifolded collectors typically use larger headers than externally manifolded collectors so that the headers can also serve as the manifold. Figure 9.31 shows the relative sizes. Expect to find $\frac{1}{2}$- to $\frac{3}{4}$-in. nipples on the nonmanifolded collector and $\frac{3}{4}$- to $1\frac{1}{2}$-in. nipples on the internally manifolded collector.

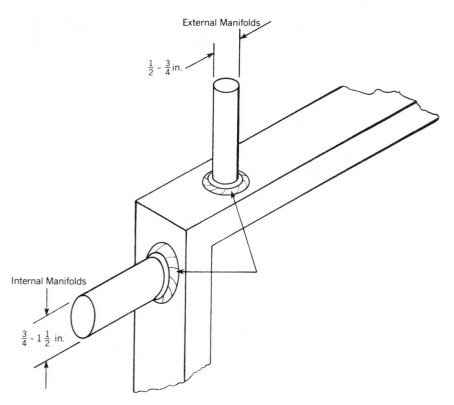

External Manifolds

$\frac{1}{2} - \frac{3}{4}$ in.

Internal Manifolds

$\frac{3}{4} - 1\frac{1}{2}$ in.

Figure 9.31 Typical stub sizes on collectors with and without external manifolds.

Thermal Expansion Field Experiences

The problem of thermal expansion in small residential systems rarely exists. Collector plates and assemblies are generally free to move within the collector enclosure enough to take care of any problems. In larger arrays, such as found in commercial systems, thermal expansion can be a major problem. Experienced solar consultants can offer assistance in solving large array problems.

Penetrating the Roof

When the array is roof mounted, piping leading from the supply and return manifolds plus sensor wiring must penetrate the roof. Such penetrations must be made very carefully.

The details of roof penetration will vary, depending on the type of roofing. However, the same techniques used for making other plumbing penetrations, such as vent pipe, are commonly used.

Both sheet metal and flexible rubber or plastic roof jacks have been used successfully. Figure 9.32 shows an example of a sheet-metal roof

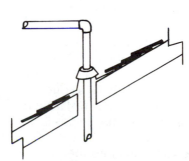

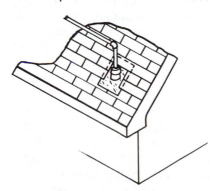

Figure 9.32 A sheet metal piping standoff. Notice the flexible pipe collar.

jack. Note the rubber cap on the top of the roof jack. Since the piping will move both horizontally and vertically, a flexible rubber cap seal is recommended.

Wiring the Control Sensors

The running of the sensor wire is of critical importance to insure proper system operation, and outdoor wiring requires great care to insure long-term reliability.

The factory-installed leads to the sensor are designed to withstand high temperatures, up to 300°F, for short periods. However, most weather-resistant sensor lead-in wiring insulation will melt at much lower temperatures, so it is best to connect the sensor leads to the weather-resistant cable under the pipe insulation and then to run the remainder of the sensor wiring outside of the insulation so that it is protected from contact with the hot pipes.

The pipe insulation surrounding the sensor and its connection to the cable should be made airtight and weather tight, using a sealant recommended by the insulation manufacturer. Often it will be necessary to use a larger inside-diameter pipe insulation over the sensor to accomodate its bulk and the bulk of the cables.

Anchor the sensor wiring every 3 ft. You may attach it to the outside of the insulation with weather- and ultraviolet-resistant cable ties or carefully staple it to wood rack members or building components. Never staple it directly to the roof.

Avoid running sensor wiring near hot boiler or furnace stacks. Take care that metal flashings do not cut or short out the wires. Also avoid any outdoor splices if possible.

Insulating the Piping

All the pipes must be insulated to prevent heat loss. The following areas need consideration.

1. The R value must be adequate (minimum R = 5).
2. The insulation must be moisture resistant.
3. It must be fire resistant.
4. It must withstand repeated thermal cycling from −20°F to +300°F.
5. It must be able to withstand environmental and solar degradation.

There does not seem to be any insulation available that meets all these requirements. Common insulation materials do not offer this complete combination of properties. The best approach seems to be to choose the insulation with the best thermal and moisture-resistant properties and to cover the insulation with either a metal or a rigid polyvinylchloride (PVC) plastic jacket; 6-mil corrugated aluminum works well. Although ultraviolet-resistant paints are available, they are expensive and must be reapplied every few years.

The four most common insulating materials used are fiberglass, flexible foam, precast rigid urethane, and precast rigid isocyanurate.

Fiberglass is prone to moisture and humidity degradation. It must be very carefully wrapped and waterproofed. Its thermal characteristics and R values are excellent, and it offers the longest life in most installations. However, the R value per inch of thickness is low, and thicker insulation is required than with foams.

Flexible foam is not generally ultraviolet resistant and must be shielded against the sun. High temperature degradation with time will most likely occur. However, it is easily applied and, when jacketed and wrapped, is acceptable. It seems to be the most popular choice.

Rigid urethane lacks both thermal and ultraviolet resistance. Expect to see urethane insulation deteriorate fairly quickly. Isocyanurate insulation, a high-temperature urethane polymer, is a better choice.

With all preformed insulations, miter the joints carefully. Seal and glue the joints as per the manufacturer's recommendation. Jacket the final result. Figure 9.33 shows some typical details. Good insulation practice requires taking one's time and applying good craftsmanship.

If the solar system you are servicing has been in operation for several years, you may have to perform some insulation maintenance.

Collector Design, Thermal Performance, and Collector Rating

Medium-temperature, flat plate collectors generally are designed either as copper and brass systems or as all-aluminum systems. *The discussion that follows does not relate to all-aluminum systems.* All-aluminum systems have different areas of consideration. The following warning, taken from the Reynold's Aluminum Collector Installation and Maintenance Manual, should be respected.

> **WARNING** Use of copper or galvanized transport and vent lines and copper heat exchangers is prohibited and voids the Reynold's warranty.

Servicepersons working with all-aluminum systems or with systems using aluminum absorber plates in the collectors should consult with the manufacturer before proceeding with maintenance. Aluminum is a highly active metal that lies near the anodic end of the galvanic series. As such, when it is improperly used in conjunction with cathodic metals such as brass, copper, tin, and bronze, it is subject to rapid galvanic corrosion under the right conditions.

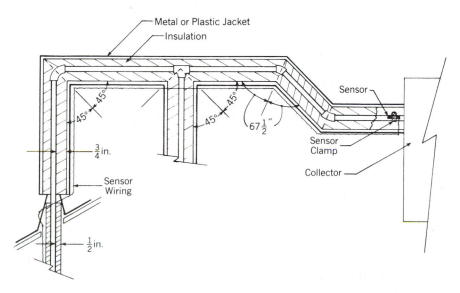

Figure 9.33 Recommended insulation practices.

In the early days of solar heating designers and installers ignored the problems posed by mixing copper and aluminum in the same system. This resulted in many failures of aluminum absorber plates. However the failures were caused by improper design and application. All-aluminum systems make good solar water-heating systems when they are properly engineered and installed.

Collector Design

A wealth of knowledge and information about the design and construction of flat plate solar collectors exists. The design of a good collector involves many tradeoffs among cost, durability, and performance.

No one construction can be said to be best, and the serviceperson cannot be expected to review the design of a collector as part of a service call. Instead, a common ground on which to base the decision as to whether the collector is constructed in a satisfactory manner must be found.

The American Society for Testing and Materials (ASTM) has an active solar committee, Committee E-44 on Solar Energy Conversion, which has set a number of test methods and standards for collector components. The current standards (December 1982) were published in July 1981 in the ASTM booklet entitled "ASTM Standards for Solar Energy (ASTM, 1916 Race Street, Philadelphia, PA 19103). Solar standards are expected to undergo some changes in the next few years, and readers should check for the latest version when ordering a copy.

One standard test method in particular, "E823-81: Standard Practice for Nonoperational Exposure and Inspection of a Solar Collector," is of interest to the serviceperson and installer. The standard involves a table of descriptions of deterioration that would cause premature failure of the collectors. Servicepersons should look for a description on the collector that states that the collector was evaluated under this standard and performed satisfactorily.

Collector Performance

ASTM Standard E823-81 includes, by reference, an American Society of Refrigeration and Air Conditioning Engineers (ASHRAE) standard, "ASHRAE Standard 93-77: Methods of Testing to Determine the Thermal Performance of Solar Collectors." This thermal performance test is usually run at the same time as exposure under ASTM E823-81.

ASHRAE 93-77 testing provides a thermal performance curve that predicts the thermal performance of the collector. This thermal performance curve allows the system designer to tailor the sizing of the system and its components correctly. Again, the serviceperson should look for the statement "ASHRAE 93-77 tested" on the nameplate of the collector.

Collector Rating

In the past several organizations have provided certified collector ratings. Florida, California, and Arizona have had such rating systems but have dropped them in favor of industry rating systems. Two industry associations, The Solar Energy Industries Association and the American Refrigeration Institute, provide certified rating systems.

ASTM E823-81 and ASHRAE 93-77 standard test methods are the methods used by the Solar Rating and Certification Corporation (SRCC), a nonprofit solar equipment-rating corporation set up by state and federal governments and the industry working through the Solar Energy Industries Association, U.S. Department of Energy, and a coalition of state energy offices. The SRCC also provides testing and rating of integral collector-storage and other specialized systems.

Check with your local state energy office to see what standards and rating systems apply within your state. Look for an SRCC or comparable rating label on the collector nameplate. When the collectors are properly tested, rated, and labeled, chances are that performance will be highly satisfactory. If none of this documentation exists, you will have to examine the collectors more closely.

Materials of Construction

Medium-temperature, flat plate solar collectors usually use:

1. Copper absorber plates and tubes.
2. Copper or aluminum fins.
3. Glass or thermoplastic glazings.
4. High-temperature fiberglass or isocyanurate insulation.
5. High-temperature flat or selective black paint, or black chrome oxide selective absorber surface.
6. Aluminum or reinforced plastic cases.
7. High-temperature rubber gaskets, grommets, and sealants.

Test methods for determining satisfactory performance of each of these materials is covered under applicable ASTM standards as outlined previously.

For further detailed information, review:

- *The Solar Decision Book,* Richard H. Montgomery with Jim Budnick, John Wiley & Sons, Inc., 1978 (Chapter 10).
- *Solar Heating Materials Handbook,* P. S. Homan and C. J. Hilliary, AnaChem Inc., 1981 (Chapters 4, 5, 6, and 8).
- *Handbook of Experience,* Solar Energy Industries Association, 1981 (Sections IIA & III).
- *ASHRAE Handbook of Experiences,* D. Ward and H. Oberi, ASHRAE 1980.

What to Look for in Examining the Collector Array

Visual examination of the collector array should include an evaluation of the glazing, absorber plate, enclosure, insulation, gaskets, seals and caulking, and connectors. The examination should be performed without any disassembly of the collectors unless a problem is found.

Glazing   The glazing should be examined for cracking, crazing, scratching, buckling, or severe clouding. You are evaluting for watertightness and ability to pass solar radiation. Dirt and dust buildup should be periodically removed.

Clouding may be due to condensed water vapor. The vapor can come from evaporated dew collected at night, a pinhole leak in the absorber plate, or from volatile chemicals out-gassed from the insulation, absorber coating, or other component within the collector box.

Absorber Plate   The absorber plate should be examined for warping or buckling, loss of bonding between tubes and fins, leakage, signs of severe corrosion, and the peeling, flaking, or blistering of the absorber's surface coating. Absorber plate degradation reduces fluid passage integrity and the ability to collect heat efficiently.

Insulation   Evaluate what you can see of the insulation for swelling, matting, slumping, discoloration, or other evidence of thermal degradation. A good test is to pass the palm of your hand over the back of the collector looking for hot spots. You are evaluating for the integrity of the insulation and to see if it has moved out of place, slumped, matted down, or lost its insulating value. Any of these faults will allow side and back heat losses to become severe.

Seals, Gasketing, and Caulking   During your inspection, examine the seals, gaskets, and caulking. The deterioration of these items starts most collector damage. Concentrate on the following points.

1. The rubber grommets around the manifold stubs must not be cracked, hardened, or softened. Hardening generally comes from exposure to solar radiation or excess thermal heat; softening usually comes from contact with organic liquids such as leaking heat transfer fluids or from excess thermal heat.
2. The seal around the glazing should not be swollen, shrunk, hardened, cracked, or softened to the point where water can enter the collector.
3. Evaluate any caulking materials in the same manner for the same problems. With caulking, also look for signs of loss of adhesion; this is one of the most common faults of caulking. Organic caulkings will be more likely to lose adhesion to glass, and silicone caulkings will be more likely to lose adhesion to metals.

Very often, minor repairs to gasketing, seals, and caulking will stop any further deterioration of the collector.

You should only consider major collector repairs when the function of the collector is severely impaired or if there is evidence that further operation of the system will gradually lead to function impairment. Minor internal visual problems such as paint discolorations, minor blistering, or minor insulation movement should be ignored if, in your judgment, they have stabilized.

Sizing Versus Load

For servicepeople, a quick rule of thumb is to count occupants and collectors, as discussed in Chapter 1. Assuming that the hot water is in the normal range and that the collectors have satisfactory thermal performance, the system should contain 20 to 25 ft² of collector per occupant.

In climatic areas that have large amounts of cloud cover or that lie at latitudes over 40°N, the system should lean toward 25 ft² per occupant. At lower latitudes or where high amounts of solar radiation occur year round, the system should lean more toward 20 ft² per occupant.

The matching of collector array size to storage tank size is also important. For each square foot of collector, there should be $1\frac{1}{4}$ to 2 gal of storage. If this rule is followed the system will perform even if the array is oversized or undersized.

Much more sophisticated techniques are used to determine the proper system size at the time of installation, as discussed in Chapter 1. Almost all manufacturers have sizing instructions in their installation manuals. ASTM has a proposed standard practice under development. When this standard is issued, it will be the most authoritative source of sizing information available.

The serviceperson should not be overly concerned about collector array size unless a serious performance problem is the purpose of the service call. Then, if the system has never performed properly, size versus load should be carefully evaluated.

Safety and Health Hazards

The safety hazards of concern in the collector array are typically either structural or thermal. An additional health hazard may also be associated with the heat transfer fluid. Health problems with the heat transfer fluid will be covered in Chapter 10.

Generally, evaluate the collector array for objectionable reflections, the correct fastening down of the array, and the building's ability to handle the load imposed by the array.

When the sun's rays hit the collector's glazing at an angle of less than 30° from the plane of the glazing (see Figure 9.34), large amounts of visible energy can be reflected. This is particularly true when a smooth glass glazing is used. This reflected energy must not present an objectionable glare for an oncoming motor vehicle, bicyclist, or pedestrian.

The rule for reflection is: *The angle of incidence equals the angle of reflection.* So any sunlight that causes glare will most likely be reflected at an angle of 10 to 30° from the plane of the collector and will occur during the early morning or late afternoon hours.

The building must be able to hold the dead load imposed by the array and any racks; environmental loads imposed by wind, snow, and ice

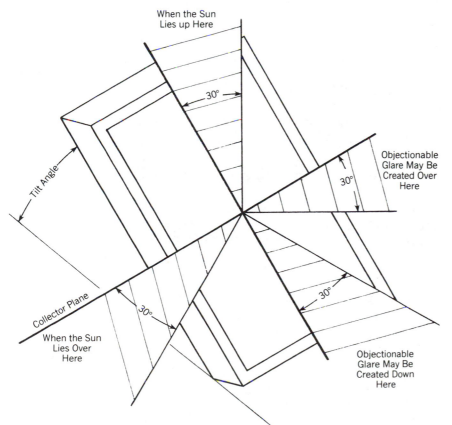

Figure 9.34 Objectionable glare may occur when the sun hits the collector at an angle of less than 30°

buildup; and live loads imposed by service personnel working on the array.

The American National Standards Institute (ANSI) publication ANSI 58.1(1979), titled "Building Code Requirements for Minimum Design Loads in Buildings and Other Structures," details structural requirements for small buildings. Section Z97.1 of the National Electrical Code gives performance specifications for safety glazing materials. The federal Department of Housing and Urban Development (HUD) has published "Minimum Property Standards for Solar Systems." Finally, the Sheet Metal and Air Conditioning Contractors Association has published standards for single- and multiple-family housing.

All these publications are incorporated by reference into ASTM E683-79, "Standard Practice for Installation and Service of Solar Space Heating Systems for One and Two Family Dwellings." Until a standard issues for the installation of solar water heaters, ASTM E683-79 should be the serviceperson's reference document.

ASTM E683-79 incorporates 17 standard practices for the collector array. It makes the following points about safety.

1. Collector supports must support the collector array under all anticipated extremes of environmental conditions and loads imposed by wind, earthquake, rain, snow, ice, and freezing temperatures without exceeding the load limitations of the structure, foundations, or soil, and the supports must not impose undue stress on the collectors.
2. Installation shall not cause water accumulation, ice dams, snow buildup, or water damage.
3. Safe access to array components for servicing shall be provided.
4. Service personnel shall be protected against injury from contact with hot surfaces.
5. Ground-mounted collectors will have guard bars or fences and warning signs to prevent thermal or other injury.
6. Ground-level glazings shall meet the safety glazing requirements of ANSI Z97.1 (shall be safety-glass glazed).
7. Collectors made of combustible materials shall not be located on or adjacent to construction required to be of noncombustible materials.

The Collector Array Service Checklist

Here is the service checklist for a complete evaluation of the collector array. Rate the array on all 37 points as good, marginal, or poor. Marginal ratings should be pointed out to the owners; poor ratings should be repaired for proper performance and safe operation.

EVALUATION POINT	GOOD	MARGINAL	POOR
1. Will structure properly support the array?	_____	_____	_____
2. Does the array meet local codes and ordinances?	_____	_____	_____
3. Was the manufacturer's installation procedure followed?	_____	_____	_____

EVALUATION POINT	GOOD	MARGINAL	POOR
4. Does the array face true south ±30°?			
5. Is the array tilted to latitude ±15°?			
6. Is the array unshaded during prime hours?			
7. Has the weather-resistant integrity of the building been maintained?			
8. Is the array properly flashed and sealed?			
9. Are mounting brackets lagged and/or bolted?			
10. Are mounting brackets sealed and caulked?			
11. Are collectors properly fastened to brackets?			
12. Are expansion and contraction allowed for?			
13. Can water, snow, or ice buildup occur?			
14. Can servicepersons gain access to the array?			
15. Are the collectors properly piped in series or in parallel?			
16. Are the nipples aligned and connected for horizontal and vertical expansion?			
17. Are the roof penetrations made through watertight roof jacks or flashing that allows for pipe movement?			
18. Is the piping well insulated and protected?			
19. Has copper or brass been added to an all-aluminum system?			
20. Do collectors meet ASTM, ASHRAE, and SRCC or ARI standards?			
21. Is the array manufactured from long-life materials?			
22. Is the collector glazing cracked, crazed, scratched, buckled, severely clouded, or dirt covered?			

EVALUATION POINT	GOOD	MARGINAL	POOR
23. Is the collector glazing watertight?	_____	_____	_____
24. Has the absorber plate warped or buckled, or have the tubes separated from the fins?	_____	_____	_____
25. Has the absorber plate coating blistered or peeled?	_____	_____	_____
26. Has the insulation moved, slumped, matted, or discolored?	_____	_____	_____
27. Does the collector box have structural integrity?	_____	_____	_____
28. Is the collector box watertight?	_____	_____	_____
29. Have array gaskets, seals, and caulking softened, hardened, cracked, or lost adhesion?	_____	_____	_____
30. Is the system properly sized for the load?	_____	_____	_____
31. Is storage size matched to collector size?	_____	_____	_____
32. Does glazing create any glare problems?	_____	_____	_____
33. If collectors are ground mounted, are appropriate guard rails, fences, and signs provided?	_____	_____	_____
34. Are ground- or vertical wall-mounted collectors properly safety glazed?	_____	_____	_____
35. Are any combustible materials improperly located?	_____	_____	_____
36. Is headroom ample where people must walk?	_____	_____	_____
37. Are there any unprotected hot spots where people could be burned?	_____	_____	_____

CHAPTER

10

Heat Exchangers and Transfer Fluids

Heat exchangers and their transfer fluids must be chosen to complement each other, or less-than-optimum results will be obtained.

Generally, the designer chooses the transfer fluid first and then matches the size and the construction of the exchanger to the fluid.

Heat Transfer Fluids

Heat transfer fluids can be divided into two general classes: **potable** and **nonpotable.** A potable fluid is suitable for human consumption; a nonpotable fluid is not.

The only two potable heat transfer fluids used in solar water heaters are drinking-quality water and fresh air. When one of these two fluids is used to transfer the heat, the design of the system is simplified, which lowers the cost. This is why thermosiphon, pumped water, draindown, and drainback systems are so popular. However, as previously discussed, these four types of systems have other design limitations, so they are not always used.

All systems using nonpotable heat transfer fluids require heat exchangers to separate the heat transfer fluid from the potable water supply. The design of the heat exchanger is strongly influenced by the toxicity of the fluid and by other fluid properties.

Transfer Fluid Toxicity

Toxic is an adjective meaning poison, for example, a toxic substance; **toxicity** is the noun corresponding to toxic and is used to say, for example, that a substance has high toxicity.

Just about all nonpotable fluids have some degree of toxicity to humans and can cause health problems if they are ingested. Heat transfer fluids are generally classified as *low, moderate,* or *dangerous* in toxicity.

Authorities differ on how toxic different substances are, and determining the toxicity of a substance to humans is complicated.

We have formulated some guidelines for servicepeople on the toxicity of different common solar heat transfer fluids. Use these guidelines with the understanding that they are only an informed opinion, not the final answer from a practicing toxicologist, who would comment on a very specific material only after very specific testing.

Low-Toxicity Fluids   We define low-toxicity fluids as fluids that, if ingested in quantity over a period of time, are unlikely to cause a serious health problem. Materials such as dimethyl silicone fluids, (poly)propylene glycol fluids, and perhaps some synthetic oils can be placed in this category.

Moderate-Toxicity Fluids   We define moderate-toxicity fluids as fluids that, if ingested in quantity over a period of time, could result in a severe illness or health problem. (Poly)ethylene glycol, some corrosion-inhibited water, and selected nonaromatic hydrocarbon oils seem to belong in this category.

Dangerous-Toxicity Fluids   Finally, we define dangerous-toxicity fluids as fluids that cause a serious illness, health problem, or death when ingested in minor quantities—once. We place aromatic hydrocarbon fluids in this category. Some authorities also place (poly)ethylene glycol in this category.

The Testing of Fluid Toxicity

As previously stated, the testing of fluids for oral toxicity is a matter for expert toxicologists. A common test method is the **lethal exposure test;** a common toxic rating system is the **Gosselin rating.**

The lethal exposure test determines the quantity of a material that will cause the death of 50% of a population of rats, mice, or other laboratory animals within 24 hours after they have been fed the material. The results are reported as grams of fluid ingested per kilogram of body weight.

The Gosselin rating system takes the results of the lethal exposure tests and estimates the amount of the material that a 150-lb human would have to eat to die. Under the Gosselin rating system, fluids are given a rating of 1 to 6, depending on the estimated human lethal dose. Here is the Gosselin rating scale.

GOSSELIN RATING	ESTIMATED LETHAL DOSE
6 Supertoxic	Less than 7 drops
5 Extremely toxic	7 drops to 1 tsp
4 Very toxic	1 tsp to 1 oz
3 Moderately toxic	1 oz to 1 pt
2 Slightly toxic	1 pt to 1 qt
1 Practically nontoxic	Over 1 qt

Table 10.1 shows the toxicity of many commercial heat transfer fluids.

We recommend that only materials having a Gosselin rating of 1 be used in residential solar water-heating systems unless an approved **double-walled heat exchanger** separates the transfer fluid from the potable water, in which case we suggest that a fluid with a Gosselin rating of 2 is acceptable. A double-walled heat exchanger is a heat exchanger constructed so that: (1) two separate walls exist between the heat transfer fluid and the storage water and (2) an open passage to the environment exists between the walls, thus allowing any leaks to be immediately detected.

There is no sound technical reason why a more toxic heat transfer fluid is required in any residential flat plate solar system and, in the opinion of

Table 10.1

TOXICITY OF SOME COMMON HEAT TRANSFER FLUIDS

HEAT TRANSFER FLUID (MANUFACTURER)	COMPOSITION	SKIN IRRITATION	LD^{50} GOSSELIN	
			g/kg	RATING
Dowtherm SR-1 (Dow Chemical U.S.A.)	Ethylene glycol	0.3	8.2	2
Ethylene glycol (Aldrich Chemical Co.)	Ethylene glycol	0.4	4.0	3
Dowfrost (Dow Chemical U.S.A.)	Propylene glycol	0.1	20.3	1
Freezeproof (Commonwealth Chemical)	Propylene glycol	0.1	>24.0	1
Nutek 835 (Nuclear Technology Corp.)	Propylene glycol	0.1	>24.0	1
Propylene Glycol USP (Union Carbide Corp.)	Propylene glycol	0.1	22.8	1
Solar Winter Ban (CAMCO Mfg., Inc.)	Propylene glycol	0.2	>24.0	1
Sunsol 60 (Sunworks, Ethone, Inc.)	Propylene glycol	0.3	>24.0	1
UCAR Food Freeze 35 (Union Carbide Corp.)	Propylene glycol	0.1	19.4	1
UCON 50-HB-280-X (Union Carbide Corp.)	Polyalkylene glycol	0	1.9	3
Solargard G (Daystar Corp.)	Glycerol/water	0.1	>24.0	1
SF-96 (General Electric)	Silicone	0	>24.0	1
Syltherm 444 (Dow Corning)	Silicone	0	>24.0	1
Caloria HT-43 (Exxon Co. U.S.A.)	Paraffinic oil	0.1	>24.0	1
Mobiltherm Light (Mobil Oil Co.)	Aromatic oil	1.2	24.0	1
Mobiltherm 603 (Mobil Oil Co.)	Paraffinic oil	0.2	>24.0	1
Process Oil 3029 (Exxon Co. U.S.A.)	Naphthenic oil	0.6	>24.0	1
Suntemp (Resource Technology)	Petroleum-based organic	1.1	>24.0	1
Dowtherm A (Dow Chemical U.S.A.)	Diphenyl plus diphenyl oxide	0.4	4.1	3
Therminol 66 (Monsanto)	Modified terphenyl	2.0	>24.0	1

Source: Adapted by permission from *Solar Heating Materials Handbook*, P. Homan and C. Hiliary, Anachem, Inc., 1981.

several state health officials, "fluids with Gosselin ratings of more than 2 represent an unacceptable risk of contaminating a public water supply as well as causing a personal injury."

Heat Transfer Fluid Flammability

Most nonpotable heat transfer fluids are flammable. In the case of propylene glycol and ethylene glycol the flammability of the fluid is reduced by diluting it with water, and the 50/50 water/glycol solutions commonly used in residental solar systems can be considered nonflammable.

Fluid flammability is also complex. *Flash point, fire point, autoignition point, smoke generation, combustion by-products, and amount of heat generated in burning* all must be considered.

Flash point is the temperature at which the fluid will burn if an outside flame is applied but at which it will self-extinguish if the flame is removed. Fire point is the temperature at which the fluid will burn if an external flame is applied and will continue to burn if the flame is removed. Autoignition point is the temperature at which the fluid will catch fire without an outside flame being applied. Smoke generation is the type and amount of smoke and ash generated during burning. This is determined by laboratory analysis of the smoke generated. Combustion by-products, also determined in the same laboratory analysis, are the chemicals manufactured during the burning process. Heat generation is a measure of the amount of heat generated per unit of fluid burned. See Table 10.2 for the flash points of typical commercial heat transfer fluids.

We recommend the following fluid flammability guidelines.

- The flash point of the fluid should exceed 325°F.
- The fire point of the fluid should exceed 400°F.

Table 10.2

FLAMMABILITY OF SOME COMMON HEAT TRANSFER FLUIDS

HEAT TRANSFER FLUID	CLOSED CUP FLASH POINT (°F)	OPEN CUP FLASH POINT (°F)
Ethylene glycol	245	250
Polyalkylene glycol	480	555
Alkylated aromatic oil	135	145
Diphenyl/diphenyl oxide	240	265
Extracted naphthenic oil	280	325
Paraffinic oil, type 1	275	305
Pafaffinic oil, type 2	325	360
Paraffinic oil, type 3	390	455
Paraffinic oil, type 4	405	465
Silicone fluid	545	605

Source: Adapted by permission from *Solar Heating Materials Handbook,* P. Homan and C. Hiliary, Anachem, Inc., 1981.

- The autoignition point of the fluid should exceed 700°F.
- The smoke should not contain gross amounts of noxious combustion by-products that could injure humans.
- The heat generated should not exceed the amount generated by the burning of a typical high-flash hydrocarbon oil.

Since small quantities of fluid, generally less than 5 gal, are usually contained in the system, smoke generation, combustion by-products, and heat generation do not need careful evaluation by the serviceperson.

The serviceperson must keep in mind that two fire sources exist: fires starting within the system from overheating or malfunction, and fires starting from an outside flame or fire. In most installations the danger of fire from an outside source is greater than the danger of fire from a system fault.

Other Important Heat Transfer Fluid Considerations

There are many other important characteristics of heat transfer fluids. The designer must consider *specific heat, temperature/viscosity slope, specific gravity, thermal conductivity, boiling and freezing points, coefficient of thermal expansion, and fluid compatability* with the other materials in the system.

Specific heat, specific gravity, and thermal conductivity will not be explained or discussed here. The serviceperson should understand that these are measures of the fluid's ability to transfer heat per unit of fluid circulated and that, because these properties differ from one fluid to another, heat transfer fluids cannot readily be substituted for each other. The system's components are matched up to operate correctly with a fluid that has a certain ability to transfer heat. If another fluid with different characteristics is substituted, the system will not operate as designed.

Temperature Viscosity Changes

Viscosity is a measure of a fluid's ability to flow under shear. Shear occurs whenever the fluid is subjected to physical deformation. Thin fluids with low viscosity, such as water, flow readily. Thicker fluids with higher viscosities, such as 30-weight motor oil, flow less readily. It takes more pumping horsepower to move them.

The viscosity of a fluid changes with temperature. The hotter the fluid, the lower its viscosity. It is said to be "less viscous." Some fluids change viscosity rapidly with small temperature changes. Other fluids change viscosity more slowly with the same temperature change. The rate of change is known as fluid *temperature/viscosity slope*. The lower the temperature/viscosity slope, the less the change in viscosity per degree of temperature change. Solar water-heater fluids should have a low temperature/viscosity slope so that they can be successfully pumped over a wide range of temperature.

Boiling Point/Freezing Point

The boiling point of a fluid is the temperature at which it changes from a liquid to a gas; the freezing point of the liquid is the temperature at which it changes from a liquid to a solid.

Heat transfer fluids in residential solar water heaters must be used above their freezing point and below their boiling point in order to pump them. In most systems this calls for the fluid to remain liquid from -20 to 300°F.

Coefficient of Thermal Expansion

All fluids expand as the temperature rises and contract as the temperature drops. The coefficient of thermal expansion is a measure of how much the fluid expands or contracts for each degree of temperature it is heated or cooled.

All closed systems that circulate a fluid between the collectors and a heat exchanger must have an expansion tank in the system so that the fluid has room to expand without rupturing the system and to cool without creating a vacuum within the system.

Expansion Tank Sizing   The size of expansion tank required in a system is a function of the volume of liquid in the system, the difference between the coldest and hottest temperatures the system will encounter, and the volume expansion characteristics of the heat transfer fluid being used. Thus, sizing is a complex procedure usually carried out by the system's designer.

Most domestic one- to four-collector solar water-heating systems will contain less than 5 gal of heat transfer fluid. These systems require only a small expansion tank. The Amtrol S-15 or its equivalent should be satisfactory in such systems. These are diaphragm-containing tanks that are normally precharged at the factory. The diaphragm separates the liquid in the system from the air in the tank and prevents the two from mixing.

Suspect too small an expansion tank or an inadequate expansion tank air charge when either fluid flow is lost in extremely cold weather or the pressure relief valve opens in extremely hot weather when the system is at rest.

Fluid Material Compatibility

The heat transfer fluid must be compatible with everything that it contacts in the system. If it is not, part of the system can deteriorate or the heat transfer fluid can degrade.

Here are two examples of how serious such incompatibility can be. In the early 1970s, an oil-type heat transfer fluid, which was incompatible with certain types of rubbers and which dissolved asphalt, was marketed widely. In several locations the fluid softened rubber hose connectors, leaked, and destroyed patches of roofing. In others it dissolved O rings in pumps, leaked, and destroyed asphalt tile or took paint off buildings.

Another oil-type fluid, which was designed to operate only in steel systems, was placed in copper/aluminum systems. These metals caused the fluids to sludge up rapidly and plug heat-exchanger tubes and collector risers. In less than a year the systems had to be drained and flushed.

Heat Transfer Fluid Evaluation

All heat transfer fluids except air should be periodically evaluated. An annual inspection is ideal, but a biannual inspection is satisfactory for most fluids.

Only a few ounces of fluid are needed for the evaluation. Two separate samples should be taken. One sample should come from a drain leg or a fill valve at the lowest possible point in the system. It should be taken after the system has been sitting for 5 to 10 minutes with the pump off. Crack the valve and drain 4 to 6 oz of fluid into a clean, dry, cappable glass sample bottle. A 1-pt mason jar works well.

This sample will contain any heavy sludge, water, or foreign material that has entered the system. The concentration of these impurities will be the highest at this low point in the system. The sample presents a picture that most likely is worse than the bulk of the fluid.

The second sample should be taken from a higher point in the system while the system is running and after the pump has been operating for 5 to 10 minutes. Use the same sampling procedure.

This second sample represents the average condition of the fluid in the system. If a laboratory analysis of the fluid is to be carried out, this sample will give the most accurate analysis. The first sample will give a high impurity analysis. Use both samples for your visual inspection at the site.

Fluid Visual Inspection

The visual inspection should include looking for

- Rust, scale, and/or water.
- Sludge and carbonaceous solids.
- Metal particles.
- Sticky, gummy, or viscous semisolids (gels).

Rust and Scale and/or Water   Large amounts of rust and/or scale indicate either deterioration of iron parts in the loop (rust) or the buildup of solids on the tubes of the exchanger and/or the risers of the collectors; the presence of water suggests a heat-exchanger or system leak.

Sludge and Carbonaceous Materials   Large amounts of sludge and/or carbonaceous materials indicate degradation of the heat transfer fluid. Eventually, this will clog the risers or plug the exchanger tubes. Fluid degradation rate increases once degradation has started, and degradation rarely can be stopped. The fluid must be drained and flushed, and new fluid must be added.

Metal Particles   The presence of solid metal particles in the sample indicates that mechanical damage to the system's fluid passages is occurring. Eventually, this will result in a perforation of the passages somewhere in the system. When this condition is seen, the source of the particles must be found.

Sticky, Gummy, or Viscous Materials The presence of these materials indicates that the rubber and plastic parts of the fluid passages are under attack by the fluid. Look for problems with seals, gaskets, valve packings, or any plastic components.

Fluid Testing

Many heat transfer fluids need periodic testing. The tests required depend on the class of heat transfer fluid. For testing purposes, the fluids can be divided into four classes: inhibited water, water/glycol mixtures, hot oils (hydrocarbons), and synthetic oils (silicone fluids).

The tests that need to be run are generally specified by the manufacturer in the instruction manual. Nothing that you read here should override the manufacturer's instructions, which apply to a specific fluid.

Inhibited Water Inhibitors are placed in water to prevent the water from attacking metal. These inhibitors are sacrificial in nature and must be periodically replaced.

Inhibitors should not be mixed. They are specific to the metals in the system. Stay with the inhibitor presently in the system unless the manufacturer specifies otherwise. Table 10.3 gives a list of common inhibitors. Note that one of these inhibitors is very toxic. Use this material *with extreme care,* even when its use is specified by the system manfacturer.

The discoloration of deionized inhibited water is normal and is not a sign that the inhibitor must be refreshed.

A good practice to follow when inhibitors need to be added to the system is to

- Completely drain the heat transfer fluid.
- Flush the system with fresh, deionized water by filling the system and circulating the water for 5 to 10 minutes.
- Make up a solution of deionized water containing the correct amount of inhibitor.
- Perform any testing on the new solution requested by the manufacturer.
- Drain out the flush water.
- Refill the system with new heat transfer fluid.

NOTE The use of deionized water is important. You can never be certain what types of minerals are present in any given water supply. At high temperature, many common waterborne minerals become insoluble and plate out on the system's fluid passages as scale. A common source of deionized water is your local grocery store, where it is sold very inexpensively by the gallon.

Table 10.3

TOXICITY OF SOME COMMON CORROSION INHIBITORS

INHIBITOR (MANUFACTURER)	COMPOSITION	SKIN IRRITATION	LD⁵⁰ GOSSELIN g/kg	RATING
Drewsol (Drew Chemical Corp.)	Polyhydroxyl organic fluid	0.1	>24.0	1
Nutek 876 (Nuclear Technology)	Concentrated inhibitors for aqueous solutions	0.1	22.9	1
Vicro Pet 300	Butoxyethyl acid phosphate diethylamine adduct	1.2	5.9	3

Source: Adapted by permission from *Solar Heating Materials Handbook,* P. Homan and C. Hiliary, Anachem, Inc., 1981.

Water/Glycol Fluids Water/glycol heat transfer fluids are generally 70/30, 60/40, or 50/50 water/glycol mixtures. If the fluid is an ethylene glycol/water mixture, the proportion of glycol can be determined with a **hydrometer.** The antifreeze checker used at your local service station is a satisfactory hydrometer. In addition to the hydrometer the serviceperson will need to carry a specific gravity chart unless the hydrometer reads directly in degrees Fahrenheit.

If the fluid is propylene glycol/water, a hydrometer cannot be used. A *refractometer* must be used instead. A refractometer is an instrument that checks the fluid's refractive index. One manufacturer of propylene glycols, Dow Chemical, recommends the use of an instrument from American Optical Company, Model 7391. The serviceperson also must have a refractive index/freeze-point chart available.

If the concentration of the water/glycol is correct, as recommended by the system manufacturer, check the pH of the system. pH is a measure of the fluid's acidity or alkalinity. The pH scale ranges from 1 to 14, as shown in the following table.

pH	ACIDITY/ALKALINITY DESCRIPTION
1	Very acid, highly corrosive to metal
3	Moderately acid, corrosive to metal
5	Slightly acid, mildly corrosive to metal
7	Neutral
9	Slightly alkaline, noncorrosive to copper, brass, and iron
11	Moderately alkaline, slightly corrosive
13	Highly alkaline, can be quite corrosive

For reference, household vinegar's pH is about 3; household ammonia's pH is about 10; Drano's pH is about 14; and drinking water ranges from about 6.5 to 8.0.

Fresh deionized water/glycol solutions, which contain alkaline inhibitors added at the factory, will have a pH of about 10. As the inhibitors are used up, the pH will gradually drop down to the acid range. When the pH reaches 7.5 to 8, fresh heat transfer fluid should be placed in the system. Again, it is wise to remove all the current material, flush the system first with pressurized tap water to remove foreign matter and then with deionized water, and fill the system with fresh deionized water/glycol solution at the recommended concentration.

Either pH paper or a pH meter may be used for these measurements. A pH meter, however, is a laboratory instrument, not a toolbox item. Usually, pH paper may be obtained from a local chemical supply house or a drugstore.

Hydrocarbon Oils

Many different hydrocarbon oils are used. They differ in their characteristics, so it is hard to describe a common practice that fits all systems. In general, oils do not need replacement unless they are sludging, have increased in viscosity, have turned acid, or are contaminated with water.

Hydrocarbon oils should not be blended or mixed. They may be incompatible with each other and will thicken or sludge when blended. To change out an oil system, proceed as follows.

- Drain the system.
- Flush the system with fresh oil. *Do not water flush.*

- Drain and strain the flushing oil.
- Refill the system with the strained oil or with a fresh batch.

Synthetic Oils   Synthetic oils, such as dimethyl silicone fluids, usually need no inspection or maintenance. The fluids are being used way below their designed extreme operating temperatures. A visual inspection of the downleg sample should be all that is needed. The oil can be drained, strained, and placed back in the system if necessary. If the oil contains water, allow the oil to stand overnight. The silicone fluid will rise to the top of the container and can be siphoned off and saved. Add any additional material needed to top off the system.

Draining and Filling the Loop

While the system is down for heat transfer fluid change, it should be tested for leaks, because leaks are much more apt to occur with organic heat transfer fluids than with water because of the organic fluid's lower surface tension. Silicone fluids are the worst offenders, but oils and glycols also can present problems. Any leaks should be repaired before the loop is refilled.

The following procedure, reprinted from *The Solar Decision Book,* will test and fill most liquid heat-exchanged loops satisfactorily.

Testing the Completed Collector Loop[1]

The collector loops discussed in *The Solar Decision Book* are not engineered for water. Nor are certain other designs for collector loops. In many cases the metals employed can be severely corroded by contact with water. As such, the collector loop should never be water tested. Exceptions, of course, would be water-based collector systems and those systems for which the manufacturer specifically recommends water testing. All other types of collector loops should be tested with air before filling.

The general procedure for testing is to charge the loop with air at a pressure 15% below the value of the system's pressure relief valve. Generally, 25 psi is adequate. The loop should be pressurized for 24 hours. No loss of air pressure over that period of time can be tolerated. If any loss occurs, a leak is indicated. The leak must be located and repaired before filling the system.

Freon- or nitrogen-testing procedures are also common. These procedures allow the use of a sniffer to determine leak locations. This is particularly useful in large, complex systems where many joints must be inspected. You should not, however, pressure-test a system that is not designed to be pressurized, nor should you pressurize a system beyond its design capabilities.

Filling the Collector Loop

The recommended systems in *The Solar Decision Book* are designed for filling with heat transfer fluid from the service room. Filling from the top of the collector array is unnecessary. Figure 10.1 shows a typical filling hookup. Filling is accomplished as follows.

- With the collector pump off (it is not designed for dry operation),

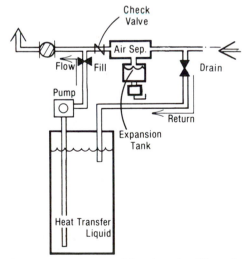

Figure 10.1   A typical hookup for filling the collector loop with heat transfer fluid. (Reprinted by permission from R. Montgomery and J. Budnick, *The Solar Decision Book,* Wiley, New York, 1978.)

[1]The sections "Testing the Completed Collector Loop" and "Filling the Collector Loop" are reprinted from R. H. Montgomery, *The Solar Decision Book* (Wiley, New York, 1978), pp. 21-12, 21-13.

start the auxilliary pump. As soon as the loop is filled, the return line will stop bubbling.

- When the loop is filled, turn off the drain valve and continue to pump until the loop pressure builds to approximately 80% of the rating of the pressure relief valve. This will usually be about 22 to 24 psi. Higher pressures are not recommended.
- Close the fill valve and stop the auxilliary pump.
- Turn on the system circulating pump.

During the first few days of operation, any air left in the system will be bled from the automatic air eliminator. At the end of that period, the system pressure should be reduced to about 15 to 30 psi. This is accomplished by cracking the drain valve to bleed excess pressure from the system. If an air lock is encountered during or after filling, the manual bleed valve on the collectors can be used to purge the system. Observation of the flow meter, if one is included in the system, will reveal the presence of air in the fluid.

The Heat Transfer Fluid Checklist

Here is a heat transfer fluid checklist that covers all the points discussed.

Fluid Data

1. Trade name _____
2. Chemical class _____
3. Potable or nonpotable _____
4. Fluid class (check one)
   _____ Drinking water   _____ Water/ethylene glycol
   _____ Fresh air   _____ Hydrocarbon oil
   _____ Inhibited water   _____ Synthetic oil
   _____ Water/propylene glycol
5. Gosselin rating _____
6. Rated flash point _____
7. Rated fire point _____
8. Rated autoignition point _____

Rate the following as good, marginal, or poor by checking the appropriate column.

	GOOD	MARGINAL	POOR
9. Does the system contain a properly sized expansion tank?	_____	_____	_____
10. Is the expansion tank operating correctly?	_____	_____	_____
11. Has the fluid softened any gaskets, hoses, or seals?	_____	_____	_____
12. Has the fluid leaked and caused any damage?	_____	_____	_____

	GOOD	MARGINAL	POOR

13. Pump leg sample evaluation.
   (a) Any rust/scale/ water? _____ _____ _____

   (b) Any sludge or carbon? _____ _____ _____

   (c) Metal particle content? _____ _____ _____

   (d) Any sticky, gummy, viscous gels? _____ _____ _____

14. Circulating system sample.
   (a) Any rust/scale/ water? _____ _____ _____

   (b) Any sludge or carbon? _____ _____ _____

   (c) Metal particle content? _____ _____ _____

   (d) Any sticky, gummy, viscous gels? _____ _____ _____

15. FLUID TEST RESULTS

Fluid test results—inhibited water

Type inhibitor _____

Concentration _____

Date last changed _____

Life expectancy _____

Estimated overall condition __ Good __ Marginal __ Poor

Fluid test results—water/glycol

Type glycol _____

Specified concentration _____ / _____

Date last changed _____

Life expectancy _____

Specific gravity _____

Refractive index _____

Actual concentration _____ / _____

pH (as tested) _____

Estimated overall condition __ Good __ Marginal __ Poor

Fluid test results—hydrocarbon oils

Type oil _____

Date last changed _____

Life expectancy _____

Estimated overall condition __ Good __ Marginal __ Poor

Fluid test results—synthetic oils

Type oil _____

Date last changed _____

Life expectancy _____

Estimated overall condition __ Good __ Marginal __ Poor

16. SERVICE RECOMMENDATION

_____ Change out fluid

_____ Fluid changeout time approaching

_____ Fluid changeout not required

Corrosion in Solar Systems

The Solar Decision Book (R. H. Montgomery and Jim Budnick, John Wiley & Sons, Inc., 1978) covered the subject of corrosion in solar systems in depth. Much of the material still applies so we have included an edited version of the material here.

Types of Corrosion

Several types of corrosion are possible in solar energy systems.

Galvanic Corrosion Galvanic corrosion is a type of corrosion that is caused by an electrochemical reaction between two or more different metals. A chemical reaction between the metals causes a small electrical current. This current erodes material from one of the metals.

Solar energy systems generally contain a number of different metals, such as aluminum, copper, brass, tin, and steel. This makes the solar system a prime candidate for galvanic corrosion. If the dissimilar metals are physically joined or if they are contacted by a common storage or heat transfer fluid, the possibility of galvanic corrosion becomes much greater.

Corrosion control thus becomes an important factor in the design and operation of a multimetal solar energy system. Without proper corrosion control, premature system failure may result.

Figure 10.2 shows an elementary setup that demonstrates the principles of corrosion. An aluminum plate and a copper plate are suspended in salt water. An external electrical connection is made between the plates. Salt water is an *electrolyte*. Electrolytes are fluids capable of carrying an electrical charge. Because the metals are different, an electrical potential or voltage will exist between them across the electrolyte. This voltage will cause a current to flow through the external circuit. The voltage acts as a force to push metal ions from the anode into solution in the electrolyte. The anode is the more electrically positive metal; the *cathode* is the more electrically negative metal.

A metal is more positive (anodic) or more negative (cathodic) than another metal because of its chemical structure. The various metals can be ranked from positive to negative in a *galvanic series*. The more positive the material, the more it will corrode. Also, the corrosive action between two metals next to each other in a galvanic series is small. The corrosion action increases to a high level between two metals widely spearated in the series. There is a greater electrical potential. A galvanic series of common metals and alloys is shown in Figure 10.3.

In the simple example of galvanic corrosion shown in Figure 10.2 the aluminum ions will go into solution. The aluminum will become corroded.

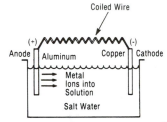

Figure 10.2 An elementary setup that demonstrates the principles of galvanic corrosion. (Reprinted by permission from R. Montgomery and J. Budnick, *The Solar Decision Book,* Wiley, New York, 1978.)

Figure 10.3 A galvanic series of common metals and alloys. (Reprinted by permission from R. Montgomery and J. Budnick, *The Solar Decision Book,* Wiley, New York, 1978.)

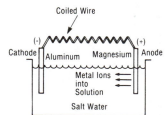

Figure 10.4 Galvanic action between magnesium and aluminum. (Reprinted by permission from R. Montgomery and J. Budnick, *The Solar Decision Book,* Wiley, New York, 1978.)

If magnesium is substituted for the copper, the reaction shown in Figure 10.4 takes place. The magnesium is more positive than aluminum in the galvanic series, so the magnesium serves as the anode and the aluminum becomes the cathode. Magnesium ions flow into solution.

The relative corrosion resistance of metals in the galvanic series can be used to the designers' advantage. If different metals are needed in a system, they can be chosen to be close in electrical charge. The corrosion will be slow. If two widely separated metals must be used, a third metal might be added to the system. This metal would be chosen as the most positive of the three. It will be sacrificed as the next most positive material is protected.

This is a typical solution for hot water tanks, as shown in Figure 10.5. The tank is carbon steel and the water lines are copper. Water is a good electrolyte, because it usually contains some amount of dissolved salts. If not protected, such a steel tank would rapidly become corroded because of galvanic action between the steel (anodic) and copper (cathodic).

To prevent this corrosion, a metal that is more anodic than carbon steel is placed in a hot water tank when it is manufactured. Magnesium is most anodic and is usually chosen. The magnesium must be consumed before the carbon steel becomes the anode and begins to corrode. This prolongs the life of such hot water tanks.

Many factors affect the rate at which the corrosion takes place: If the electrolyte is agitated, increasing the flow, if the temperature is increased, if more oxygen is introduced, or if the electrolyte is made more active, the rate of corrosion generally will increase.

The relative size of the anodic metal to the cathodic metal is also very important. If the anode is extremely small and the cathode is extremely large, corrosion will be rapid. If the cathode is small and the anode is large, corrosion will be slower.

In the solar energy system the most anodic metal must be protected to prevent premature system failure.

Pitting Corrosion   In this type of corrosion the metal ions leave localized areas, causing a pitted surface or uneven corrosion such as shown in Figure 10.6. When heavy metal ions such as iron or copper plate out on a more anodic metal such as aluminum, a small local galvanic cell can be formed. This will cause a local "pit" or corrosion spot. As the anode is "capped" by the cathodic plating, pitting will usually continue until the metal is perforated.

Heavy metal ions can either come as a natural impurity in a water-mixture heat transfer fluid or from corrosion of other metal parts of the solar system. Pitting corrosion can also be aggravated by the presence of

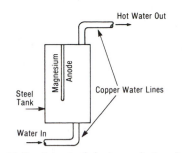

Figure 10.5   A typical design solution for hot water tanks. A magnesium rod in the tank serves as a sacrificial metal to prevent rapid corrosion of the steel tank. (Reprinted by permission from R. Montgomery and J. Budnick, *The Solar Decision Book,* Wiley, New York, 1978.)

Figure 10.6   An exaggerated example of pitting corrosion on a metal surface. (Reprinted by permission from R. Montgomery and J. Budnick, *The Solar Decision Book,* Wiley, New York, 1978.)

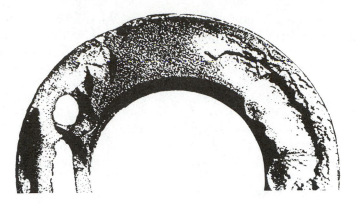

Figure 10.7 An example of crevice corrosion on a flange under a leaky gasket. (Reprinted by permission from R. Montgomery and J. Budnick, *The Solar Decision Book,* Wiley, New York, 1978.)

chloride or other chemicals that can be part of the water mixture or a contaminant from solder fluxes.

Aluminum is very susceptible to pitting corrosion while copper generally is not, except on rare occasions.

Crevice Corrosion Like pitting corrosion, crevice corrosion occurs in localized areas. Figure 10.7 shows a typical example of crevice corrosion on a flange under a leaky gasket. Crevice corrosion is generally associated with bad gasketing, poorly fitting joints, internal blockages, scale deposits from hard waters, and poor design. Both aluminum and copper are susceptible to crevice corrosion.

Erosion/Corrosion A metal placed in the atmosphere quickly develops a protective coating because of its interaction with oxygen. A surface metal oxide is formed. High-velocity liquid flow, particularly when coupled with air bubbles or abrasive debris, can mechanically remove this protective film and cause the metal to erode/corrode. An example of such erosion/corrosion is shown in Figure 10.8.

Guidelines for Corrosion Prevention

A practical solar energy system cannot be built using a single type of metal throughout. All solar energy systems incorporate many types of metals. Thus, corrosion prevention is important in all types of systems.

Galvanic corrosion and localized pitting have caused the most failures in solar collector systems. Problems with crevice corrosion and erosion/corrosion have been minor, but these types of corrosion should be guarded against. There are other forms of corrosion but they have not appeared in solar energy systems.

Since aluminum is usually the most anodic metal in the system, it requires the most corrosion protection. It offers too many advantages not to be used. It is light in weight, high in strength, and easily fabricated. Possibly most important, aluminum is an economical material that has a higher thermal conductivity.

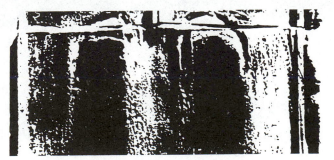

Figure 10.8 An example of erosion corrosion caused by localized turbulence. (Reprinted by permission from R. Montgomery and J. Budnick, *The Solar Decision Book,* Wiley, New York, 1978.)

Aqueous Systems When water is used as the collector heat transfer medium in a closed loop (recirculating system), the water should be deionized, demineralized, and neutralized. A "getter"[2] should be installed in the loop as shown in Figure 10.9. It should be galvanically isolated from the panels and directly before them.

A getter is very simply a column or cartridge containing an active metal, such as aluminum, that will be sacrificed to protect some other metal in the system. Any heavy metal ions developing in the system will plate out on the "getter" metal and corrode it. Getters are not 100% effective, and they must be periodically replaced.

In addition to the getter, inhibitors should be added to the water. Anodic inhibitors, which are primarily oxidizing agents that halt corrosion, can be used.

Anodic inhibitors that form insoluble films can also be used. Phosphates, arsenates, and silicates are effective. Filming anodic inhibitors can also aggravate localized corrosion when used in too-small quantities.

The cathodic inhibitors normally used with iron and steel are generally not used in the solar collector loop. They are not effective for aluminum.

Aqueous/Antifreeze Systems An aqueous system must be drained down during freezing weather to prevent mechanical damage. Otherwise, an antifreeze must be added. Ethylene and propylene glycols are the most commonly used antifreezes. When either of these are mixed 50/50 with water, freeze protection down to −30°F is provided, and the boiling point is raised to about 230°F.

The addition of glycols to the collector heat transfer loop presents a new corrosion problem not seen in aqueous systems. At temperatures close to the boiling point, glycols can break down to form glycolic acid. Further breakdown to oxalic acid may also occur. The formation of these acids reduces the pH of the heat transfer fluid. This makes the fluid corrosive to most all metals, including aluminum, copper, and steel. In automotive applications the normal practice is to add pH buffers such as borates to slow the acidification of the heat transfer medium. However, solar collectors may see temperatures well over 300°F, while automotive systems operate under 230°F. At these high solar system temperatures, breakdown is more rapid and the buffer is more quickly depleted.

Anodic inhibitors are still required in the system. However, oxidizing inhibitors such as chromates are not suitable. They promote rapid degradation of the glycol. Extreme care must be exercised in selecting inhibitors for water/glycol systems. Toxicity can be a consideration, particularly with ethylene glycol. Again, a getter column would provide additional corrosion protection.

The strict schedule of maintenance mentioned for aqueous systems is also required for water/glycol systems. However, because of the buffer depletion and subsequent acid formation problem, the system must be monitored for pH changes continually and drained and refilled as soon as the pH begins to turn acid.

Nonaqueous Fluids Fluids that are nonelectrolytes seem to be a practical solution to the multimetal corrosion problem. Current practice indicates that there are several promising candidates of this type in the marketplace. These include silicone heat transfer fluids. To date, no corrosion problems have been seen with this fluid.

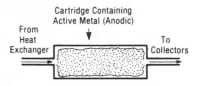

Figure 10.9 A "getter" for corrosion protection in the collector loop. No longer in common use. (Reprinted by permission from R. Montgomery and J. Budnick, *The Solar Decision Book,* Wiley, New York, 1978.)

---

[2]Most systems in use today avoid the use of aluminum, and "getter" cartridges are not in general use.

Other Considerations "Internal corrosion" within the collector loop fluid passages is not the only area for prevention measures. There are two other places where corrosion prevention must be carefully practiced.

- Between solar storage and the heat load.
- On the external surfaces of the solar system metals.

Solar storage is usually accomplished with water held in large tanks. These tanks cycle from 45 to 200°F and can be made from a variety of materials. Solar storage water must be properly inhibited and maintained. The corrosion problem differs, depending on the metal combinations, and therefore cannot be generalized. A local water-treatment company should be consulted for details.

For the external surfaces of the solar collector loop, the following corrosion-prevention techniques should be practiced. Regardless of what medium is chosen for internal heat transfer, external failure is generally much less likely to occur.

- Two widely dissimilar metals such as copper and aluminum should never be directly connected. A nonconducting joint or flange should be used. An example would be a high-temperature flexible rubber hose. This, too, should be carefully chosen for compatibility with the heat transfer agent and the temperature conditions. Generally, steel-to-brass, and copper-to-brass connections do not cause serious problems.
- Moisture from condensation or leakage should be carefully avoided. Without an aqueous medium present, corrosion rarely occurs. Contamination from salt spray and atmospheric pollutants such as sulphur dioxide must be avoided.
- Collector surface paints and platings must be chosen and applied with care to prevent galvanic effects in crevices, and the rear of the panel must be protected.

Corrosion in Other Systems (Non-Heat-Exchanged)

Aluminum is no longer used in most solar water-heating systems, so steel becomes the most anodic metal in the system and must be protected against corrosion.

New steel water-storage tanks are protected against corrosion by lining the tank walls with either ceramic frits (glass lined) or cementacious materials (stone lined) and suspending a magnesium anode rod in the tank.

This presents few problems in completely new water-heating systems, but it can create problems in coupling new solar systems to old auxiliary water heaters. In older water heaters galvanic corrosion may have consumed the anode rod and begun to attack the tank walls.

Coupling a new copper-plumbed solar water-heating system to such an existing system will accelerate the tank corrosion and may cause rapid tank failure. Replacement of the anode rod is usually difficult or impossible because of corrosion around the anode's threaded fitting.

This potential problem exists wherever copper pipes are joined to steel plumbing. The corrosion of the steel speeds up, particularly around the locations where the two dissimilar metals contact each other. Corrosion in these joints can actually fill in the pipe bore and block water flow.

The usual defense against this type of corrosion is the use of dielectric unions in locations where the dissimilar metals join. Unlike conventional unions, dielectric unions contain a rubber washer and a plastic collar that insulate the pipes from each other. This insulation prevents the formation of a galvanic circuit in the piping. Although the metal ions can freely travel through the water, the electric potential that draws them from the steel to the copper is broken. Even though some corrosion will occur, the rate of corrosion will be slowed.

Steel-to-brass and copper-to-brass connections are less likely to create galvanic corrosion problems. However, it is improper to substitute a brass union or fitting for a dielectric union in a copper and steel piping system.

In drainback systems water and oxygen (air) are constantly mixed, creating good conditions for corrosion. Most manufacturers recommend the use of deionized (distilled) water with a nontoxic inhibitor added in the drainback tank to insure long system life and optimal performance.

Heat Exchangers

Chapter 5 described the six common classes of heat exchangers used in solar water heaters. A seventh type, the air-to-liquid exchanger, was described in Chapter 6. To review, these exchanger types are as follows.

- Internal-flued tank.
- Plate around tank.
- Tank around tank.
- Cascade over tank.
- Coil in tank.
- External separate exchanger.
- Air collector-to-liquid storage exchanger.

Chapter 5 also explained what heat-exchanger effectiveness is, why it is important, and how to determine it.

Heat-Exchanger Effectiveness—General Method

Since heat-exchanger configurations vary from system to system, a further discussion of a general method of determining exchanger effectiveness in collector-to-storage loops follows.

$$\frac{\text{Collector outlet temp} - \text{collector inlet temp}}{\text{Collector outlet temp} - \text{bottom tank temp}} \times 100 = \% \text{ effectiveness}$$

Or, when referring to Figure 10.10,

$$(T1 - T2)/(T1 - T3) \times 100 = \%E$$

When evaluating the exchanger configurations that only have one liquid circuit such as a coil-in-tank or internal-flued system, use the tank water temperature as measured from or at the drain valve near the bottom of the tank. Always check effectiveness during full sunlight hours under the test conditions shown in Chapters 6 and 7.

Evaluating the Heat Exchanger

Heat-exchanger evaluation consists of checking for

- Exchanger effectiveness.
- Tube plugging.
- Physical damage.

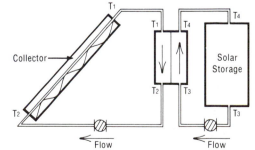

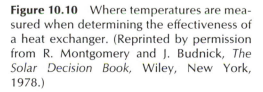

Figure 10.10 Where temperatures are measured when determining the effectiveness of a heat exchanger. (Reprinted by permission from R. Montgomery and J. Budnick, *The Solar Decision Book,* Wiley, New York, 1978.)

- Internal leaks.
- Structural mounting.
- Suitability of physical location.

Exchanger Effectiveness Exchanger effectiveness under maximum solar radiation conditions with a storage tank temperature under 90°F should be a minimum of 50%. If the effectiveness of the exchanger is lower, the design may be poor, the flow rate may be wrong, the exchanger may be partially plugged, or the system may be airbound. In any case, the reason for low effectiveness should be isolated.

Chapter 6 indicated which designs would be apt to have low effectiveness. If the system uses a cascade-over-tank, tank-around-tank, or double-wall plate-around-tank system, most likely the design is at fault and system modification would be needed to increase exchanger effectiveness.

Tube Plugging With coil-in-tank, air-to-liquid, and external exchangers, the small tubes carrying the liquids may be plugged with sludge, rust, or scale.

When this occurs, the tubes must be flushed and cleaned. Procedures will vary, depending on the exchanger design. Some exchangers, like the Doucette exchanger shown in Figure 10.11, have a removable tube head that allows cleaning with a stiff wire brush. Other exchangers, like the coil in tank, or the coiled tube in shell, do not provide access for mechanical cleaning.

For tubes that cannot be mechanically cleaned, two alternative procedures can be considered: treatment with chemical solutions and blowing out the tubes with compressed air.

Chemical treatment of the fluid passages of a solar water-heating system should never be undertaken unless the manufacturer has specifically recommended a chemical treatment. Any chemical aggressive enough to remove scale will also attack many of the metals used in the system. Most manufacturers' warranties will be voided if chemical treatment is undertaken.

Compressed air can be safely used as long as the pressure limits of the system are not exceeded. Look for the equipment's pressure limits on the manufacturer's nameplate.

Physical Damage A visual check will generally reveal any physical damage to the system. Look for dents, crushed pipes, and damage caused by thermal expansion.

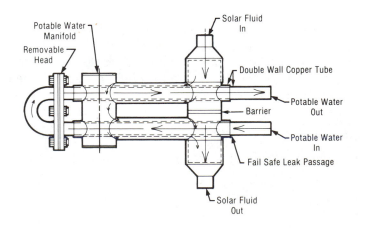

Figure 10.11 Some exchangers have removable heads that allow the tubes to be mechanically cleaned. (Reprinted by permission from R. Montgomery and J. Budnick, *The Solar Decision Book,* Wiley, New York, 1978.)

Internal Leaks   The system can be quickly evaluated for internal leaks by one of three different methods.

- Check the pressure of the fluid on the two sides of the exchanger. The storage tank side of the exchanger should be at normal water pressure: 30 to 60 psi. A pressure gauge mounted to the drain valve (flush before installing) or elsewhere in the storage tank system will quickly give the water tank pressure. Most collector loops have a pressure gauge installed. This should read in the 5- to 25-psi range. If the loop does not contain a pressure gauge, one may be mounted to the drain or fill valve. If the system shows a 5-psi or greater differential between the storage side of the exchanger and the collector side of the exchanger, the system most likely does not have an internal leak.
- In oil systems check the exchanger fluid for water. The collector side of the exchanger generally runs at lower pressures than the storage tank side. The presence of water in the exchanger fluid *may* indicate an internal exchanger leak.
- As a final check before removing a heat exchanger that is suspected of leaking, check the manufacturer's pressure rating of the exchanger; disconnect the tubes of the exchanger, mount a pressure gauge and an air fitting, and pressurize the exchanger tubes to rated pressure. The tubes should hold this pressure without leakage. A 5- to 10-minute test should reveal any leaks.

Structural Mounting   When the heat exchanger is an integral part of the storage tank, the mounting of the exchanger to the tank will be done by the manufacturer. In this case the serviceperson should evaluate the piping and the components attached to the exchanger to determine if

- Too much weight has been added to the mounting structure.
- Thermal expansion has been properly allowed for.
- Vibration, caused by the pump or other apparatus, has loosened the mountings or fatigued the metal in the exchanger.
- Separate exchangers are correctly mounted to a firm foundation or wall in a manner that insures they will not come loose.

Physical Location   All heat exchangers should be carefully insulated and placed in physical areas where they cannot freeze or be subjected to other adverse environmental conditions that would create substantial energy losses from the outer shell or otherwise interfere with the exchanger's operation.

Heat-Exchanger Evaluation Checklist

Here is a checklist for evaluating heat exchanger condition and installation.

EVALUATION POINT	GOOD	MARGINAL	POOR
1. Type _____			
2. Is the exchanger a recommended type?	_____	_____	_____
3. Is the exchanger effectiveness satisfactory?	_____	_____	_____

EVALUATION POINT	GOOD	MARGINAL	POOR
4. Are the tubes scaled or plugged?			
5. Is there any physical damage?			
6. Are there any internal leaks?			
(a) Pressure differential?			
(b) Water in transfer fluid?			
(c) Air pressure test?			
7. Is the exchanger securely mounted?			
(a) Ability to support the weight?			
(b) Thermal expansion allowance?			
(c) Vibration or fatigue problems?			
8. Is the physical location suitable?			
(a) Properly insulated?			
(b) Protected against freezing?			
(c) Any adverse environmental factors?			

CHAPTER
11

Pumps and Controllers

All solar water-heater systems except ICS, thermosiphon, and gravity return phase-change systems contain a pump and a controller. The pump circulates the heat transfer fluid between the collectors and the storage tank; the controller turns the pump on when the collectors are hotter than the storage tank and turns it off when they are not.

Pumped water, draindown, drainback, and heat-exchanged systems that have the heat exchanger as an integral part of the tank, such as coil-in-tank systems usually use one pump. Systems that use separate heat exchangers, such as the shell-and-tube exchanger, usually require two pumps. Even in dual-pump systems, only one controller is generally used.

Pumps

The ability to pump at the correct flow rate and without premature pump failure requires some basic knowledge about pump design, pump performance curves, and pumping pressure drop.

Pump Design Types

Residential solar water-heating systems use **centrifugal pumps,** which contain a series of curved vanes on the face of, or within, a rapidly rotating impeller. The liquid enters at the center of the impeller and is "slung out" the discharge port of the pump by the vanes.

There are three major types of centrifugal pumps: the **external motor mechanical shaft seal pump,** the **magnetic-drive pump,** and the **canned wet-rotor pump.**

Mechanical Shaft Seal Pump   Figure 11.1 shows an exploded view of a typical centrifugal pump with an external motor and a mechanical shaft seal. The pump consists of the

- Motor.
- Pump frame.
- Seal seat.
- Shaft seal.
- Seal spring.

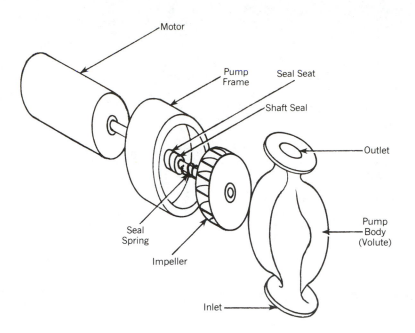

Figure 11.1 An exploded view of a mechanical shaft seal pump.

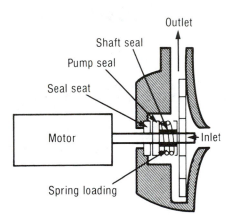

Figure 11.2 Cross section of a mechanical shaft seal pump. (Reprinted by permission from R. Montgomery and J. Budnick, *The Solar Decision Book,* Wiley, New York, 1978.)

- Impeller.
- Pump casing with liquid inlet and outlet.

This type of pump is used most often to pump water. When other heat transfer fluids are used, a magnetically driven or a canned wet-rotor pump is chosen.

Figure 11.2 is a cross-sectional view. When the motor starts up, the shaft seal, the shaft seal spring, and the impeller turn together as a unit; the frame, the seal seat, and the casing remain stationary. Water coming in the center inlet contacts the rapidly turning impeller, which contains a number of curved vanes (see Figure 11.3), and is slung out through the outlet. Water is prevented from leaking out around the motor shaft by the spring-loaded shaft seal, which rides against the seal seat.

Magnetic-Drive Pump Figure 11.4 shows a magnetic-drive pump. In this type of pump the interior of the pump is mechanically isolated from the motor, so that no pump seal is needed. A set of strong, permanent

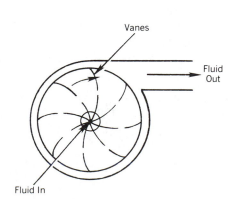

Figure 11.3 The vanes on the rotating impeller sling the fluid out the discharge port.

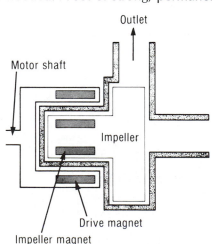

Figure 11.4 A magnetic drive pump. (Reprinted by permission from R. Montgomery and J. Budnick, *The Solar Decision Book,* Wiley, New York, 1978.)

drive magnets is located on the end of the motor shaft. These drive magnets surround a similar set of magnets located in the impeller. When the motor turns, the rapidly turning drive magnets attract the impeller magnets, so that the impeller turns with the motor. The elimination of the shaft seal allows the pumping of liquids that might leak out through a seal.

Canned Wet-Rotor Pump Figure 11.5 shows a canned wet-rotor pump. In this type of pump one set of motor windings (the stator winding) is built into the pump frame, and a second set (the rotor winding) is built into the pump impeller. When the motor is energized, the rotor and impeller turn as a unit, with the rotor supported on internal bearings and running in the liquid. Again, the elimination of the seal allows many different liquids to be pumped.

Pump Motor Start-Up Alternating-current motors need an initial spin to get them started. Most fractional horsepower pumps used in solar water-heating systems use a *starting winding wired in series with a starting capacitor* to provide this spin. A starting winding is a separate small winding in the motor. A starting capacitor is an electrical device that stores up energy and discharges it into the starting winding.

Because there is a time delay in charging the capacitor, its electrical charge reaches the motor out of phase with the charge in the main winding and creates an initial spin that starts up the motor. Such motors are known as *permanent split capacitor motors.*

If the capacitor is corroded from exposure to water, damaged by excess heat, or develops a leak, the capacitor needs to be replaced. When an energized pump hums but does not start up, the starting capacitor may need to be replaced. If the pump has an exposed shaft, cautiously give it a spin by hand. If it runs, the problem is isolated to the starting circuit.

New capacitors should be of the rating and type suggested by the manufacturer.

Pump Motor Overload Protection Most fractional horsepower pump motors are *type A* motors. Type A motors are engineered to run at ambient temperatures of 104°F or less. Higher ambient temperatures may cause premature failure because of overheating.

The windings in the pump may reach 176°F during normal operation. Higher temperatures than this can cause damage, so most pump motors are designed either to dissipate the excess heat or to turn off at excessive temperatures. Most magnetic-drive and canned wet-rotor pumps are designed for heat dissipation. These are known as *impedance-protected* motors. Most mechanical shaft seal pumps contain a thermal overload switch. These are known as *thermally protected* motors.

Even with overload protection, repeated overheating can warp the bearings or damage the windings. This overheating generally occurs for one of six reasons.

- The pump is located in a room with an excessively high ambient air temperature.
- Air cannot circulate freely around the pump.
- The line voltage is less than required.
- Water cannot flow freely through the pump.
- The pump impeller is jammed and cannot rotate.
- The pump is airbound.

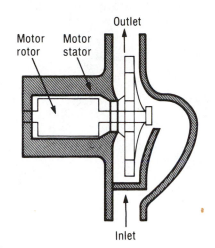

Figure 11.5 A canned wet-rotor pump. (Reprinted by permission from R. Montgomery and J. Budnick, *The Solar Decision Book*, Wiley, New York, 1978.)

The serviceperson must determine why the pump failed before replacing it if a second pump failure is to be avoided.

Other Considerations   Centrifugal pumps will not provide suction to lift water, and they are not self-priming. The pump must always be filled with liquid for it to operate. Therefore, the pump must be located low in the system, where it will not become airbound, and it must be supplied with a column of liquid above the pump inlet at all times.

The bearings in many centrifugal pumps are lubricated by the liquid being pumped. If the pump is run dry, the bearings will be ruined quickly.

Some pump motors have bearing oil cups that must face up. Usually, the pump body can be rotated around the shaft to different positions to accommodate a horizontal or vertical discharge pipe while still maintaining the correct positioning of the motor's oil cups. Check the manufacturer's guidelines before changing the position of the body.

Pump motors must always be electrically grounded to prevent any possible electrical shock. The pump is generally piped with conductive metals, and a short to the motor frame could create an electrical hazard throughout the entire solar water-heating system.

Most pumps used in solar water-heating systems are of a *drip-proof* design. A drip-proof design is not intended to be immersed in liquid and should be installed in a dry environment. Do not install these designs outdoors or in an excessively damp environment.

Pump Sizing

A pump converts the electrical energy provided to the motor into the mechanical energy required to circulate the liquid.

The pump's capacity to do work is expressed as the amount of liquid it will circulate against a given resistance to flow in a given unit of time. Normally, flow is expressed as gallons per minute (gpm), and resistance is expressed as feet of vertical head (ft of water).

Figure 11.6 shows a performance graph for a hypothetical pump. Note the multiple scales. Scales A and C are used with a single pump and scales B and D are used with two pumps.

Let us first look at a single pump (scales A and C). The graph shows that if a 20-ft-high vertical pipe were erected (scale A = 20) and pumping was attempted, the liquid being pumped would just reach the top of the column (scale C = 0). No overflow would result, since the pump is only able to do enough work to raise the liquid for 20 ft. You should also realize that under this no-flow condition, the diameter of the pipe under test makes little difference. The pipe could be 1 in. or 1 ft in diameter and the same result would be seen. It would merely take longer for the pipe to fill.

If the pipe were only 15 ft high, 10 gpm of liquid would be pumped out the top of the pipe; if the pipe was 10 ft high, 18 gpm would flow; and if there were no vertical pipe, 31 gpm would flow.

The Pump Performance Curve   Pump performance curves of this type are generated experimentally by the pump manufacturer for the liquids that will be pumped. A different performance curve is needed for each liquid because the viscosity and specific gravity of liquids vary, and these variations change the flow rate. In a pumping curve for water the vertical axis (head) reads in feet of water. However, in a pumping curve for

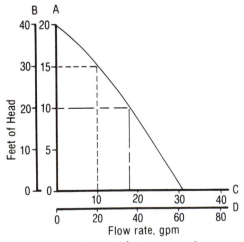

Figure 11.6 A typical pump performance curve. (Reprinted by permission from R. Montgomery and J. Budnick, *The Solar Decision Book*, Wiley, New York, 1978.)

50/50 glycol/water, the vertical axis should read in feet of 50/50 glycol/water.

Once a series of pumping points such as the four just discussed has been experimentally established, a smooth curve like the one in Figure 11.6 can be used to connect the points. This is called the pump performance curve. It allows one to predict what flow will occur at any head shown on the scale. It is a good design practice to choose and operate pumps near the middle of their performance curves instead of near either end. This gives the most efficient use of the pump.

Staging Pumps   When one pump will not provide the needed flow, pumps may be staged by connecting them together to change the flow rate at a given head. Look at Figure 11.7, which shows pumps staged in series and in parallel.

Assume two hypothetical pumps were connected in series, with the inlet of the second pump connected to the discharge of the first pump. The liquid entering the second pump would already be moving at the velocity imparted by the first pump. The second pump would do an equivalent amount of work and double the height to which the liquid would move. The two pumps will move the same amount of fluid against twice the head (scales B and C).

Now assume that the two pumps are connected side by side in parallel. Against a given head, the two pumps would pump twice as much liquid (twice the flow rate), because each pump will do an equivalent amount of work (scales A and D).

It is sometimes cheaper for a designer to stage two or more pumps in series, parallel, or a combination of series and parallel than it is to install a larger pump. The same is true for the serviceperson. A second pump can sometimes be staged into the system to increase flow or overcome head at less cost than removing the undersized pump and replacing it with a larger one. However, a system containing two pumps may need more service than a system containing only one pump.

The Pump Performance Rules   Remember these points.

- Two pumps in series will pump at a given flow rate against twice as much head as a single pump.
- Two pumps in parallel will provide approximately twice the flow against a given head as a single pump.
- It is sometimes cheaper to stage two pumps than it is to purchase one larger one.
- The pump performance curve must be generated for the liquid that will be pumped, and the vertical scale should read in feet of the liquid under test.

Resistance to Flow Through the System

The pumping head characteristics that you just studied referred to only one factor of system flow resistance, **static head.** Static head is the resistance to flow caused by the need to pump the liquid vertically against the force of gravity.

The pump must also overcome the friction caused by moving the liquid through the piping and the components in the system. This resistance to flow is known as **dynamic head.**

The total resistance to flow is known as **system pressure drop.** System pressure drop is the sum of the dynamic head and the static head.

Pressure drop is also called ΔP. ΔP can be stated in feet of liquid or in

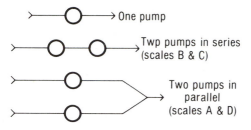

Figure 11.7   Pumps can be staged together in both parallel and series. (Reprinted by permission from R. Montgomery and J. Budnick, *The Solar Decision Book,* Wiley, New York, 1978.)

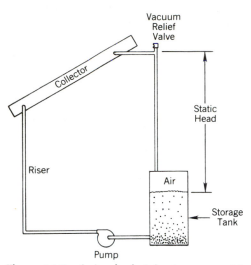

Figure 11.8 A simple drainback system with a static head.

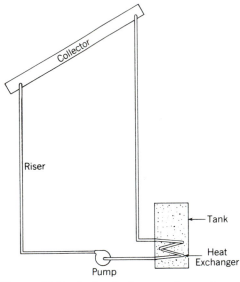

Figure 11.9 A simple heat-exchanged system with no static head.

pounds per square inch. Here is a correlation between feet of water and pounds per square inch that you should remember:

1 psi = 2.3 ft of water

Static Head Static head is created when a liquid has to be pumped vertically up a riser and no continuous, closed, fluid-filled downcomer exists to counterbalance the weight of the fluid that must be lifted. Static head generally only exists if air is present in some part of the system. The magnitude of the static head does not change with fluid velocity, hence the name.

Figure 11.8 shows a simple drainback system that has such a static head. Water from the storage tank is pumped to the top of the collectors and returns by gravity to the top of the tank. Because air is free to enter at a high point in the system, no siphon effect is created by the falling fluid in the return pipe. The size of the static head in this system would be the difference in vertical feet between the top level of liquid in the tank and the highest point to which the liquid must be pumped in order to return. The overall length of the pipe running between these two points does not affect the size of the static head. It is only the difference in height that is counted. If that vertical distance were 10 ft, the system would have a static head of 10 ft of water, or 4.35 psi (10/2.3 = 4.35).

Figure 11.9 shows a simple heat-exchanged system that has no static head. In this system all the liquid is held in a closed loop. The weight of the fluid in the downleg counterbalances the weight of the fluid that must be pumped up the riser. Only the head created by friction (dynamic head) needs to be considered. Generally, any system that is completely filled with fluid when the pump is not operating has no static head and will siphon.

Figure 11.10 shows a simple drain down system. This system also has no static head. As long as there is an automatic air vent at the high point in the system, it will refill when the valves close, becuase the water is under city or well pressure.

You may occasionally encounter an improperly designed draindown system that does not include an air vent at the system's highest point. In these systems the incoming water may encounter an air plug at the top

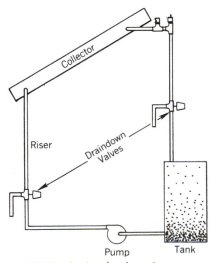

Figure 11.10 A simple draindown system. Properly designed draindown systems have no static head.

of the system. The air plug will resist displacement, and the water must force its way around the plug until it drags the air, in the form of bubbles, slowly down into the storage tank. Hopefully, the designer has included an air vent at the tank, which would eliminate the air.

Flow rates on the order of 1 to 2 ft/s are needed to eliminate these air plugs. That flow rate cannot be provided by the average pump used in a system that supposedly has no static head. The correct solution is the installation of properly located air vents, not the installation of a larger pump.

The serviceperson must be able to recognize the difference between air blockage and excessive static head so that he or she can take the appropriate action. Air plugs are not unique to draindown systems. They can occur in any type of liquid-filled system.

Dynamic Head Dynamic head is the friction head created by the flow of liquid through the system. Every foot of pipe, every fitting, every valve, and every component in the system offer resistance to fluid flow, and the dynamic head of a system changes with the rate of flow. The faster the fluid flows, the higher the resistance offered by the system. Hence the name dynamic head.

The ΔP caused by dynamic head in a system is a function of the velocity at which the fluid is moving, the diameter of the pipe through which it flows, the number and type of fittings and components through which the fluid passes, and the type of liquid flowing.

Figure 11.11 is a graph of pressure drop for water per 100 ft of pipe. The vertical axis shows the pressure drop in 100 ft of pipe as feet of water; the horizontal axis shows the flow rate in gallons per minute. Feet of water can be converted to pounds per square inch on the vertical axis by dividing the feet of water by 2.3 to obtain pounds per squre inch.

Note the five solid lines slanting upward to the right labeled $\frac{1}{2}$ in., $\frac{3}{4}$ in., 1 in.,$1\frac{1}{2}$ in., and 2 in. Each line represents a different pipe size. Now note the two dashed lines slanting downward to the right labeled 1 ft/s and 3 ft/s. These lines indicate the velocity of the liquid flowing through the pipe at the flow rates shown on the horizontal axis. Good engineering practice in the past in hydronic heating systems has called for a fluid velocity of 1 to 3 ft/s, but solar designers have found that 4 ft/s is an acceptable upper limit, while 1 ft/s may be too low for some draindown and drainback systems.

If the flow rate in a system was 5 gpm and the system used 100 ft of $1\frac{1}{2}$-in. pipe, the pressure drop through the pipe (ΔP) would be 0.3 ft of water. ΔP would be only half as much in 50 ft of pipe and twice as much in 200 ft of pipe.

If the system used 100 ft of 1-in. pipe, the ΔP would be about 1.5 ft of water. In $\frac{3}{4}$-in. pipe, ΔP would be about 5 ft of water per 100 ft.

The velocity of the liquid through $1\frac{1}{2}$-in. pipe would be less than 1 ft/s, so $1\frac{1}{2}$-in. pipe should not be used. The use of either $\frac{3}{4}$-in. or 1-in. pipe would place the velocity in the correct range.

Let us assume that our system contains 400 ft of pipe and that a flow rate of 5 gpm is desired. The pressure drop graph shows that with $\frac{3}{4}$-in. pipe, $\Delta P = 20$ ft of water. With 1-in. pipe, $\Delta P = 6$ ft of water.

Go back to Figure 11.6, the pumping curve. The pump shown in that curve will only move 5 gpm at 18 ft of water, so $\frac{3}{4}$-in. pipe cannot be used with a single pump. However, if two pumps were staged in series (scales B and C), the system would pump up to 18 gpm at a 20-ft head, and the throttle valve could be used to reduce that flow to 5 gpm.

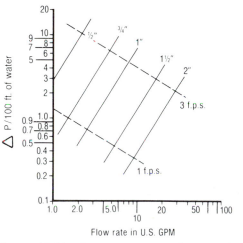

Figure 11.11 A water pressure drop graph. (Reprinted by permission from R. Montgomery and J. Budnick, *The Solar Decision Book,* Wiley, New York, 1978.)

FITTING SIZE INCHES	EQUIVALENT LENGTH OF PIPE IN FEET					
	STD 90°	ELLS 45°	90° Tee STRAIGHT RUN	PIPE COUPLING	GATE VALVE	GLOBE VALVE
3/8	0.5	0.3	0.15	0.15	0.1	4
1/2	1.0	0.6	0.3	0.3	0.2	7.5
3/4	1.25	0.75	0.4	0.4	0.25	10
1	1.5	1.0	0.45	0.45	0.3	12.5
1 1/4	2	1.2	0.6	0.6	0.4	18

Figure 11.12 A table of equivalent pipe lengths for copper fittings and brass valves. (Reprinted by permission from R. Montgomery and J. Budnick, *The Solar Decision Book,* Wiley, New York, 1978.)

If the system were 1-in. pipe, only 6 ft of head would exist and one pump would move up to about 25 gpm. Again, the throttle valve could be used to reduce the flow to the correct amount.

Both approaches require heavy throttling. A more efficient solution would be to substitute a pump better matched to the system.

Pressure Drop in Fittings and Valves Additional friction is created in fittings and valves. This results in increased pressure drop. This pressure drop is calculated by converting the fittings and valves to **equivalent lengths of pipe.** An equivalent length of pipe is the length of pipe that would create the same pressure drop as the fitting or the valve. Figure 11.12 shows a table of equivalent pipe lengths for copper fittings and brass valves. This table shows, for instance, that a 1-in. globe valve has the same ΔP as $12\frac{1}{2}$ ft of 1-in. pipe or that a $\frac{1}{2}$-in., 90° elbow has the same ΔP as 1 ft of $\frac{1}{2}$-in. pipe.

Pressure drop in fittings and valves is calculated by converting the fittings and valves to equivalent lengths of pipe and adding the total length to the piping in the system. Here is an example that uses the supply piping in Figure 11.13.

ITEM	FEET OF $\frac{1}{2}$-IN. PIPE
$\frac{1}{2}$-in., 90° elbow (3)	3
$\frac{1}{2}$-in. ball valve (fully open)	1.5
30 ft of $\frac{1}{2}$-in. pipe	30
Total	34.5

The equivalent length of pipe from the bottom of the heat exchanger to the lower manifold is 34.5 ft.

Calculating System Pressure Drop

The total system pressure drop consists of the static head pressure drop plus the dynamic head pressure drop. The static head pressure drop is calculated by measuring the vertical height that the pump must raise the water and expressing the result in feet of water.

The dynamic pressure drop through the system consists of two separate pressure drop components: pressure drop caused by friction through the pipes, fittings, and heat exchanger, and pressure drop through the collector array.

Figure 11.13 will be used as the first example to illustrate pressure drop

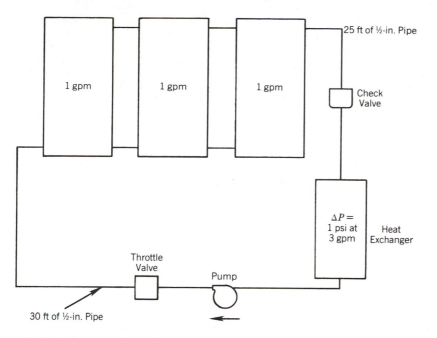

in a closed system where there is no static head. This system consists of three collectors piped in parallel, 55 ft of $\frac{1}{2}$-in. pipe, and a number of components. The flow rate through the pump is 3 gpm.

PIPING PRESSURE DROP

ITEM	EQUIVALENT PIPE LENGTH	ΔP IN FT OF WATER 3 gpm FLOW
55 ft of $\frac{1}{2}$-in. pipe	55	
Throttle valve	1.5	
Check valve	3.0	
Total pipe and fittings	59.5	12.0
Heat exchanger		2.3
Total pressure drop through pipe and exchanger		14.3

The pressure drop through the collectors is highly dependent on whether they are piped in series or in parallel. When collectors are piped in parallel, the flow is divided among the collectors, and the total pressure drop is equal to the pressure drop across a single collector. *The pressure drops are not additive.*

When collectors are piped in series, the flow rate through each collector is the same as the flow rate in the piping, and a pressure drop must be calculated for each collector. *The total collector pressure drop is the sum of the pressure drops across each collector.*

In Figure 11.13 the collectors are connected in parallel, and 1 gpm flows through each collector. As shown Figure 11.13, the collector has a pressure drop of 0.1 psi or 0.23 ft of water at a 1-gpm flow rate, so the total collector pressure drop is 1 × 0.23 ft of water, or 0.23 ft of water.

PARALLEL COLLECTOR ARRAY PRESSURE DROP

ITEM	FLOW RATE (gpm)	ΔP IN FT OF WATER 1gpm/COLLECTOR
Collector 1, 2, or 3	1	0.23
Total		0.23

The total system pressure drop is 14.3 + 0.23 = 14.5 ft of water.

This example shows a very typical head for a solar water-heating system, and the pump shown in the pump performance curve would be adequate.

A glance through the table shows that a 3-gpm flow rate through $\frac{1}{2}$-in. pipe is a high flow rate. The velocity of the fluid exceeds 3 ft/s. *Pressure drop increases as the square of velocity increase.* At 3 ft/s, the pressure drop is 9 times as large as it is at 1 ft/s, and at 4 ft/s it is 16 times as large as at 1 ft/s.

In Figure 11.14 the three collectors are now piped in series, so the flow rate through each collector is 3 gpm and the pressure drop across each panel is 0.9 psi, or 2.1 ft of water.

SERIES COLLECTOR ARRAY PRESSURE DROP

ITEM	FLOW RATE (gpm)	ΔP IN FT OF WATER
Collector 1	3	2.1
Collector 2	3	2.1
Collector 3	3	2.1
Total		6.9

The total system pressure drop = 14.3 + 6.9 = 21 ft of water.

The best source for pressure drop information in collectors and heat exchangers is the manufacturer of the equipment who has determined the pressure drop as part of the design work.

As a general rule, follow these sizing guidelines to avoid excessive pressure drop in small residential systems.

PIPE SIZE (IN.)	MAXIMUM FLOW RATE (gpm)
$\frac{1}{2}$	2–3
$\frac{3}{4}$	4–5
1	8–9

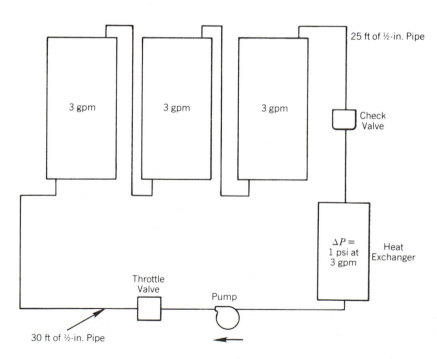

Figure 11.14 An illustration of the example in the text for determining pressure drop, when there is no static head and the collectors are connected in series.

When Pressure Drop Is a Concern   Servicepeople only need to be concerned with calculating pressure drop when the system is circulating too slowly. This will show up as too high a temperature differential across the collectors. In cases where no other reason can be found for the temperature differential, calculate the pressure drop and determine if the pump is undersized for the job. If so, change the pump or correct the piping size.

Pump Installation Considerations

There are a number of pump installation considerations, as shown in Figure 11.15. The pump should

- Be installed low on the storage tank.
- Run in the correct direction.
- Be properly secured and mounted.
- Have valves on both the inlet and outlet sides so that it may be removed for service.
- Be electrically connected to a fused circuit.
- Be electrically grounded.
- Have no piping constrictions on the inlet side.
- Be protected by a strainer.
- Be followed by a throttle valve.
- Be mounted so that the motor shaft is horizontal, with any oil cups pointing up.
- Have the motor left uninsulated for proper cooling.
- Be protected against adverse environmental conditions.

Low Installation Location   The circulating pump should be installed low in the system, because centrifugal pumps are not self-priming. A column of liquid must always be present above the pump inlet, and the pump should not be run dry or the bearings may be ruined. Always install the

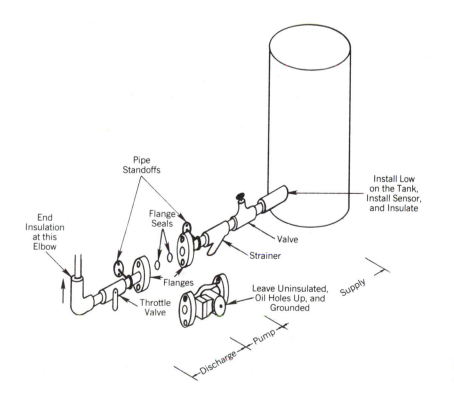

Figure   11.15   The   pump   installation considerations.

pump below the lowest fluid level seen in the storage tank. Follow the manufacturer's directions. A 3-ft head above the pump suction inlet is a good rule of thumb. In a system operating at line water pressure, like a pumped service-water or draindown system, the line pressure primes the pump and pump location is less critical.

Proper Flow Direction   The pump should be mounted so that the flow of the liquid is from the bottom of the storage tank to the bottom supply manifold of the collectors. If the pump is installed backward, the check valve in the downcomer will close and the liquid will not circulate. A temperature gauge mounted on the downcomer will quickly reveal this condition, because the hot liquid from the collectors will not flow and the downcomer temperature will not change when the pump is energized.

Proper Mounting   The pump and its associated piping are subject to differential expansion, pump motor vibration, and the weight of the pump. Generally, small fractional horsepower pumps rely on the piping to provide the support. The pump flanges usually make an excellent structural mounting point on many pumps. Pipe standoffs should be located just above and below the flanges.

Larger pumps may require a vibration-dampening rubber mounting block bolted to the floor or the wall.

Isolation Valves   Pumps periodically require service. Therefore, they should be installed so that they can be isolated from the rest of the system. Flanges containing isolation valves can be used, or a positive-seal butterfly or ball valve can be used on the discharge side as a throttle valve while a gate valve can be placed on the intake side. If the gate valve precedes the strainer, cleaning the strainer is simplified.

Electrical Connections   All pumps should be connected to a properly fused circuit. Pumps of $\frac{1}{6}$ hp and less can be placed on a 15-A circuit, because they draw less than 3 to 4 A. In any case, follow the National Electrical Code. Some small circulator pumps plug into the controller with a three-wire conductor, the third wire being a ground. This is usually very satisfactory if the controller plug is grounded. Some localities do require that the pump be hard wired with no plug.

Electrical Grounding   All pump motors should be wired with a three-wire system where one wire leads to ground. It is also wise to protect the supply-and-return piping coming to and from the collectors independently against lightning strikes.

Piping Constrictions   There should be no piping constrictions between the outlet on the tank and the inlet of the pump. Ideally, the piping should be the same size as or larger than the pump inlet opening. Any valves should be gate valves, not globe or flow-constricting ball valves. Constrictions in the supply piping can cause the pump to be starved for liquid to pump.

Strainer Protection   The use of a strainer on the supply side of the pump prevents any foreign particles from entering the pump or from being pumped up to the collectors. Large particles in the pump could cause the pump to bind and seize up, which could burn out the motor; large particles lodging in the small-diameter risers of the collectors could block the risers.

However, there are two schools of thought on the location of the pump strainer. One advises locating the strainer on the suction side, as shown throughout this book; the other advises placing the strainer on the discharge side so that a plugged strainer will not lead to pump cavitation. Either location is generally satisfactory.

Throttle Valve The pump outlet should be followed by a throttle valve so that the flow of the liquid can be adjusted to the correct point. A gate valve should not be used in this location; it will freeze up with time, so that it cannot be adjusted. Use a butterfly or ball valve.

Horizontal Motor Mounting Generally, centrifugal pump motors are manufactured without vertical thrust bearings. If the motor is mounted vertically, premature wear on the bearings and the pump's shaft seal can result. When oil cups are present and are not positioned correctly, the motor bearings cannot be properly lubricated.

Leave Motor Uninsulated Electric motors generate heat as they run. If the motor cage is insulated, the heat generated will build up and cause motor failure.

Adverse Environmental Conditions Pumps used for water systems must be located where they will not freeze. The pump motor must be protected against water entering the motor and shorting out the windings. The ambient temperature of the air surrounding the pump must be low enough so that the pump motor will run cool. This temperature is usually stated on the motor's nameplate.

Pump Service Checklist

Here is a checklist for the pump. Rate the pump on all 19 points.

EVALUATION POINT	GOOD	MARGINAL	POOR
1. Has the proper pump type been used? (a) Mechanical shaft seal? (b) Magnetic drive? (c) Canned, wet rotor?	_____	_____	_____
2. Is the pump located low in the system where it will be self-priming?	_____	_____	_____
3. Will the bearings be lubricated at all times?	_____	_____	_____
4. Do any oil holes or cups face up?	_____	_____	_____
5. Is the motor electrically grounded?	_____	_____	_____
6. Is the pump sized correctly for the load?	_____	_____	_____
7. If pumps are staged, is the staging correct? (a) Series? (b) Parallel? (c) Combined series/ parallel?	_____	_____	_____
8. Is the flow velocity between 1 and 4 ft/sec?	_____	_____	_____

EVALUATION POINT	GOOD	MARGINAL	POOR
9. Does the pump run in the right direction?	_____	_____	_____
10. Is the pump properly secured and mounted?	_____	_____	_____
11. Does the pump have a shutoff valve on the inlet?	_____	_____	_____
12. Does the pump have a shutoff valve on the outlet?	_____	_____	_____
13. Is the pump connected to a fused circuit?	_____	_____	_____
14. Are there any piping restrictions on the inlet side?	_____	_____	_____
15. Is the pump inlet protected with a strainer?	_____	_____	_____
16. Is the pump followed by a throttle valve?	_____	_____	_____
17. Is the motor shaft horizontal? If not, does it have a thrust bearing?	_____	_____	_____
18. Will the motor cool properly?	_____	_____	_____
19. Is the motor protected against adverse environmental conditions?	_____	_____	_____

Controllers

Any system that contains a pump or electrically operated valves must also contain a **controller.** A controller is a device that automatically turns the pump off and on and/or opens and closes the valves as conditions within the system signal for these actions.

Solar water-heater system controllers are *temperature-signaling controllers* that take action based on temperature changes within the system. A temperature-signaling controller is a device that reads the required temperatures and tells the pump motor or the electrical valves what action to take.

Solar system controllers are also **differential temperature controllers (differential thermostats).** A differential temperature controller is a controller that compares two or more temperatures and takes action based on the difference between those temperatures.

An interesting exception to these controller guidelines is the photovoltaic solar power module. This recently developed device uses solar electric photovoltaic cells to provide power to a small pump when the sun is shining on the collectors and PV cells. In cloudy weather the solar intensity is reduced and the PV module acts like a proportional controller. Pump speed is reduced in accordance with the available light. When solar intensity drops below a threshold level, the pump shuts off.

Photovoltaic power modules offer an alternative to differential temperature solar pump controllers. However, their efficiency and reliability have not been adequately tested in the field to justify detailed coverage here. Manufacturer's instruction manuals can provide information on servicing specific products.

Servicepeople used to dealing with the average heating-system controls will find a major difference in solar controls. The average heating-system control, such as a thermostat or a stack control, uses bimetallic mechanical elements to signal changes; the more sophisticated solar control uses electronic circuits and variable resistances to signal change. Such controls are more complicated and require different servicing procedures.

What the System Controller Does

The typical solar water-heater controller measures the temperature at

- The top of the collectors.
- The bottom of the storage tank.
- The top of the storage tank.
- Strategic points on the collectors and along the outdoor piping.

It first compares the temperature of the top of the collector to the temperature of the bottom of the storage tank. When the collector temperature is a few degrees hotter than the bottom of the storage tank, usually 8 to 18°F, it turns the pump on. When the temperature at the top of the collector drops down to $2\frac{1}{2}$ to 6°F of the temperature at the bottom of the storage tank, it turns the pump off.

However, this differential temperature decision can be overridden. If the temperature at the top of the tank has reached a preset limit, usually between 160 and 180°F, it overrides the pump turn-on circuit and prevents the pump from operating. In some draindown systems it also opens the collector dump valves and drains the collectors at 160 to 180°F. These actions assure that the system does not overheat and build up pressure.

In a pumped service-water system, when the outdoor pipe temperature reaches 40 to 44°F, the system may turn the pump on to circulate warm water from storage. This prevents freezing. At 50 to 54°F, the controller shuts the pump down and reverts to operating from the temperature difference between the collector and the storage tank.

In a drainback system, when the outdoor pipe temperature reaches 40 to 44°F, the controller overrides the pump turn-on circuit and prevents the pump from operating, which might cause a freeze-up.

Also, in a draindown system, when the outdoor temperature reaches 40 to 44°F, the controller both overrides pump turn-on and opens the drain valves to dump the water out of the collector array.

The Controller Logic

Logic is the path of reasoning that the controller travels in making its decisions to turn on and off pumps and to open and close valves. The logic varies from system type to system type. Figures 11.16 to 11.19 show logic diagrams for the five major types of solar water-heating systems.

Recirculation or Drainback Systems   Assuming that the pumped service-water system allows for freeze recirculation and high-limit cutoff, the controller follows the logic path shown in Figure 11.16.

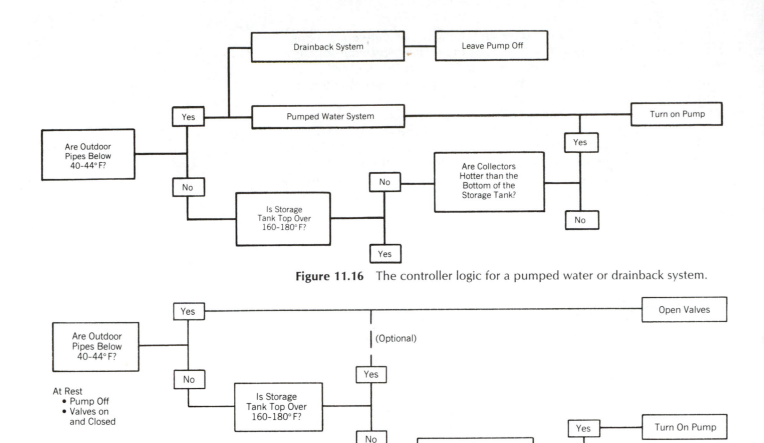

Figure 11.16 The controller logic for a pumped water or drainback system.

Figure 11.17 The controller logic for a draindown system.

In a recirculation system, if the outdoor pipes are below 40 to 44°F, the controller turns on the pump, regardless of the tank-top temperature or the storage/collector differential. At 50 to 54°F, the controller turns the pump back off.

If the outdoor pipes are above 40 to 44°F, the tank-top temperature is below 160 to 180°F, and the collectors are 8 to 18°F warmer than the storage tank; then the controller turns the pump on. The pump turns back off when the tank-top temperature rises to over 160 to 180°F or when the collectors are only 2.5 to 6°F hotter than the bottom of the storage tank.

If the outdoor pipes are above 40 to 44°F and the storage tank temperature is over 160 to 180°F, the pump cannot turn on, regardless of the storage/collector differential.

The drainback system works in the same manner, except that the pump stays off if the outdoor pipes are below 40 to 44°F.

Draindown Systems The controller in a draindown system has to perform two functions. It has to turn the pump off and on and it has to energize and deenergize the freeze dump valves. Before the start of solar collection, the pump is off and the freeze dump valves are energized to close them. If the power to the controller is lost, its fail-safe logic drains the collector array automatically to prevent freezing. Figure 11.17 shows such a control system's logic.

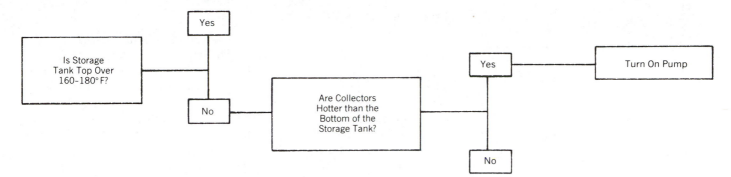

Figure 11.18 The controller logic for a heat-exchanged system.

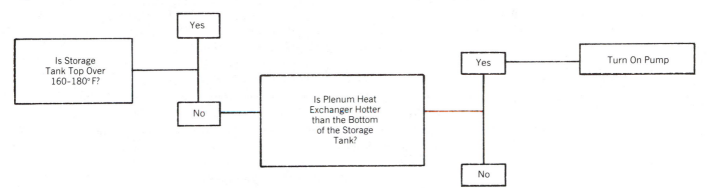

Figure 11.19 The controller logic for a collector-to-storage loop in an air-exchanged system.

If the outdoor pipes are below 40 to 44°F, the valves open to drain the collectors. This action usually overrides all the other actions. The valves reclose at 50 to 54°F.

If the outdoor pipes are above 40 to 44°F and the temperature at the top of the tank is over 160 to 180°F, the freeze dump valves may or may not open, depending on how the system is designed. The pump cannot turn on, regardless of what storage/collector differential exists. The valves reclose when the temperature at the tank top drops below 150 to 170°F.

If the outdoor pipes are over 40 to 44°F and the temperature at the top of the tank is below 160 to 180°F, the pump turns on if the collectors are 8 to 18°F hotter than the bottom of the storage tank.

Heat-Exchanged Systems Figure 11.18 shows the control logic for a heat-exchanged system. There is no freeze control function necessary, which simplifies the system.

If the temperature at the top of the tank is over 160 to 180°F, the pump cannot turn on.

If the temperature at the top of the tank is below 160 to 180°F, the pump turns on if the collectors are 8 to 18°F hotter than the bottom of the storage tank.

Air Heat-Exchanged Systems This controller may be part of a larger controller that controls the balance of the system. Its operation is identical to the heat-exchanged system except that the temperature at the plenum heat exchanger is measured instead of the temperature at the top of the collectors. Figure 11.19 shows the controller logic.

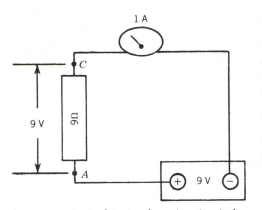

Figure 11.20 In this simple series circuit the current is 1 A and the voltage drop across the resistor is 9 V.

How Controllers Work

Solar water-heater controllers work by sensing temperature change with an electronic sensing device called a **thermistor.** This device electronically regulates the flow of electrical current to relays that open and close to turn the pump off and on and to open and close the system's valves. A thermistor is a resistor that changes in resistance with temperature.

To understand this action, you need to know some elementary things about the flow of electricity. The flow of electricity through a circuit is governed by **Ohm's law,** which states that the amount of current flowing through a circuit is equal to the electrical potential of the circuit divided by the resistance in the circuit. When the current is stated in amperes, the potential is stated in volts, and the resistance is stated in ohms.

The formula for Ohm's law is stated as $I = E/R$, where I = current in amperes (A), E = potential in volts (V), and R = resistance in ohms (Ω).

Figure 11.20 shows a very simple circuit containing a 9-V battery, a 9-Ω resistor, and an **ammeter.** An ammeter is a meter that shows how much current is flowing. This ammeter would read 1 A, because $I = 9/9 = 1$.

If the voltage between points A and C were measured, the voltmeter would read 9 V, the same as the battery terminals. This change is known as **voltage drop.** Any time that electric current flows through a resistance, the voltage drops.

Figure 11.21 shows the same circuit, but two $4\frac{1}{2}$-Ω resistors, R1 and R2, have been substituted for the 9-Ω resistor. The ammeter still reads 1 A because, when resistors are connected in series as these are, the total resistance is equal to the sum of the individual resistances.

$$R1 + R2 = RT \quad (RT = \text{total resistance})$$

The voltage drop measured from points A to C is still 9 V. However, if you were to read the voltage drop across either individual resistor (points

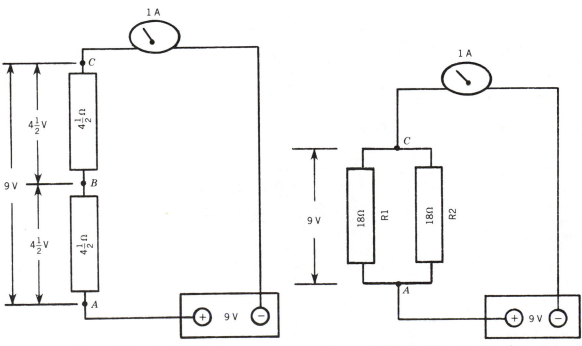

Figure 11.21 In this series circuit the current is still 1 A and the voltage drop across each resistor is $4\frac{1}{2}$ V.

Figure 11.22 In this circuit, where two resistors are connected in parallel, 1 A of current still flows.

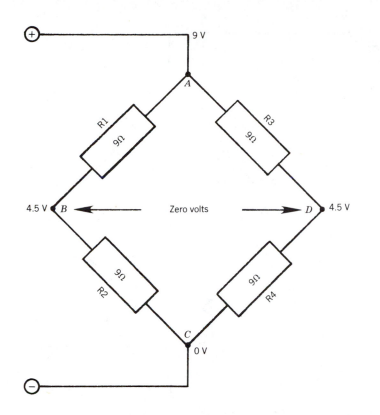

Figure 11.23 This is a series-parallel circuit similar to Figure 11.22. The battery has been left out of the illustration.

A to *B* or points *B* to *C*), the voltage drop across one resistor would be only $4\frac{1}{2}$-V. When resistances are connected in series, the voltage drop across any one resistor is equal to that resistor's value in ohms divided by the total resistance of all the resistors and multipled by the total voltage drop.

$$(R1/RT) \times \text{voltage} = R1 \text{ voltage drop}$$

Figure 11.22 shows the same circuit with two 18-Ω resistors connected in parallel in the circuit. The ammeter still reads 1 A because, when resistances are connected in parallel as these two resistors are, the total resistance is calculated by the following formula.

$$1/RT = 1/R1 + 1/R2 \ldots (+ 1/R3 + 1/R4, \text{etc.})$$

The voltage drop across either or both resistors (points *A* to *C*) would be 9 V.

Now look at Figure 11.23. In this illustration only the battery posts have been shown and the resistors have been drawn in a diamond pattern so that more devices can be attached to the circuit as the discussion goes on. In this circuit there are four resistors. They are connected in a combination of series and parallel. R1 + R2 makes up an 18-Ω resistive leg that is parallel to the 18-Ω resistive leg formed by R3 + R4. However, because each leg is formed by two resistors in series, there is an intermediate point in each leg that can be used to make connections or to measure voltages.

The voltage drop across any of the four resistors is $4\frac{1}{2}$ V (points *A* to *B*, *B* to *C*, *C* to *D*, and *A* to *D*). If the voltage between points *B* and *D* were measured, the voltmeter would read zero, since both points are at a potential of $4\frac{1}{2}$ V.

Figure 11.24 shows the circuit with the same two 18-Ω parallel legs, but now the resistances in series within these legs have been changed to

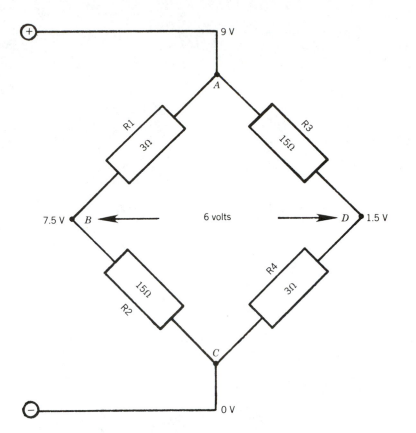

Figure 11.24 By relocating the resistors, a voltage potential can be developed between points B and D.

3 Ω and 15 Ω. The position of the resistors has also been changed around so that in the left leg, the current flows through a 3-Ω resistor first and in the right leg the current flows through the 15-Ω resistor first. As a result, the voltages at points B and D are now different and, if a voltmeter were placed across points B and D, the voltmeter would read 6 V.

This voltage existing between points B and D can be used to control the pump and valves. To understand how, look at Figures 11.25 and 11.26. Figure 11.25 shows a **switching relay.** A switching relay is an electromagnetic switch in which a low-current, small-voltage source can be used to open and close a higher-powered circuit.

In an electromagnetic switching relay, a coil of wire is wound around

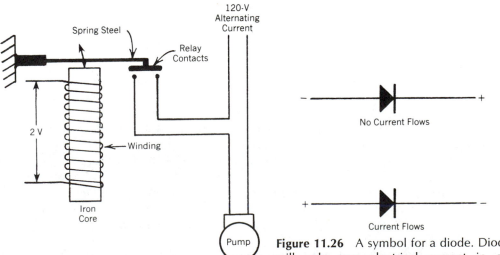

Figure 11.25 A switching relay.

Figure 11.26 A symbol for a diode. Diodes will only pass electrical current in one direction.

an iron core. When a voltage of a minimum size, in this case 2 V, is applied across the coil, the iron core is turned into a magnet strong enough to pull a switching mechanism against the core. The switching mechanism is connected to a higher-powered circuit (in this case, a pump). When the switch closes, the pump turns on.

When the voltage is removed from the wire coil, the core is no longer magnetized and a spring in the switching mechanism opens the switch. This turns the pump off.

The switch in the pump circuit could have been wired to be either on or off when the relay was energized. This is one of the advantages of using a relay. For freeze dump valves, the switch might have been wired to open when the relay closed.

For this reason, the word *energized* is used to describe the relay when it is closed. This prevents any misunderstanding in speaking of the relay opening and closing instead of the circuit switch that it controls. An open relay is described as *deenergized*.

Figure 11.26 shows a symbol that represents a semiconductor called a **diode.** A diode is a device that, when connected into a circuit, allows electrical current to flow from negative to positive. When the end to which the symbol's arrow points is at a higher voltage (more positive) than the other end, current cannot flow. When the end to which the symbol's arrow points is at a lower voltage (more negative) than the other end, current can flow.

Go back to Figure 11.24. Lightly pencil in the diode between points B and D so that the arrow points to D. Current would now flow from point B to point D. Turn the diode around so that the arrow points to B. Current cannot flow from point B to point D, because the diode will not pass electricity in that direction. The action is the same as if the two points were not connected.

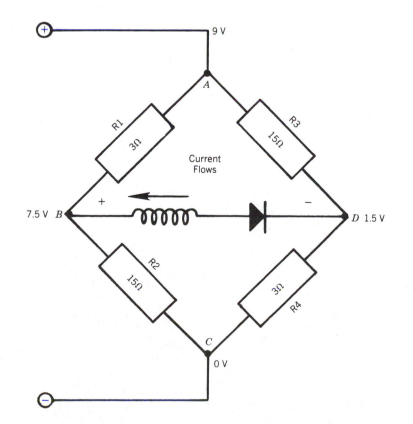

Figure 11.27 A switching relay in series with a diode has been connected between points B and D in a circuit where the relay will energize.

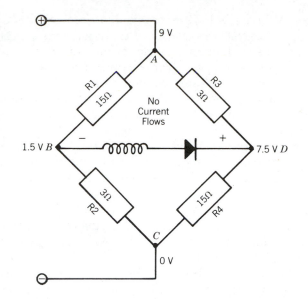

Figure 11.28 A switching relay in series with a diode has been connected between points *B* and *D* in a circuit where the relay will not energize.

In Figure 11.27, the switching relay's wire coil is connected in series with a diode across points *B* and *D* so that the diode points to *D*.

A 6-V potential exists between points *B* and *D* so, when the circuit is turned on, the relay will energize.

In Figure 11.28, the resistors have been changed in position so that point *B* is $1\frac{1}{2}$ V and point *D* is $7\frac{1}{2}$ V. The relay cannot energize, because current cannot flow backward through the diode.

If there were a way to switch these resistors around automatically with temperature changes, an automatic control circuit would exist.

As previously noted, there are devices called thermistors whose resis-

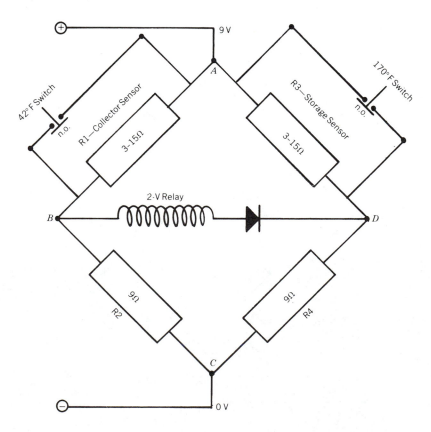

Figure 11.29 A circuit that simulates the action of a solar water-heater controller.

TABLE 11.1

OPERATING CONDITIONS AND CIRCUIT ACTIONS FOR FIGURE 11.29

CONDITION	R1	V_B	R3	V_D	V_{B-D}	RELAY
1. Hot collector/cold tank	3	6.8	15	3.4	+3.4	Energized
2. Hot collector/hot tank	3	6.8	3	6.8	+0.0	Deenergized
3. Cold collector/hot tank	15	3.4	3	6.8	−3.4	Deenergized
4. 42°F switch closed	0	9	≥3	≥3.4	≥2.2	Energized
5. 170°F switch closed	0	≥3.4	0	9	≤−2.2	Deenergized

tance changes in magnitude with temperature. When the thermistor is designed so that its resistance decreases as temperature increases, the thermistor is known as a **negative coefficient thermistor.** The sensors used in most solar water-heater controllers are negative coefficient thermistors. They drop in resistance as temperature increases, and they do this in a very precise manner.

Some solar controllers use positive coefficient thermistors. Positive coefficient thermistors increase in resistance as the temperature increases. Obviously, negative coefficient thermistors cannot be used with positive coefficient controllers, and vice versa.

Figure 11.29 shows a circuit that stimulates the action of a typical solar water-heater controller. R1 is a thermistor that can vary from 3 to 15 Ω located at the top of the collectors. It runs in series with R2, a 9-Ω resistor. A normally open switch that automatically closes when the temperature drops to 42°F is connected across R1.

R3 is a similar thermistor located at the bottom of the storage tank. It is connected in series with R4, a 9-Ω resistor. A normally open switch that automatically closes at 170°F is connected across R3.

A 2-V switching relay is connected in series with a diode between points B and D.

Table 11.1 shows the resistance of R1 and R3, the voltages at points B and D, the difference in potential between points B and D, and the condition of the relay under five typical operating conditions.

With the two temperature switches open and under condition 1, where R1 is small and R3 is large, representing a hot collector and a cold storage tank, the potential between points B and D is +3.4 V, so the relay energizes.

With the two temperature switches open and under condition 2, where R1 and R3 are both large, representing a hot collector and a hot tank, the potential between points B and D is 0 V, so the relay remains deenergized.

With the two temperature switches open and under condition 3, where R3 is small and R1 is large, representing a cold collector and a hot tank, the potential between points B and D is −3.4 V, so the relay remains deenergized.

In condition 4, with the 42°F switch closed (freezing conditions), R1 is shorted out of the circuit. Point B becomes 9 V. The potential between points B and D can vary from +2.2 to +5.6 V, depending on the storage tank temperature. Under all conditions, the relay is energized.

In condition 5, with the high-temperature switch closed, R3 is shorted out of the circuit and point D is always 9 V. Point B can never be more than 9 V. Therefore, the relay cannot energize and the pump cannot start up.

The actual controller circuits look nothing like these simple circuits.

The actual circuits use switching transistors and integrated circuits to give very low current drain and high reliability. The theory of using this bridge circuit to control the system remains the same. As you troubleshoot controllers, keep this bridge circuit in mind and the troubleshooting procedures will make good sense to you.

The Control System

The control system consists of three parts.

- The controller.
- The power circuits.
- The sensors.

The controller is the electronic circuit board and its parts. These are located in the control box.

The power circuits are the wiring from the power source to the pumps and valves in the system. They include the switch(es) inside the relay(s). The relay is located in the control box, and the wiring is external.

The sensors are located at various places throughout the system where the temperature must be measured. Long wires connect them to terminals on the controller circuit board.

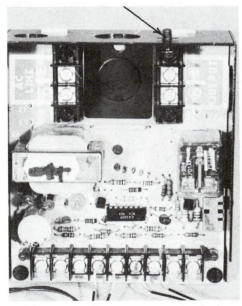

Figure 11.30 A solar water-heater controller.

Controller Installation Considerations

Most solar water-heater controllers are designed for indoor installation. They usually come in a UL-approved enclosure such as the C-30 controller from Independent Energy, Inc., shown in Figure 11.30

For safe and satisfactory operation, the controller enclosure must be

- Electrically grounded.
- Environmentally protected.
- Firmly secured to a solid mounting surface.
- Near a properly fused and grounded 120-V electrical service.

Grounding The enclosure must be grounded through the 120-V electrical supply in accordance with the local code or the National Electrical Code. Full instructions are generally included with all models of controllers. Figure 11.31 shows a close-up of the C-30 controller with the cover removed. The arrow points to a grounding screw at the rear of the box. The ground wire of the incoming electrical circuit, which is generally either a wire with green insulation or a bare copper conductor, is used to make this connection.

The enclosure can also be grounded by attaching it to a grounded surface, but this type of grounding should not be substituted for electrical circuit grounding. If the circuit that you are working with does not have a ground wire included, run a number 14 or larger copper wire from the grounding screw to a clamp firmly fastened to either an approved grounding rod or to the cold water piping where it exits from the home. (This assumes that the cold water piping is metallic.)

Figure 11.31 The C-30 controller with the cover removed. The arrow points to the grounding screw.

Environmental Protection The controller should be mounted indoors where it will not be exposed to water, rain, or snow. Most controllers will operate satisfactorily at wide extremes of temperature, but it is best not to place the controller where the electronic components will become overheated, because this shortens their effective life.

Mount the controller where the surrounding air temperature is between 32 and 105°F, where the air contains less than 95% relative humidity and is free to circulate. Controllers typically contain a transformer to step the incoming voltage down and a direct-current power supply that converts the alternating current to 12- or 24-V direct current. This process builds up heat in the controller box that must be dissipated to the air surrounding the controller.

The relays within the controller are mechanical devices. They are sealed to prevent dust and dirt from entering the switches and should have a long life. However, it is wise not to place the controller where it would be exposed to excess dust and dirt accumulation. Dust and dirt can enter the relays, build up on the printed circuit board, and provide a path for moisture to short-circuit the board.

Mounting The controller must be securely mounted so that it can support the wiring from the power circuits and the sensors. It can be mounted on a separate board or on the equipment, or it can be attached to a device that it is controlling. Look back at Figure 11.31. In the upper center of the controller between the alternating-current line and output terminals, there is a box knockout on the rear of the box. This knockout can be used to mount the controller directly onto the pump's electrical box. However, because the pump produces heat during operation and can leak, it is better not to mount the controller directly on the pump. There are also box knockouts on either side of the controller that can be used for pump mounting or to mount the controller on the side of an electrical switchbox. Controllers are delicate. Handle them carefully.

Fused and Grounded Electrical Service The purpose of the controller is to control the flow of 120-V electricity to pumps and valves. These power circuits should be properly fused and grounded in order to meet the local code that prevails in your area or the National Electrical Code in the absence of a local code. Therefore, the controller must have access to this electrical service.

In small systems such as the residential solar water-heater system, where the power required is small, plug-in installation is often used and can be adequate. Some solar controllers come with an attached three-prong electrical cord that provides power both to run the controller and to run the valves and pumps. To meet code requirements in most locations, this cord must be plugged into a properly fused and grounded electrical circuit. Some locations have electrical codes that require the pump be directly wired instead of plug in. Check the local code if there is any question.

Electronic Circuit Board Considerations Figure 11.32 shows the electronic circuit board contained in the C-30 controller. The manufacturer of the C-30 has designed the controller so that the board may be easily removed and replaced. This facilitates servicing. The arrow on the left points to the relay switch, and the arrow on the right points to the power transformer. The solid-state electronic components lie behind the transformer and the switch; the power circuits connect to the terminals in the foreground. The tank sensor and the collector sensor have been connected to a terminal board in the background to show where the connections are located.

Servicepeople should not attempt to troubleshoot the electronic circuits. Such troubleshooting requires sophisticated electronic equipment and a knowledge of solid-state circuits. Troubleshooting of the transformer primary circuit, sensor signal input, and relay output can be readily accomplished and will be covered in Chapter 17.

Figure 11.32 The C-30 controller's electronic circuit board. The left arrow points to the relay; the right arrow points to the transformer.

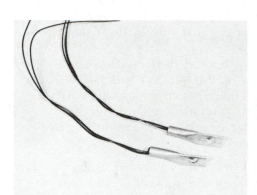

Figure 11.33 Two 10-k thermistor sensors.

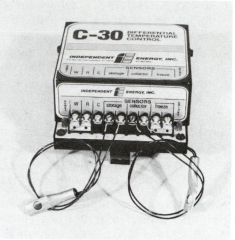

Figure 11.34 Where the thermistor sensors attach to the controller.

The Power Circuits

Controllers require very small amounts of power and do not place a large load on the power circuits. One ampere or less will run most all solar water-heater system controllers. The pumps and valves that are being controlled use larger amounts of power, so the circuit should provide for the use of about 10 A of 120-V current in most applications. A 15-A fused standard house circuit is generally adequate if the same circuit is not also running other large devices within the home. If the circuit is overloaded, a large voltage drop can occur that will hurt the valve and pump performance. In such cases a separate power circuit should be installed.

Check the electrical code carefully for your particular area. It may require a separate circuit for this electrical application.

The Sensors

Most controllers use two types of sensors: temperature-dependent, variable-resistance sensors (thermistors) and temperature-dependent switching sensors (snap switches).

The thermistor varies in resistance with temperature change in a precise manner, while snap switches open and close at a predetermined temperature level.

Thermistor Sensors   Figure 11.33 shows two thermistor sensors. These are 10,000-Ω thermistors enclosed in a preformed copper housing. Figure 11.34 shows these two thermistors attached to the proper teminals on a C-30 controller. Of course, in actual practice, the thermistors would usually be located many feet away from the controller, and long wires would be used to make the connections.

Of the many different thermistors manufactured, two types are commonly found in solar control applications: the 3-kΩ thermistor and the 10-kΩ thermistor. In the electronic industry k stands for 1000. So a 3-k thermistor is a 3000-Ω thermistor, and a 10-k thermistor is a 10,000-Ω thermistor.

A thermistor is identified by its resistance at 25°C or 77°F. Figure 11.35 shows how the resistance of these two types of thermistors changes with temperature. This is a graph of temperature in degrees Fahrenheit versus resistance in thousands of ohms. The dashed lines show the resistance

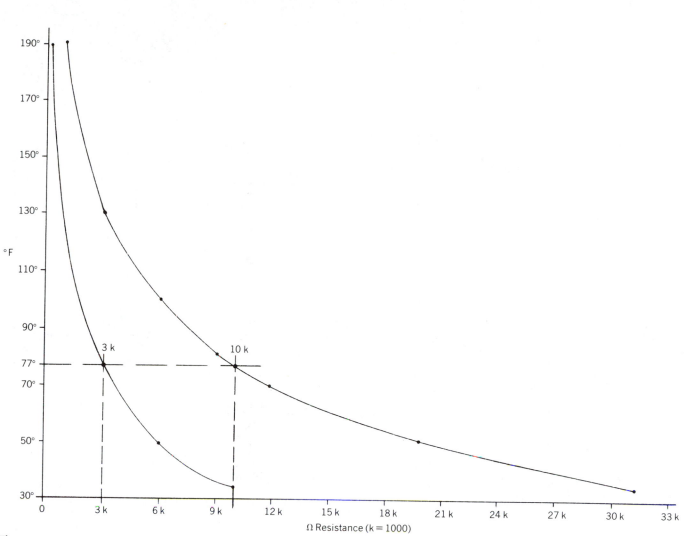

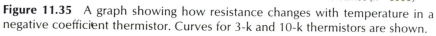

Figure 11.35 A graph showing how resistance changes with temperature in a negative coefficient thermistor. Curves for 3-k and 10-k thermistors are shown.

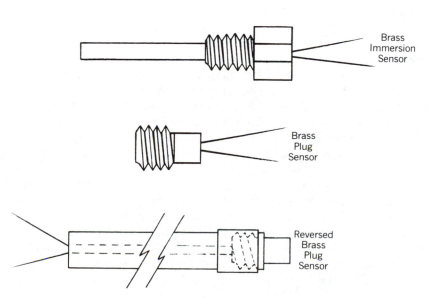

Figure 11.36 The thermistor can be mounted in several different types of housings.

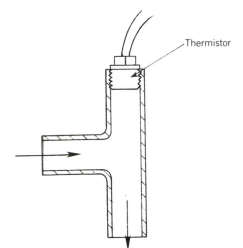

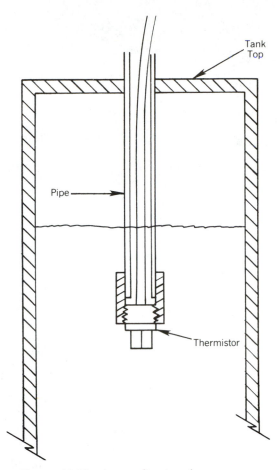

Figure 11.37 An application for a brass plug-housed thermistor.

Figure 11.38 An application for a reverse plug-housed thermistor.

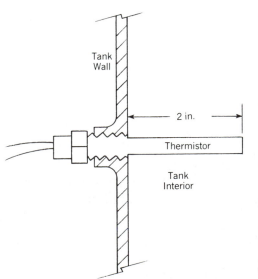

Figure 11.39 An application for a immersion plug-housed thermistor.

of the two sensors at 77°F. Follow the curve for the 10-k sensor and you will see that it has a resistance of about 32 k at freezing (32°F) and a resistance of about 1 k at 190°F. The curve for the 3-k thermistor shows a resistance of about 10-k at freezing and about 300 Ω at 190°F.

You can obtain the resistance at any temperature by reading the curve. Upon request, the manufacturer of the controller will supply you with a table of thermistor resistances at any temperature.

Note also that it is not possible to say the thermistor changes value a certain number of ohms for each degree of temperature change. If that were possible, these curves would be straight lines. At the lower temperatures, there is a much larger change in resistance per degree of temperature then there is at the higher temperatures.

It should also be apparent that it is not possible to use a 10-k thermistor in a circuit designed for a 3-k thermistor, or vice versa. If you were to put a 10-k thermistor into a 3-k circuit, when the point being measured was 77°F (10-k), the controller circuit would think it was seeing a temperature of about 30°F (10-k).

The thermistor sensor can also be obtained in different types of housings. Three common types of housings are the brass plug housing, the reverse brass plug housing, and the brass immersion sensor. Figure 11.36 shows these three thermistors. The brass plug thermistor is designed to be placed in a fitting such as a threaded tee. The reverse brass plug thermistor is designed for immersion into a nonpressurized tank at the end of a dip tube. The brass immersion thermistor is designed to give deep

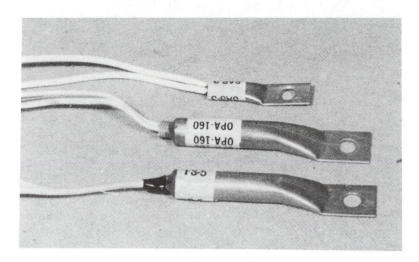

immersion into a pressurized tank or a collector header. Figures 11.37, 11.38, and 11.39 show typical applications for these housing styles.

Switching Sensors Switching sensors are sensors that open and close at specific temperatures. They are used to open or close a circuit within the controller at a given temperature. Figure 11.40 shows a photograph of a thermistor sensor and two different switching sensors. The thermistor sensor is the sensor at the top of the photograph. Note that the switching sensors are much larger in size.

The switching sensor can be either a **normally open** (n.o.) or a **normally closed** (n.c.) switch. A normally open switching sensor closes in response to a predetermined condition, such as an exceptionally high or low temperature. A normally closed switching sensor opens in response to an exceptionally high or low temperature. In Figure 11.29 the two switches used to simulate the circuit action were normally open switches.

Switching sensors open and close, or close and open, within a temperature range. Usually, a normally closed freeze circuit switching sensor will open when the temperature drops to 42°F ± 5°F. It will reclose at 52°F ± 5°F.

Switching sensors may be wired in a number of different configurations. For details, consult the manufacturer's technical information.

Other Temperature Sensors New sensor types are introduced. The serviceperson may encounter the resistance temperature device, the diode temperature sensor, the constant-current temperature sensor, or the multisensor probe. These sensors operate differently, and their use should be studied by the serviceperson prior to servicing the control system.

Sensor Mounting The standard sensors, such as the ones shown in Figure 11.33, are designed to be surface mounted on a pipe or a metal surface. When they are mounted on a flat metal surface, a screw or bolt through the hole in the sensor generally is used to assure a good thermal connection. Many pressurized solar water tanks are fitted with a stud for this connection. When the sensor is mounted on a pipe, hose clamp is used to obtain this thermal connection. The locations of the screw or bolt and the hose clamp are shown in Figure 11.41.

Overtightening the sensor's screw or bolt or the hose clamp is not of particular concern according to one manufacturer, Midwestern Compo-

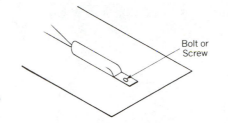 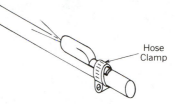

Figure 11.41 How sensors are mounted to a pipe or a flat metal surface.

nents, Inc., of Muskegon, Michigan. The flat extension of the sensor housing is very rugged and not subject to deformation. Be certain that the connection is tight, but do not exert excess pressure or attempt to place the hose clamp around the main tubular housing, because the thermistor is enclosed in a glass bead in the tubular portion that can be crushed under compression.

Midwestern Components also recommends the use of a thermally conductive grease or compound between the housing and the mounting surface. This compound will ensure a good thermal bond and fast response to temperature changes. Any good thermally conductive grease may be used. The heat sink compounds used to mount power transistors are excellent and can be obtained from any electronics parts supplier. The grease is nonsetting, and the sensor may be easily removed. A typical material would be Dow Corning 340 Heat Sink Compound.

All sensors *must* be protected from the elements and *must* be tightly insulated from the air surrounding the device. Otherwise, the device will be measuring the temperature of the air as well as, or instead of, the temperature of the pipe. The devices may be placed under moisture-tight insulation or you can **pot** them. Potting consists of pouring a rubbery material around the device and surface to which it is mounted. The potting material must be nonconductive and free of moisture. An available and easy-to-use potting compound is a one-part silicone sealant found in most hardware stores. One precaution must be taken. Many one-part silicone sealants contain small amounts of acetic acid as a curing agent and need some moisture from the air to cure. Therefore, at least one surface of the potting material should be left exposed for 24 hours to be certain that the acetic acid is driven off and the material has cured.

When the device is connected to a pipe, the potting compound or the insulation should surround the entire device, the pipe, and the hose clamp. On a flat surface, there should be a minimum of $\frac{1}{4}$ in. of potting compound over the entire device.

NOTE Often there is a need to attach temporarily a sensor to a surface for troubleshooting. This can be accomplished easily with an aggressive, heat-resistant, pressure-sensitive tape such as duct tape.

Generally, standard number 18 or 20 wire is used to connect the sensors to the controller. Smaller-gauge wires can also work. Note, however, that copper wire does add resistance to the circuit. In runs longer than 100 ft, wire below 20 gauge may cause the controller to read the sensor temperature incorrectly.

Wire nuts can be used to make the connections. Wire nuts used outdoors should be potted with a one-part silicone sealant. Installation of the wire nuts should take place in two steps. First, the wire nut should be used to make a firm connection *without* any potting compound being present. Then the wire nuts should be removed; the potting compound

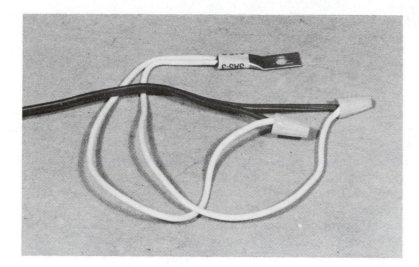

Figure 11.42 An sensor connected to an 18-gauge lamp cord. Notice how the connections are offset.

should be placed in the nuts; and the nuts should be reinstalled. This procedure is used to insure that a high-resistance connection is not made. Silicone sealants are good insulators. They could interfere with making a good connection unless the wires are first firmly connected.

3M makes a superior connector, the Scotch UY connector, which connects and pots the wires in a single step.

Figure 11.42 shows a sensor connected to number 18 lamp cord (zip cord). The use of lamp cord has some limitations. Lamp cord cannot be exposed to the elements. It must be protected, expecially from high heat. High heat will embrittle the insulation in a very few years, necessitating wire replacement. It is best not to use this type of wire outdoors. Instead, choose a wire that will take 140°F for long periods without embrittling, such as PVC- or Teflon-coated wire. These have good weathering and ultraviolet resistance. Run the sensor wires outside the insulation to avoid hot pipes that can melt the insulation and short out the wires.

Whenever possible, use twisted-pair sensor wires, especially if long wire runs are required. Twisted pairs are inherently resistant to "noise" caused by stray voltages or radio frequency signals that can give a false signal to the controller.

Note that the ends of the lamp cord are slightly offset in Figure 11.42. This insures that the sensor cannot inadvertently be shorted out during installation.

Remember that copper is a good conductor of heat as well as electricity and that if the wires are not well insulated, the cold can be transmitted to the sensor, giving a low collector temperature reading. Similarly, when the wires connecting the cold storage sensor are installed, they should not be run against a hot surface. The heat from the wires will be transmitted to the storage sensor and heat it up. This will cause a high storage tank temperature reading.

Sensor Placement All temperature-dependent sensors must be very carefully installed at the proper places in the system so that they are located where they will measure the proper temperature and so that they actually measure the temperature of the piping containing the fluid and not the temperature of the air surrounding the piping.

The collector sensor should be placed on the upper manifold of the highest unshaded collector, as close to the collector enclosure as possible. Generally, it is attached with a hose clamp. Occasionally, the sensor may come premounted on the collector absorber plate fin inside the collector box. If this sensor fails, replace it with an external sensor instead of opening up the collector, unless the manufacturer insists that the internal sensor be replaced.

Note that drainback systems generally perform most efficiently when the collector sensor is mounted on an absorber plate fin, to minimize daily pump cycling.

Attach the sensor firmly. It is best to use a heat sink compound between the pipe and the sensor body to get a good thermal bond and fast response time to temperature changes.

Pot and/or insulate the sensor and the connections to the lead-in wires so that the connections are weathertight and the sensor is not exposed to ambient air temperatures or to the sun.

The storage tank sensor should be attached to either the tank wall under the insulation close to the bottom of the tank or to the pipe where the liquid exits back to the collector. Use a hose clamp if the sensor is to be pipe mounted. Bolt the sensor to the threaded tank stud if one is provided. Otherwise, use a thin layer of setting compound such as an epoxy resin to mount the sensor on the tank wall. Some installations use a pressure-sensitive tape. This tape can age and lose its adhesion unless it is very carefully chosen, and it is usually a poor method of mounting sensors. Insulate and pot the sensor and its connections. Run the lead-in wires where they will not heat up from contact with hot air or hot surfaces.

The freeze sensors are mounted outdoors in locations where the collectors and pipes are most likely to be the coldest. Generally, at least two collector freeze sensors are installed for reliability: one at the collector array inlet and one at the collector array outlet as close to the collectors as practical. In areas where the chances of freeze damage are high, it is a good investment to install an additional freeze sensor inside the collector box on an absorber plate tube about one-third of the way up from the lower header, near the center of the collector. Studies have shown this to be the first part of a collector to freeze.

Solar system piping may freeze, too. Consider the temperature differentials in the collector array, the shading of the array, the wind direction and velocity in the colder months, the effectiveness of the piping insulation, and the length and position of the exposed runs in determining which parts of the piping will freeze first. Install additional freeze sensors in any critical locations. Keep freeze sensors clear of roof penetrations, where attic heat may cause localized warm spots in pipes.

Recirculation freeze-protection sensors must always be mounted on a header or absorber plate tube, never on a fin. This will minimize storage heat loss due to recirculation.

The high-limit tank temperature sensor must be mounted at the top of the solar storage tank. It can be mounted to the tank wall under the insulation or it can be mounted on the hot water pipe close to where it exits the tank. An epoxy resin can be used to mount the sensor on the tank wall, and a hose clamp will securely fasten the sensor to the hot water exit pipe. The lead-in wires from this sensor should not be mounted on cold water pipes, because this could result in unintentional thermal cooling of the sensor.

Controller Installation Checklist

Here is the controller installation checklist. Evaluate the installation against each of the 23 points.

EVALUATION POINT	POOR	MARGINAL	GOOD
1. Is the controller encased in a UL-approved box?			
2. Is the controller electrically grounded?			
3. Is the controller mounted indoors, where it is protected against the environment?			
4. Is the ambient air around the controller free to circulate and between 32 and 105°F?			
5. Can dust and dirt enter the controller box and damage the relay or short the circuits?			
6. Has the controller been securely mounted using either the box mounting clip or the box knockouts?			
7. Does the controller have easy access to a fused and grounded 120-V, 15-A circuit?			
8. If the system is plug in instead of hard wired, are the controller, valve, and pump connector cords all three-wire grounded cords?			
9. Does the electrical hookup of the power circuits meet local codes and the National Electrical Code?			
10. Have the proper sensors been used?			
11. Are the sensors located in the right places?			
12. Are the sensors properly mounted?			
13. Are the sensors protected from the elements?			
14. Are the sensors properly insulated?			

EVALUATION POINT	POOR	MARGINAL	GOOD
15. Are the lead-in connections properly made and potted?	_____	_____	_____
16. Is the lead-in wire at least 20 gauge?	_____	_____	_____
17. Are the lead-in wires a twisted pair?	_____	_____	_____
18. Is the lead-in wire insulation cracked or embrittled?	_____	_____	_____
19. Can the lead-in wires to any of the sensors cause unintentional thermal heating or cooling of the sensors?	_____	_____	_____
20. Does the sensor wire touch hot pipes?	_____	_____	_____
21. Is the sensor wire insulation melted or snagged on a sharp surface?	_____	_____	_____
22. Has the sensor wire been anchored every 3 ft?	_____	_____	_____
23. Is the sensor wire abraded or broken where it passes through metal flashings?	_____	_____	_____

CHAPTER 12

Solar Storage and Auxiliary Water-Heater Tanks

Most solar water-heating systems contain a solar storage unit and an auxiliary water heater. The two may be combined into one tank, or the system may contain two separate tanks, one for solar-heated water storage, and one for auxiliary water heating.

The systems described in Chapters 1 to 8 showed solar storage and auxiliary water heating taking place in separate tanks. In almost all of these systems a single tank could have been substituted if an electrically fired auxiliary water heater was acceptable.

However, it is more costly to heat water with electricity than with natural gas or propane. A two-tank system often has lower operating costs but a slightly higher initial investment.

In Chapters 1 to 9 the subject of **tempering valves** was not discussed. A tempering valve is a valve that automatically adds cold water to the hot water stream when it is hotter than 120 to 140°F. In this chapter you will learn that all solar water-heater systems should contain a tempering valve.

The Separate Solar Storage Tank

When a separate solar storage tank is used, there are many possible tank types to consider. These tanks break into two separate categories: the pressurized tank and the unpressurized tank. This book will not cover unpressurized tanks. *The Solar Decision Book* shows the various unpressurized tank possibilities and methods of construction.

The Pressurized Tank

The pressurized tank normally runs at service-water pressure of 15 to 80 psi. Occasionally, a system will run at 100 psi or more, and such a system should contain a pressure-reduction valve in the cold water lines. The tanks are generally cylindrical, with ends that are concave or convex. The most popular sizes are 65, 82, and 120 gal.

Pressurized solar water tanks are generally made of steel. They are either hot dip galvanized (not generally recommended), coated on the inside with an epoxy resin, coated with a fused ceramic frit (glass lined), or lined with spun concrete (stone lined) to cut down on the corrosion caused by the use of multimetal systems where the collectors are generally copper, a more noble (cathodic) metal than steel. To delay corrosion further within the glass-lined tank, a sacrifical anode of magnesium is generally suspended in the tank. The magnesium rod corrodes first, thus protecting the steel tank until the anode is used up.

Pressurized solar water tanks may be supplied with or without exterior insulated jackets. Uninsulated tanks must be site insulated with fiberglass or urethane foam insulation. Factory-insulated storage tanks are fiberglass or foam insulated and include a decorative steel jacket to contain the insulation.

The solar storage tank may or may not include a built-in heat exchanger as part of the tank. These heat exchangers can be plate-around-tank, tank-around-tank, internal-flued tank, or coil-in-tank systems, as previously shown.

All solar storage tanks must make provision for

- Mounting temperature and pressure safety relief valves.
- Mounting a tank drain valve.
- Piping in the cold service water at or near the bottom of the tank.
- Piping out the heated water at the top of the tank.
- Piping up the heat exchanger to the collector loop or piping the service water to the collector loop.
- Mounting and insulating a storage tank temperature sensor and a high-limit cutoff sensor.

The layout and the location of the fittings will vary with the type of system. Most tanks have their hot and cold water inlets and outlets located at the top of the tank; a plastic dip tube extends from the cold water inlet down to the bottom of the tank, but some tanks have their cold water inlet near the bottom of the tank and are equipped with a diffuser plate to keep the cold water at the bottom of the tank.

The fittings for the solar collection loop may be added as tee fittings to standard tanks, or tanks may be purchased that have separate fittings especially designed for collector loop addition. The separate fittings are advisable.

Codes and Safety Devices

Solar-heated storage tanks are subject to the same plumbing codes and safety considerations as electrically heated or fossil fuel-heated water tanks. Codes differ from location to location and require the serviceperson and the system's installer to be well versed in the local rules and regulations. Here are some typical considerations.

- Pressure relief valves shall be installed for all equipment used for heating or storing hot water. The rate of discharge of such a valve shall limit the pressure rise for any given heat input to 10% of the pressure at which the valve is set to open.
- Temperature relief valves or energy shutoff devices shall be installed for equipment used for heating or storing hot water. Each temperature relief valve shall be rated as to its British thermal unit capacity. At 210°F, it shall be capable of discharging sufficient hot water to prevent any further rise in temperature. As an alternative, an energy shutoff device may be used that will cut off the supply of heat energy to the water tank before the temperature of the water in the tank exceeds 210°F.

- Separate or combined temperature and pressure relief valves must be approved by, or meet the specifications of the American Gas Association or the National Board of Casualty and Surety Underwriters.
- There shall be no shutoff valve or check valve between the relief valve and the tank.
- Temperature relief valves or combined temperature/pressure relief valves shall be situated so that the valve stem or thermal element extends into the tank and is in direct contact with the hot water flow. In no case shall the valve be located more than 6 in. down from the top of the tank or more than 3 in. away from the tank. There shall be no direct closed connection of the discharge from the valves to a closed water or waste system.
- All tanks used for the heating or storage of hot water shall contain a suitable drain-and-flush valve.
- All service-water supply valves shall be nonrestrictive, full-flow valves (gate or ball valves).

Installation Considerations

The tank must be located on a foundation designed to carry the weight. Generally, the average $3\frac{1}{2}$-in. concrete basement floor is adequate. If the tank is located on a wood floor above a basement, the adequacy of the foundation should be examined and a metal drip pan that contains a downward sloping drainpipe to a storm sewer or outside the building should be added under the tank.

A sanitary sewer or storm drain should be available to handle any flushing operation, safety valve overflows, or leaks from the tank.

In northern latitudes the tank should be adequately protected against freezing.

Insulation Requirements

Electrically fueled or fossil-fueled tanks were poorly insulated until a few years ago. Fuel was relatively inexpensive, and competition for water-heater sales was high. Such a condition cannot be tolerated in solar storage tanks. Solar heat fluxes are low. The times when solar energy is available are unpredictable, and there is a need to store the energy for much longer periods.

The value of insulation can be seen by examining an 80-gal tank with 22 ft^2 of radiating surface filled with 130°F water sitting in a 60°F room. If the tank were uninsulated, the water would cool to body temperature in 13 hours. With R-5 insulation, the tank would take 65 hours to cool to body temperature; with R-10 insulation, it would take 130 hours; and with R-20 insulation it would take 250 hours.

This analysis clearly shows that R-10 to R-20 insulation is essential for solar storage tanks. Servicepeople evaluating the performance of a system should pay close attention to how well the tank is insulated. The evaluation should include evaluating the base of the tank. Many tanks are not insulated on the bottom and sit on cold floors. Slipping one inch of rigid insulation under the tank, or blocking the tank up off the floor and placing insulation under it, will lower heat losses.

Standard tanks with foamed insulation or combined foam/fiberglass insulation having R-8.3 to R-16.7 are now standard industry equipment and should be used in almost all systems.

Coping with Water Supply Problems

The water supply in many municipalities or private wells contains calcium salts, iron salts, sulfur compounds, or has an acid pH. When the water supply is acid or contains large amounts of any of these products, water conditioning ahead of the water-heating system should be considered.

Hard water, caused by an excess of calcium salts, is handled with ion-exchange resin water softeners. Iron salts are removed by feeding controlled amounts of potassium permanganate. Sulfur odors are removed with charcoal filtration, and acid pH is generally neutralized by the addition of food-grade phosphates.

Calcium salts build up in the system on the collector and heat-exchanger surfaces. The action is hastened at high temperatures, because the salts become less soluble as the water heats up. These salt deposits lower the heat-exchange rates and hurt the performance of the system.

Iron salts can build up on the heat-transfer surfaces, can contaminate the ion-exchange resins used to remove the calcium salts, and can cause staining of plumbing fixtures and clothes during washing.

Sulfur compounds, while not particularly harmful to the system, give a rotten and swampy smell to the water. Water with an acid pH will hasten corrosion of the system.

These water conditions, unless controlled, will quickly degrade solar storage systems. Solar storage systems are expensive to service and replace, so it is important to give strong consideration to the installation of water-conditioning equipment, which should be located between the cold water supply and the water-heating system.

Storage Tank Sizing

For maximum efficiency, the solar storage tank should provide 1.2 to 2 gal of storage per square foot of collector. The following chart provides typical guidelines.

NUMBER OF COLLECTORS	GALLONS OF STORAGE PER COLLECTOR			
	20 ft^2	25 ft^2	32 ft^2	40 ft^2
1	25–40	30–50	40–64	50–80
2	50–80	60–100	80–128	100–160
3	75–120	80–150	120–192	150–240
4	100–160	120–200	160–256	200–320

Multiple-Tank Hookups

In pumped water and draindown systems, where the service water is passing through the collectors, a thermostatic three-way valve may be used to increase solar storage without making any major changes in the current solar storage tank.

A thermostatic three-way valve is a valve that sends the water stream out one outlet if it is above a certain temperature and sends it through a second outlet if it is below that temperature. In Figure 12.1 the incoming water stream enters the valve through inlet A. An internal sensor, set by an adjusting screw on the top of the valve, directs the water out the proper port. Assume the device is set for 130°F. Water less than 130°F would exit through outlet C, and water hotter than 130°F would exit

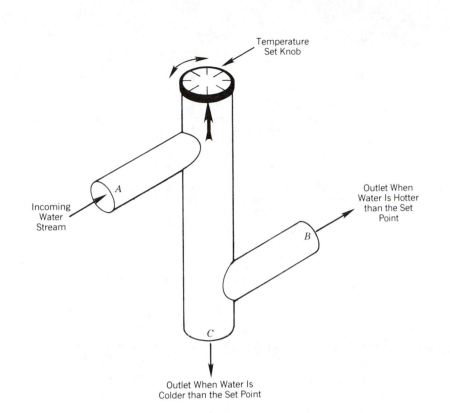

Figure 12.1 The thermostatic three-way modulating valve.

through outlet *B*. There is about a 10°F range where water exits through both outlets.

If the solar storage tank currently in the system is undersized, the additional solar storage may be obtained in the auxiliary water heater instead of changing to one larger tank. Figure 12.2 shows how to hook up this tank using a thermostatic three-way valve.

In heat-exchanged systems, a second tank can be added either in parallel or in series with the first tank. Figure 12.3 shows a series connection. Tank B, which is closest to the water user, gets the hottest collector fluid.

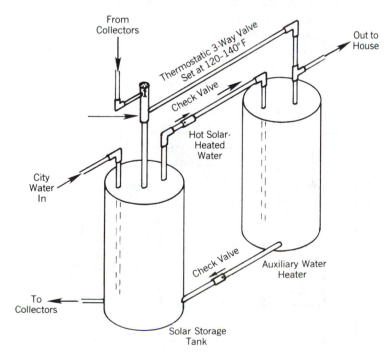

Figure 12.2 How the auxiliary water heater can be connected as a second solar storage tank in systems using water as the heat transfer medium.

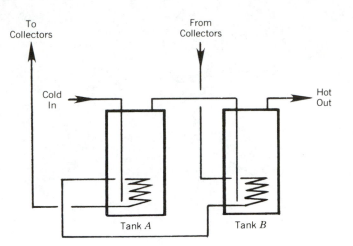

Figure 12.3 How two solar storage tanks with two heat exchangers can be connected in series.

Tank A removes more heat from the collector fluid and utilizes tank A as a preheater for tank B.

In Figure 12.4 the two tanks are piped in parallel. Cold water flows to both tanks. The collector fluid flow is divided across the tanks, and hot water for the house is drawn from both tanks at the same time.

Parallel configurations require identical tanks, exchangers, and reverse-return plumbing. They offer large quantities of warm water. Series connection can use dissimilar tanks, and no reverse returns are needed. They offer one tank hotter than the other.

Solar Storage Tanks Containing Electric Heaters

It is very common to see one-tank systems where the solar storage tank contains electric elements to provide auxiliary heat. One element may be provided in the upper third of the tank or two elements, one in the upper half and one at the bottom of the tank, may be provided.

True solar storage cannot exist where the entire tank can also be heated by active electric elements. *The electric element at the bottom of the tank must be disconnected, and any thermostats controlling the element(s) must be located above the higher element. Any tank space located above the remaining active element must not be counted as solar storage.*

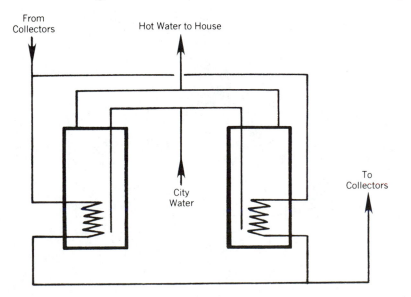

Figure 12.4 How two solar storage tanks with two heat exchangers can be connected in parallel.

A 120-gal electric heater with an element in the center connected is a 60-gal solar storage tank coupled with a 60-gal auxiliary heater tank.

The addition of a separate auxiliary heater tank is strongly advised if the solar system does not have at least 1.2 gal of true solar storage per square foot of collector surface.

The Auxiliary Heater

Auxiliary water heaters are standard water heaters and need few special considerations. In the event that the solar system is large and well designed, a minimum-sized auxiliary heater as small as 10 or 20 gal can be considered. The use of this size auxiliary instead of a 40- to 65-gal tank cuts standby losses drastically.

Instantaneous gas or electric booster heaters can be used to replace an auxiliary heater at the point of use. Such heaters have no standby losses but are expensive to install.

In large homes with extended pipe runs, it may also be possible to gain efficiency by using two or more auxiliary heaters located close to the point of use.

Feed the preheated solar water to the cold water piping of the auxiliary heater. Install a heat trap on the hot water piping, and consider the use of hot water pipe insulation if the home contains long runs of hot water piping through unheated areas. Make certain that the auxiliary heater has at least R-10 insulation.

Teach the Customer to Look for Heat Gain, Not Temperature

Conceputally, consumers using solar water heaters for the first time tend to continue to think in terms of final temperature instead of in terms of heat gained. Hence, many complaints about performance hinge around the fact that the solar system is not heating water to the 120 to 140°F temperature that the customer wants delivered.

The serviceperson should explain to the customer that heat gain from 60 to 100°F topped by auxiliary heat to the final 120 to 140°F is a more efficient use of solar collection than attempting to use the solar system to provide 120 to 140°F water. Use a collector performance curve to make this point and show the customer how collector performance drops off as the collectors run hotter.

This will clear up many misunderstandings about the role of the average solar water heater.

System Hookups and Connections

In Chapter 1 you learned that the system plumbing should allow either the solar heater or the auxiliary heater to be removed from the water-heating system without affecting the use of the other heater. For simplicity, only one simple hookup was shown in the first section of the book.

The system shown in Figure 2.15 for bypassing either the solar or the auxiliary heater used six valves. It is possible to make this hookup with fewer valves if two three-way ball valves are used. Native Sun Energy

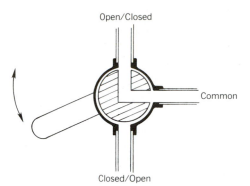

Figure 12.5 The three-way ball valve.

Systems, Inc., of Petaluma, California, has allowed us to show one way they make these connections.

Figure 12.5 shows a cross section of a standard three-way ball valve. The ball is the crosshatched section shown with a 90° hole drilled through it. The opening on the right is a common opening that is connected to one of the other two openings, depending on which way the handle on the left is positioned. The handle is down and the valve is open to the top outlet. If the handle were moved up, the common inlet would be open to the bottom outlet. Any of the three ports may be used as the inlet, so the valve is truly multipurpose.

Figure 12.6 shows a complete hookup. Bypass is allowed for, and a tempering valve has been added. For the moment, we will ignore the tempering valve and concentrate on the bypass system, which consists of valves V3, V4, and V5. V3 and V4 are the two three-way valves in the system; V5 is a standard gate valve.

Cold service water feeds the system through V2. Tempered hot water leaves the system through tempering valve V1.

Figure 12.7 shows how the system's valves are set when both solar and auxiliary systems are in use. Cold supply water feeds through V4 to the cold water dip tube of the solar storage tank. The water in the storage tank is warmed by solar energy and is returned to the auxiliary heater through open gate valve V5, to V3, to the cold water inlet of the auxiliary tank. The auxiliary heater does any necessary temperature topping off and feeds the hot water out to the house through the tempering valve, which is not shown.

Figure 12.8 shows how solar storage is bypassed. V4 is reversed so that cold supply water feeds to the cold water pipe of the auxiliary heater. V3 is set in either position. The gate valve, V5, is closed to prevent any flow to or from the solar storage tank. Now the auxiliary heater operates just as if there were no solar system.

Figure 12.9 shows how the auxiliary heater is bypassed. V4 passes cold water to the bottom of the solar storage tank. The water is warmed by solar energy and returned through the open gate valve, V5, passes

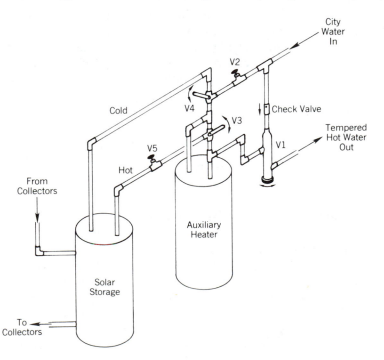

Figure 12.6 A complete two-tank hookup with three-way by pass valves and a tempering valve.

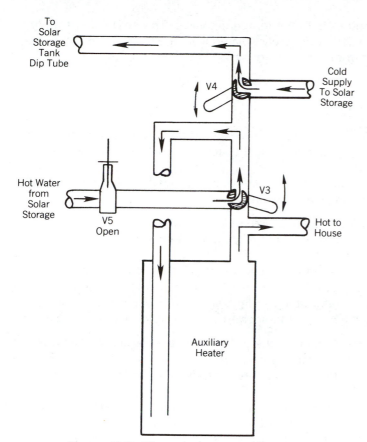

Figure 12.7 The valve settings when both solar and auxiliary heaters are in operation.

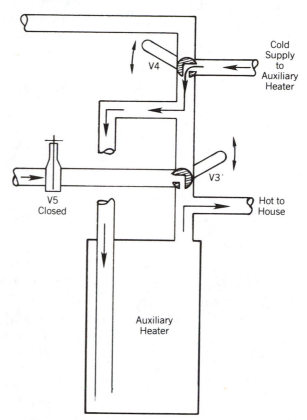

Figure 12.8 The valve settings when solar storage is being bypassed.

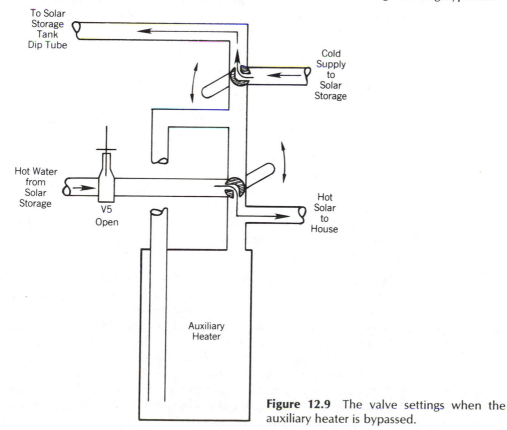

Figure 12.9 The valve settings when the auxiliary heater is bypassed.

through V3, and exits to the tempering valve. The cold water supply to the auxiliary tank is completely isolated from the system, which prevents any flow to or from the auxiliary tank.

Tempering Valves

All solar systems should contain a tempering valve to prevent scalding water from entering the house. In fact, most codes will insist on the valve's inclusion.

Consider the nature of solar energy and you will realize why the inclusion of a tempering valve is so essential.

- Solar energy is not available on demand, so it cannot be turned on and off at will. The owner takes what he or she can get when he or she can get it.
- Solar systems can build up heat to 160 to 180°F before the controller responds and turns the system off. In a thermosiphon system or a system without a high-limit switch the water can actually build up to 210°F before a temperature or pressure relief valve will vent.

Thus, fine control of the high-limit temperature is just not possible in solar systems. Any temperature over 140°F poses a danger to human safety. Water-heater temperatures over 180°F pose severe dangers and are prohibited by building codes.

The tempering valve can be placed between the solar storage tank and the auxiliary tank if desired. However, when it is placed in front of the final tank outlet going to the house, as in Figure 12.6, any overheating caused by the auxiliary heater is also remedied.

Plumb the tempering valve so that it is located below the tank's hot water outlet. Isolate the valve from the cold water inlet with a check valve.

Water Hammer

Water hammer occurs when a shock wave, created by closing a valve, creates a slamming or banging noise in the pipes. The antithermosiphon valves located in many solar storage tanks can create water hammer where it did not exist before the solar system was installed.

Water hammer can be eliminated by installing a shock-absorbing device in the water lines. One such device is a rubber-seated spring check valve. Another is a length of capped, air-filled pipe plumbed vertically upward from a tee fitting in a horizontal pipe line. A third device is a small expansion tank specifically designed to eliminate water hammer noise.

Where to Place the Device   Often the water hammer shock wave travels from the cold water taps through the cold water supply valve to the solar storage tank, where a closing check valve creates noise heard throughout the cold water system. Open and close the solar system isolation valves and trigger the hammer to identify the noise location.

It is then a simple matter to plumb the device between the faucet that triggers the shock wave and the check valve that creates the noise.

In some cases you may have to replace the storage tank check valve with a rubber-seated spring check valve. Make sure that the replacement valve does not impede normal flow through the system.

Dielectric Unions

All dissimilar metals, such as copper piping and steel water tanks, should be isolated from each other with dielectric unions, as described in Chapter 10. Figure 12.10 shows a typical dielectric union.

As you saw in reading about corrosion, metal corrosion is caused by galvanic action. Isolating the dissimilar metals will retard any corrosion in the system.

Common locations where dielectric unions are required are at the inlet and outlet of the collector piping to the water tanks and at the hot and cold water outlet and the inlets leading to and from the house.

Dielectric unions also cut down on the rate of heat transfer from one piece of piping to another or from the piping to a valve. Automatic air vent valves commonly contain plastic and/or rubber parts. When a "bake" controller strategy is used, these valves can overheat and malfunction. A dielectric union between the piping and the valve will often solve this vexing problem.

Heat Traps

The final hot water line leading from the last storage tank in the system to the service lines for the house should contain a *heat trap* to prevent thermosiphon heat loss from the service lines.

A heat trap is merely the inclusion of a U shape in the line at the top of the tank, such as the one shown in Figure 12.11. Hot water is lighter than cold water. Hence, it will not travel downward through the U shape and up the service piping where the heat can be dissipated.

Solar Storage and Auxiliary Water-Heater Installation Checklist

Here is the solar storage installation checklist. Evaluate the installation against each of the 35 points.

Figure 12.10 A dielectric union.

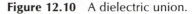
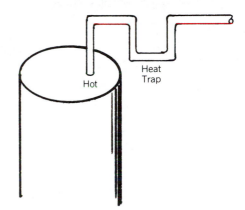

Figure 12.11 A heat trap at the top of the storage tank lowers heat losses in the house plumbing, because hot water cannot flow down around the U-Shape.

EVALUATION POINT	POOR	MARGINAL	GOOD
1. Does the pressurized tank meet ASME specifications?	_____	_____	_____
2. Is the water pressure between 15 and 80 psi?	_____	_____	_____
3. Does the tank have a suitable interior coating to prevent corrosion?	_____	_____	_____
4. If the tank is glass lined, does it contain a magnesium anode?	_____	_____	_____
5. Has the tank suffered any physical damage that may have injured the lining?	_____	_____	_____
6. Are the fittings on the tank suitable for the metals in the system?	_____	_____	_____

EVALUATION POINT	POOR	MARGINAL	GOOD
7. Is the tank insulated to at least R-10?	_____	_____	_____
8. Are the safety valves an approved type?	_____	_____	_____
9. Are the safety valves mounted correctly?	_____	_____	_____
10. Is there a suitable drain valve in the proper location?	_____	_____	_____
11. Does the cold service water enter at or near the bottom of the tank or through a dip tube?	_____	_____	_____
12. Is the hot water piped out at the top of the tank?	_____	_____	_____
13. Does the final storage tank hot water exit pipe contain a heat trap?	_____	_____	_____
14. Is the heat exchanger or the service water properly piped up to the collector loop?	_____	_____	_____
15. Are all plumbing code standards met?	_____	_____	_____
16. Is the temperature relief valve located within 6 in. of the top of the tank and no more than 3 in. away from the tank?	_____	_____	_____
17. Is the temperature relief valve element located in the hot water flow?	_____	_____	_____
18. Is the discharge from the temperature and pressure relief valves adequate to prevent further heat or pressure buildup?	_____	_____	_____
19. Are there any shutoff or check valves located between the safety valves and the tank?	_____	_____	_____
20. Are the supply valves to the tank full-flow valves (gate or ball valves)?	_____	_____	_____
21. Is the tank foundation adequate? Is a drip pan installed and piped if required?	_____	_____	_____

EVALUATION POINT	POOR	MARGINAL	GOOD
22. Is a sanitary sewer or storm drain available to carry away any overflows or leaks?	_____	_____	_____
23. Is the tank protected against freezing?	_____	_____	_____
24. Is the bottom of the tank insulated?	_____	_____	_____
25. Does the water supply need conditioning?	_____	_____	_____
26. Does the customer understand the difference between British thermal unit gain and final temperature?	_____	_____	_____
27. Is the auxiliary fuel the most cost-effective choice?	_____	_____	_____
28. Is the size of the auxiliary heater correct?	_____	_____	_____
29. Are the hot water pipes insulated?	_____	_____	_____
30. Is the auxiliary heater heat trapped?	_____	_____	_____
31. Is the auxiliary heater insulated to at least R-10?	_____	_____	_____
32. Is the lower electric element of the auxiliary heater disconnected?	_____	_____	_____
33. Does the final outlet contain a tempering valve?	_____	_____	_____
34. Can either solar storage or auxiliary heater be bypassed?	_____	_____	_____
35. Is any water hammer noise heard when faucets are opened or closed?	_____	_____	_____

CHAPTER
13

Piping, Valves, and Gauges

The piping that connects all the solar system components together, the valves that control the flow of the liquids through the system, and the gauges that monitor the condition of the system have some special considerations in solar water-heating systems. They must be chosen to perform their designed function, engineered to withstand their particular environmental conditions, and built to have long life within the system.

The Piping

There are three classes of materials used in solar water-heater piping: aluminum, copper, and plastic. Generally, the serviceperson will find copper piping in most of the systems he or she is called on to repair, but aluminum and plastic are common enough so that they should be mentioned.

Aluminum Piping

Aluminum is a highly anodic, active metal that, when improperly used, is subject to a high degree of galvanic corrosion. That corrosion tends to manifest itself as pitting, which rapidly leads to perforation of the piping.

Aluminum piping should only be used in all-aluminum systems where the manufacturer of the system mandates aluminum to protect the heat exchanger and the collector's liquid passages. The substitution of copper or galvanized pipe in those situations can result in component damage.

It is possible to substitute chlorinated polyvinyl chloride (CPVC) or polybutylene (PB) plastic piping when absolutely necessary. However, the environmental conditons must be carefully considered to be certain that the plastic piping will not rapidly deteriorate.

Joining aluminum pipe through soldered connections is a special art that requires the use of an inert gas blanket. This is hard to accomplish on the job unless a special torch is available. The serviceperson is usually better off to consider nonelectrolytic rubber or plastic hose connections where repairs must be made.

In all cases relating to all-aluminum systems, consultation with the manufacturer's technical representatives is strongly advised prior to making any piping repairs or changes.

Plastic Piping

Four types of plastic piping are used in the construction field; rigid ABS (acrylonitrile-butadiene-styrene), rigid PVC (polyvinyl chloride), rigid CPVC (chlorinated polyvinyl chloride), and flexible or rigid PB (polybutylene). Of these four types, only CPVC or PB are suitable for use in pressurized solar systems.

ABS and PVC piping have a top temperature use of 100°F under pressure conditions. They should not be used in solar water heating systems.

Both PB and CPVC have a top temperature use of 180°F under normal household water-pressure conditions. They can be used in exchanger storage loops, in air bleed lines, in dump valve discharge tubes, and in piping storage tanks to auxiliary heaters.

NOTE As of July 1983, the use of PB piping in water systems was being questioned by several regulatory agencies. Check the latest code information before using it.

The use of plastic pipe should be limited to systems containing only water. Various plasticizers are used to give the piping its flexibility and hardness. Organic heat-transfer agents will often leach out the plasticizers with time and cause the pipe to be weak and embrittled.

Choose any plastic pipe that you use with care. Quality varies from manufacturer to manufacturer. There is little advantage in price of plastic piping over copper piping, but the piping may be solvent joined with valves and fittings at room temperature, which cuts labor costs on the job. However, many of the valves used in solar systems are not available in plastic.

As a rule, keep plastic piping out of the collector loop and use it indoors, where it is protected from the environment. If faced with the need to repair an aluminum-piped system, choose CPVC or PB over copper or galvanized steel pipe.

Copper Piping

The standard piping of the solar water-heating business is copper. Generally, rigid piping is used but soft, annealed, flexible copper tubing also has a place in systems for making wide, sweeping changes in direction and for handling differential collector array expansion.

Rigid copper piping and flexible copper tubing are supplied in three grades: types M, L, and K. The grades differ by wall thickness, with type M having the thinnest wall and type K having the thickest wall. A fourth type, ACR tubing, is similar to type L but comes in rolls of small-diameter tubing suitable for drainback air bleed lines.

Most solar collectors use type M copper tubing, which is most satisfactory for connecting components *if the local codes do not specify the use of type L*. The use of type K piping is generally restricted to use in areas where it would be prohibitively expensive to replace the piping, such as in buried trenches, inside finished walls, and under concrete floors.

Joints in Copper Piping   Three types of joints are used with copper piping: compression ring fittings, swaged fittings, and soldered fittings.

Any of these joints are satisfactory with water systems, although the compression fitting suffers when subjected to gross expansion and contraction with heating and cooling. However, when considering permanent installations, high thermal expansion, and the use of organic fluids, the soldered copper joint has proven to be best.

The following solder joint discussion gives the general techniques.

Solder Joints

Making good solder joints is extremely important. The general technique is worth reviewing.

In the fabrication of soldered joints, a satisfactory joint begins with selection of the proper flux and solder. The flux should be a nonaggressive material free of chlorides and highly acidic or alkaline materials. A water-based flux is generally more satisfactory than an animal-fat-based flux. *Under no circumstances should a 50/50 tin/lead solder be used. Such a material is not satisfactory for the collector loop temperature ranges.* The minimum satisfactory soldering material is 95/5 tin/antimony solder. A brazing alloy melting at or above 1000°F could also be used.

There are 12 simple steps to make a good solder joint. These are shown in Figure 13.1, and are as follows.

1. Measure the length of tube.
2. Cut the tube square.
3. Ream the cut end.
4. Clean the tube end.
5. Clean the fitting socket.
6. Apply flux to the tube end.
7. Apply flux to the fitting socket.
8. Assemble the joint.
9. Remove the excessive flux.
10. Apply heat.
11. Apply solder.
12. Allow the joint to cool.

These steps are basically simple, but they make the difference between good and poor joints. They normally take less time to do than to describe. For good results, none should be omitted.

Cutting The tube must be cut to the exact length with a square cut. Tube cutters should generally be used for sizes up to 1 in. Alternate methods for larger pipes are abrasive saws or hack saws. With these methods, the joint must be filed square following cutting.

Reaming Either the tube cutter or the abrasive saw will leave small burrs on the end of the tubing. These must be removed to prevent localized turbulence, which will result in excessive corrosion. Often, particularly with soft-annealed tubing, the tube will be out of round and must be brought to true dimension and roundness with a sizing plug and ring.

Cleaning The surfaces to be joined must be completely clean and free from oil, grease, and heavy oxide. The end of the tube must be cleaned for a distance slightly more than is required to enter the socket of the fitting. Fine crocus cloth (00), cleaning pads, or special wire brushes may be used. Rub hard enough to remove the surface film or soil, but not hard enough to remove metal. Take extreme care so that particles of material do not fall into the tube or the fitting. The socket of the fitting

Figure 13.1 The 12 steps to a good solder joint. (Reprinted by permission from R. Montgomery and J. Budnick, *The Solar Decision Book,* Wiley, New York, 1978.)

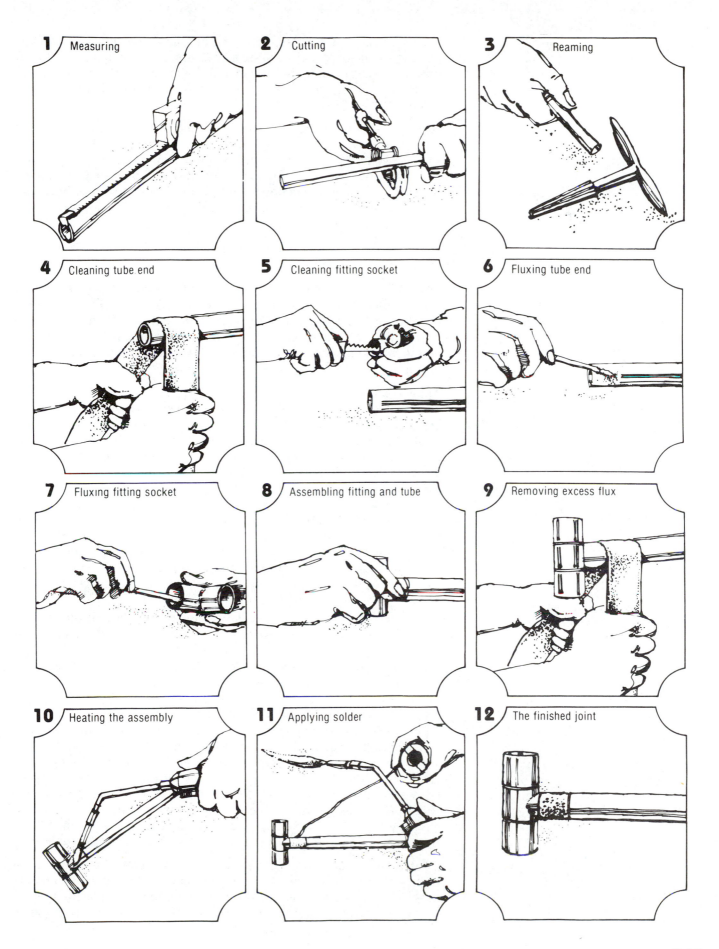

1 Measuring

2 Cutting

3 Reaming

4 Cleaning tube end

5 Cleaning fitting socket

6 Fluxing tube end

7 Fluxing fitting socket

8 Assembling fitting and tube

9 Removing excess flux

10 Heating the assembly

11 Applying solder

12 The finished joint

should be similarly cleaned, observing the same precautions. Tubes and fittings are made to close tolerances. Abrasive cleaning should not remove significant amounts of metal. If too much metal is removed during cleaning, the capillary space may become so large that a poor joint will result.

Fluxing As soon as possible after cleaning, the surfaces to be joined should be covered with a thin film of flux. Fluxes must be mildly corrosive and generally contain zinc and ammonia chlorides. At all cost, avoid highly aggressive fluxes and tallow bases. Stay with only mildly aggressive fluxes in water-based compounds. Apply the flux with a small brush or a clean rag. Be careful not to insert the flux into the fitting to the point where it remains in the loop.

NOTE For safety, avoid using fingers. Flux accidentally carried to the eyes can be very harmful.

Assembling Assemble a joint by inserting the tube into the fitting. Make sure the tube is firmly against the end of the socket. Give the tube a small twist to help spread the flux over the two surfaces. Then remove any excess flux from the outside of the pipe. The joint is now ready for soldering. Soldering should take place within 1 or 2 hours of fluxing. The flux assembly should never be left overnight before soldering.

Apply Heat and Solder Heat is usually applied with a propane, butane, air-acetylene, or oxy-acetylene torch. The flame is played on the fitting and moved to heat as large an area as possible. The flame must not be pointed into the socket. This will overheat and burn the flux, destroying its effectiveness. Once the flux has been burned, solder will not enter the joint. The joint must be opened, recleaned, and refluxed. Overheating is apt to cause joints to crack, particularly with cast fittings, and must be avoided.

You can tell whether the joint is hot enough by bringing the solder in contact with the tube with the flame removed. If the solder does not melt, more heat should be applied. With a correctly heated joint, capillary action will draw solder over the entire joint. Pretinning is not necessary and is not recommended. It will interfere to some extent with the capillary action of drawing the solder into the joint.

Soldered joints depend on capillary action to draw free-flowing molten solder into the narrow clearance between the fitting and the tube. The flux acts as a cleaning and wetting agent. It permits uniform spreading of the molten solder over the surfaces to be soldered. Capillary action is most effective when the spaces between the surfaces to be jointed are between 0.002 and 0.005 in. A certain amount of looseness of fit can be tolerated, but this will cause difficulty with large-size fittings, particularly when the joints are horizontal. Oversized fittings and horizontal joints usually result in too large a gap at the top and a resultant leak.

Good support and alignment contribute to good joints. Wherever possible, joints should be made as subassemblies in the vertical position before installation. Wherever copper pipe is to be joined to copper valves and other materials, the manufacturer's instructions should be carefully followed. These usually include keeping the valve in the full-open position before applying heat and applying heat to the tube only.

Cooling The joint must cool naturally for some time. This is particularly true if cast fittings are used. Too-rapid cooling cracks fittings and contributes to poor solder joints. Do not test solder joints by placing them in water or placing water in them. If the joints have been made as described, there should be no leaks, and final air testing of the collector loop will reveal this.

Brazed Joints

Brazed filler metals, sometimes referred to as "hard solders" or "silver solder," call for much different techniques than for standard solders. Contractors doing brazing should consult with brazing manufacturers.

Technical Assistance

The Copper Development Association, Incorporated, has published a book called *The Copper Tube Handbook for Plumbing, Heating, Air conditioning and Refrigeration.* This book is generally available from your local plumbing or heating supply house. If not, you can obtain a copy of this book by writing the Copper Development Association at the following address:

Eastern Regional Office
Copper Development Association, Incorporated
Greenwich Office Park 2
Box 1840
Greenwich, CT 06836-1840

Nothing is more frustrating than to build a solar collector loop and then find that less-than-perfect soldering or brazing is causing leaks. The system must then be drained and cleaned and the leaks repaired—a very costly experience for the contractor. Thus, manufacture sample joints at the bench and saw them apart to insure that the particular procedure used is doing the job.

Valves

Solar water-heating systems use a number of different types of manual and automatic valves. These valves must be carefully chosen so that they properly perform their function, withstand the operational conditions, have good life in the operating environment, and are compatible with the system's components.

Valves can be divided into manually operated valves and automatically operating valves. The automatically operating valves can be further divided into valves requiring outside electrical power to operate them and valves operated by system conditions. The following list separates the solar system's valves into these three types.

Manual Valves

- Balancing valves.
- Ball valves.
- Boiler drain valves.
- Gate valves.

System Condition-Operated Automatic Valves

- Automatic air vent valves.
- Automatic fill valves.
- Backflow prevention valves.
- Check valves.
- Pressure-actuated safety valves.

- Pressure-reduction valves.
- Temperature- and pressure-actuated safety valves.
- Tempering valves.
- Thermostatic bleed valves.
- Thermostatic three-way modulating valves.
- Vacuum relief valves.

Electrically Operated Automatic Valves

- Motorized valves.
- Solenoid valves.
- Spool-type drain valves.

The Quality of Valves

Cast bronze valves are normally used in solar systems. Valves are made in light- and heavy-duty types. Light-duty types are designed to withstand 125 psi water pressure at 200°F; heavy-duty valves are designed to withstand 125 psi water pressure at 350°F.

Obviously, light-duty valves are pushed to their limits in solar water-heater systems. For the small difference in price, the serviceperson should use heavy-duty valves.

Valve Choices and Considerations

There are a number of important considerations in choosing and installing the valves used in each part of the system.

Isolation valves

Isolation valves are used to shut off parts of the system. They are normally seen at the cold water service entrance, between the pump and the tank, and at the outlet to the auxiliary heater or to the hot water service. A gate or ball valve such as the ones shown in Figures 13.2 and 13.3 should be used in shutoff locations. Stop valves, stop-and-waste valves, or globe valves are not recommended and, in many areas, do not meet the plumbing code. In general, always choose a valve that provides unrestricted flow when open and positive, nonleak shutoff when closed.

Never isolate the collector loop unless it contains a pressure relief valve. The fluid in the collectors can expand when hot, create a high pressure, and rupture the lines or the fluid passages.

Throttling or Flow Control Valves

A throttling valve for controlling collector loop flow rate is generally found on the discharge side of the pump. A ball valve or a butterfly balancing valve, such as the one shown in Figure 13.4, should be used in this location. Gate valves should be avoided because they are apt to lock up or chatter and break and are not made for throttling.

Drain and Fill Valves

Drain and/or fill valves are found in three locations: at the bottom of storage tanks, at the bottom of the pump drain leg, and in the collector loop fill system. It is not necessary to use gate valves in these locations. Boiler drain valves like the valve shown in Figure 13.5 will operate very

Figure 13.2 A gate valve.

Figure 13.3 A ball valve.

Figure 13.4 A butterfly balancing valve.

Figure 13.5 A boiler drain valve. **Figure 13.6** An HVAC hot water system balancing valve.

satisfactorily and also make it possible to attach a fill or drain hose to the valve.

Balancing Valves

Balancing valves are found where the pump feeds more than one solar collector array. They provide a means of balancing the rate of flow between the arrays. A regular hydronic hot water system balancing valve like the one in Figure 13.6 can be used, or a ball or butterfly valve similar to the throttling valve may be used at these locations. Do not use cast iron valves—they are subject to rapid corrosion in most solar applications.

Diversion Valves

Three-way ball valves, such as the one in Figure 13.7, are commonly used as manual diverting valves in heater bypass plumbing.

Vacuum Relief Valves

Draindown and open drainback systems require a vacuum relief valve (see Figure 13.8) at the top of the collectors. These valves are critical. Only the highest-quality valve should be installed. The vacuum relief valve should be separate from the automatic air vent valve. Plastic combination valves are not advised. Protect the valve from the elements. Frost and ice entering the valve can cause it to freeze up and fail.

Automatic Air Vent Valves

Automatic air vent valves (see Figure 13.9) are located at the top of pressurized or open drainback systems and at the air eliminator in a heat-exchanged system. The typical valve is a needle-and-float valve, where the valve opens when air replaces the liquid in the valve. Again, only the highest-quality valves should be installed, because a leaking valve can cause roof damage. They must be installed in a vertical position, and they should be protected against freezing.

Figure 13.7 A three-way ball valve.

Figure 13.8 A vacuum relief valve.

Figure 13.9 An automatic air vent valve.

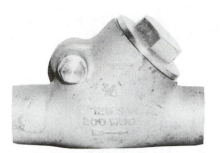

Figure 13.10 A horizontal flapper check valve.

Figure 13.11 A vertical spring-loaded check valve.

Any additional air vent valves installed as added protection to the array or the tank should be the same high quality.

Automatic air vent valves are available in two pressure ranges, 0 to 45 psi and 0 to 125 psi. Choose the valve to match the system conditions.

Check Valves

A check valve is normally located in the hot downcomer piping from the collectors. Its purpose is to prevent reverse thermosiphoning of hot fluid from the storage tank to the cold collectors during nonsunlight hours.

There are swing check valves designed to be mounted in a horizontal pipe (Figure 13.10) and spring check valves designed to be mounted in a vertical pipe (Figure 13.11). The normal vertical pipe check valve is usually a spring-loaded ball valve that creates a high pressure drop but is more positive acting than the horizontal check valve, which employs a swinging gate action. The horizontal valve is more prone to leaking, but both valves should be regarded as nonpositive, leaking valves. Both types can freeze up when used outdoors and should be installed in heated areas. A different type of vertical check valve is used in thermosiphon systems, where high pressure drops cannot be tolerated. This is the sinking/floating plastic ball check valve. Unlike the normal vertical check valve, this type is not spring loaded shut.

Temperature and Pressure Relief Valves

Each solar storage tank and each auxiliary heat storage tank must have its own temperature and pressure relief valves. The most common practice is to use one valve that combines both temperature and pressure relief, such as the one in Figure 13.12. The valve must be located within 6 in. of the top of the tank, not more than 3 in. away from the tank, have its temperature element inserted in the hot water flow, and have no shutoff or check valve between the valve and the tank. The overflow from the valve must be large enough to prevent any further temperature or pressure buildup, and the valve must be set to operate at or below 210°F. The code in some locations may mandate that the valve be at the top of the tank and not side mounted.

Pressure Relief Valves

Many collector arrays can be isolated from the storage tanks by closing valves on the riser and the downcomer. If an array were closed when it was cold, extremely high pressure would be built up when the collectors heated and the liquid expanded. Therefore, it is good practice to place a 125-psi pressure relief valve like the one in Figure 13.13 at some point

Figure 13.12 A temperature and pressure relief valve.

Figure 13.13 A pressure relief valve.

in the collector array to prevent accidental rupture of the collectors or the piping.

Most closed drainback and liquid heat-exchanged collector loops operate under low pressure, usually around 10 to 30 psi. This is in contrast to thermosiphon, pumped service-water, and draindown systems, which operate at service-water pressure. The pressure relief valve, usually rated between 30 and 60 psi, protects the collector loop against high pressure.

Always check the pressure rating of a pressure relief valve before installing a replacement.

Automatic Fill Valves

Automatic fill valves connected to the service water should not be used in any heat-exchanged system. A leak between the service water and the antifreeze side of the heat exchanger could result in the dilution of the antifreeze and subsequent freeze-up of the system.

Pressure-Reduction Valves

When city water pressure is in excess of 60 to 80 psi, a pressure-reduction valve (Figure 13.14) should be placed in the supply line where the city water enters the house to reduce the pressure in the entire water system to 40 to 60 psi.

Backflow Prevention Valves

When the guidelines relating to heat transfer fluids outlined in Chapter 10 are followed, chances are that no backflow prevention device will be needed between the city water service and the solar water heater. If the code demands one, follow local plumbing codes in its installation. Make certain that the water system contains an adequately sized expansion tank downstream from the backflow prevention valve.

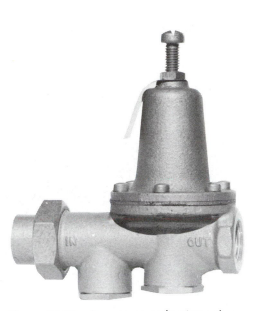

Figure 13.14 A pressure-reduction valve.

Freeze Dump and Heat Dump Valves

Electrically operated freeze dump and heat dump valves are a problem in draindown solar systems. Many different types of electrically actuated two-way and three-way valves have been tried, but most have failed.

The highest rate of failure from collector freeze-ups seems to come with the electrically actuated three-way valves. Three-way valves are generally not completely tight and, in a freezing situation, water can leak back into the collectors and freeze.

The Spool-Type Collector Draindown Valve

A new type of valve known as a spool-type collector draindown valve has been offered to help solve most of the problems related to freeze dump and heat dump valving. An example is the Sunspool Collector Draindown Valve Model No. 356, which is shown in Figure 13.15.

This valve may be used with collector arrays of up to 150 ft². When the valve is energized, it automatically fills the collector array with pressurized city water and then permits normal pumped operation. When deenergized by the controller signal or by a power failure, the valve isolates the collector array from the city water supply and drains the collectors. Figure 13.16 shows the valve partially disassembled; Figure 13.17 shows a cross section of the valve's mechanism.

Figure 13.15 The Sunspool draindown valve.

Figure 13.16 The Sunspool draindown valve disassembled.

A review of Figure 13.17 shows that the power-off mechanism is a stainless steel spring; the valve's spool continuously oscillates while the valve is in the fill position, thus preventing bind-up; the mechanism is Teflon-coated brass or polysulfone plastic; and the entire valve can be quickly disassembled in the field for checkout and cleaning. Field reports on spool-type valves indicate that their use solves many of the problems associated with electrically operated freeze and heat dump systems.

Tempering Valves

The tempering valve (see Figure 13.18) is a three-way valve containing a thermostatic element. Hot water and cold water are supplied to the valve inlet ports. The two mix, and warm water exits from the valve. A dial mechanism allows the maximum exit temperature to be set. Normally, tempering valves in solar systems are set between 120 and 160°F.

In Chapter 12 you learned where the valve can be placed in the system and what configuration the piping should assume to maximize the life of

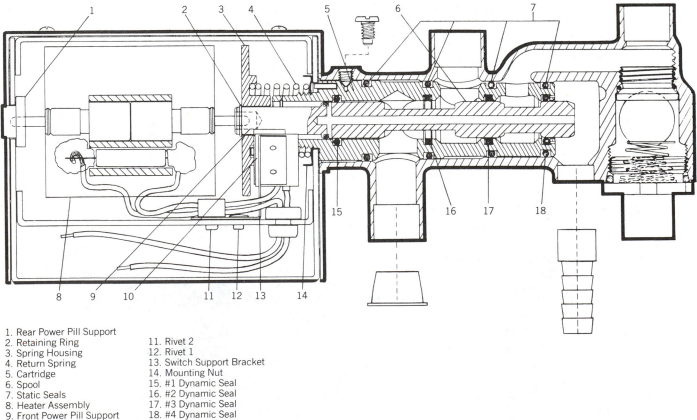

1. Rear Power Pill Support
2. Retaining Ring
3. Spring Housing
4. Return Spring
5. Cartridge
6. Spool
7. Static Seals
8. Heater Assembly
9. Front Power Pill Support
10. Switch
11. Rivet 2
12. Rivet 1
13. Switch Support Bracket
14. Mounting Nut
15. #1 Dynamic Seal
16. #2 Dynamic Seal
17. #3 Dynamic Seal
18. #4 Dynamic Seal

Figure 13.17 A cross section of the Sunspool draindown valve.

the valve. Here are the three major points to be remembered about tempering valves.

- The valve should be placed below a heat trap on the outlet pipe from the auxiliary heater to extend the life of the thermostat and provide maximum protection for the user.
- A check valve should be placed in the cold water feed to prevent hot water from backing up into the cold water line.
- Always remove the thermostatic element from the valve before soldering the valve in place. Excess heat can ruin the element.

Thermostatic Three-Way Valves

The thermostatic three-way valve was shown in Chapter 12. It is used as a diversion valve to change the flow of a stream of incoming water from one outlet to another based on the temperature of the incoming water.

Themostatic Bleed Valves

A valve, popularly known in the trade as a dribble valve, has been developed for additional protection of thermosiphon and pumped water systems. The dribble valve uses either a bimetallic element, a temperature-sensitive wax, or the pressure decrease of Freon gas to sense the onset of freezing temperatures. Figure 13.19 shows a typical Freon-powered valve.

The valve is attached to the collector array at the highest point in the array, and a postive-sealing check valve is plumbed into the collector outlet piping. When the temperature drops to 40 to 44°F, a small port in the valve opens and allows a small amount of water to flow from the collector inlet piping, through the collectors, and out to a drain. Even cold city water is generally over 50°F in the moderate climates, where thermosiphon and pumped water systems are operating in freezing weather, so the dribble valve continues to provide freeze protection even though the hot water in the tank is used up. Since the valve is not electrically powered, it will protect recirculation systems even during power failures as long as city water pressure is maintained.

The main limitations of dribble valves come with extreme temperatures. In very cold weather the slow water velocity may cause ice to plug the orifice in some dribble valves. At stagnation temperatures, wax-filled valves may be permanently damaged.

Figure 13.18 A tempering valve. (Photo courtesy of Watts Regulator Company, Lawrence, Mass.)

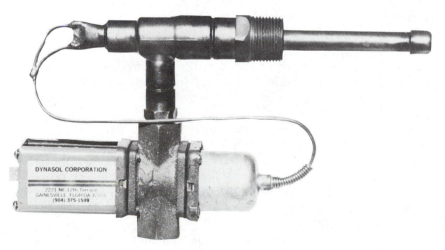

Figure 13.19 A Dynasol Corporation thermostatic bleed valve.

Gauges

The solar water heater may or may not contain gauges that report on the system's conditon. These gauges can include pressure gauges, temperature gauges, flow meters, and elapsed runtime meters.

Pressure Gauges

The pressure gauge is usually seen on the heat-exchanged system and is used as a monitoring device for heat transfer fluid leaks. When the system is charged, the pressure is run up to about 5 to 10 psi for oil systems and 20 to 30 psi for glycol/water systems. When the collectors are hot, the pressure can rise to 25 to 30 psi in oil systems and 30 to 40 psi in glycol/water systems. A properly operating oil system will show a gauge pressure of 5 to 30 psi, and a glycol/water system will show 20 to 40 psi.

Lower pressures might indicate a leak between the heat transfer fluid and the atmosphere; higher pressures could indicate either overheating of the collectors or the existence of a leak between the two sides of the heat exchanger. The difference can be determined by waiting until the collectors have cooled. If high pressure is still seen with cool collectors, most likely the storage water is leaking into the collector loop.

No special pressure gauges are required. The normal gauges used in water systems will work fine. The range of the meter should be 0 to 100 psi for most systems. A typical gauge is shown in Figure 13.20.

Figure 13.20 A pressure gauge.

Temperature Gauges

The pressure gauge can be combined with a temperature gauge into one instrument, as shown in Figure 13.21, or a separate temperature gauge can be installed. Normally, a temperature gauge is located just before the pump on the outlet of the tank or the heat exchanger. It reads the temperature of the fluid that is supplied to the bottom of the collectors. Often, a second temperature gauge is located on the other side of the exchanger or at the point where the hot collector downcomer enters the storage tank, so that the heat transfer taking place within the tank can be estimated by taking the difference between the two temperatures. It is also common to place a temperature gauge on the hot water pipe exiting solar storage and feeding the auxiliary heater to record the temperature of the water being supplied by solar energy.

Many different styles and types of gauges and thermometers are used for these applications. Most work very satisfactorily.

Sophisticated system controllers also often contain special circuits that read the resistance of the various sensors and report the resistance as degrees of temperature with a digital readout device. This is a very accurate and convenient method of reporting temperatures.

Figure 13.21 A combined temperature/ pressure gauge.

Flow Meters

Some systems incorporate a flow meter of one type or another into the piping. These flow meters have calibration problems and should be viewed only as general indicators of flow. They are designed to operate with water in a narrow temperature range. When they are used outside this temperature range or are used to check the flow of other liquids, their accuracy is poor.

However, such a flow meter is valuable for telling whether flow is taking place and for determining relative differences in flow rate with system adjustments. A typical flow meter is shown in Figure 13.22.

Elapsed Runtime Meters

The elapsed runtime meter is an electrical meter inserted in the output circuit of the controller wired in parallel with the pump. It keeps track of the time that the pump runs.

Piping, Valves, and Gauges Checklist

Here is the piping, valves, and gauges checklist. Evaluate the installation against each of the 33 points.

Figure 13.22   An in-line flow meter.

EVALUATION POINT	POOR	MARGINAL	GOOD
1. What is the general condition of the piping?	_____	_____	_____
2. Is the piping copper or does it contain aluminum, CPVC, or PB?	_____	_____	_____
3. Does the system contain any black iron or galvanized fittings that could cause corrosion?	_____	_____	_____
4. Are the fittings compression ring, swaged, or soldered? Any leaks?	_____	_____	_____
5. Are there threaded joints? Any leaks?	_____	_____	_____
6. Are the valves lightweight or heavy duty?	_____	_____	_____
7. Are all isolation valves either gate or ball valves?	_____	_____	_____
8. Does the collector array contain a pressure relief valve?	_____	_____	_____
9. Is the throttling valve a butterfly balancing or a ball valve?	_____	_____	_____
10. Are the drain and fill valves either boiler drain valves or gate valves?	_____	_____	_____
11. Are the valve stems of all manually operated valves located so that sediment cannot settle into the stem?	_____	_____	_____
12. Are array balancing valves either ball or standard HVAC balancing valves?	_____	_____	_____
13. Is the vacuum relief valve separate from the automatic air vent valve?	_____	_____	_____

14. Is the vacuum relief valve mounted in a vertical position, protected from the elements, and protected against freezing?   _____ _____ _____

15. Is the automatic air vent valve mounted in a vertical position, protected from the elements, and protected against freezing?   _____ _____ _____

16. Have the proper types of check valves been installed?   _____ _____ _____

17. Can horizontal swing check valve hinges freeze?   _____ _____ _____

18. Is the check valve being used as a positive leakproof control valve?   _____ _____ _____

19. Is the pressure/temperature relief valve on each storage tank located to meet the local codes?   _____ _____ _____

20. Is the presssure/temperature relief valve an approved type?   _____ _____ _____

21. Does the system contain an automatic fill valve?   _____ _____ _____

22. Is the system properly engineered for automatic filling (not recommended for heat-exchanged systems)?   _____ _____ _____

23. Is there a pressure-reduction valve and is it working properly?   _____ _____ _____

24. Does the system contain a backflow prevention device and does it work properly?   _____ _____ _____

25. Does the system contain freeze dump or heat dump valves?   _____ _____ _____

26. Are the dump valves two-way or three-way electrically actuated valves or a spool valve?   _____ _____ _____

27. Are any of the dump valves binding or sticking?   _____ _____ _____

28. Do the dump valves cycle easily when the controller is switched off and on?   _____ _____ _____

29. Are all gauges working correctly?   _____ _____ _____

30. Does the system contain a tempering valve?   _____ _____ _____

EVALUATION POINT	POOR	MARGINAL	GOOD
31. Is the tempering valve properly installed and operational?	_____	_____	_____
32. Are any thermostatic bleed valves working properly?	_____	_____	_____
33. Are any three-way modulating valves working properly?	_____	_____	_____

Repair
The
System

Solar water-heating systems contain up to six major components: collectors, heat exchangers, heat transfer fluids, pumps, controls, and tanks. After the serviceperson has identified the system and evaluated the installation, the system must be repaired.

Chapters 14 to 17 show how to repair or replace the major components; Chapters 18 and 19 show what tools, equipment, and instruments to carry to the job.

At the end of these six chapters, readers should be able to

- Arrive at the job prepared with the essential repair parts, tools, equipment, and instruments.
- Determine which of the system's components need repair or replacement.
- Properly repair or replace the components.
- Restore the system to good operating condition.
- Modify the system when necessary.

Repair
the Collector

The flat plate solar collector is a simple device that can readily be repaired. The easiest way to see how to repair a collector is to take one apart.

Preparation

Solar collector glazings are usually breakable materials that come in large sheets and are difficult to handle. This is particularly true of glass glazings. The absorber plate, while representing less of a problem than the glazing, is also difficult to handle and can be easily bent or otherwise damaged during handling. Thus, the solar water heater serviceperson should prepare a jig for holding and handling both the glazing and the absorber plate. The jig should be suitable for transporting a glazing or an absorber plate from the point of purchase to the installation point.

Handling Glazings

Solar collector glazings should be safety glazings. Safety glazings are glass or thermoplastic materials that conform to the Consumer Product Safety Commission's "Safety Standard for Architectural Glazing Materials," 16 CFR 1201. This standard mandates the use of certified safety glazing materials in sliding patio doors, storm doors, interior and exterior doors, tub and shower doors, and other critical locations. The willful violation of this standard is considered a criminal offense. Thus it is prudent to check the local codes and laws in your particular area carefully to see if solar collectors are covered.

Tempered Glass

A special grade of glass known as tempered glass is commonly used in solar collectors. Tempered glass is the only grade of glass that qualifies as a safety glazing material under 16 CFR 1201.

Tempered glass is fabricated by subjecting standard annealed glass to a special heat-treating process. The surface of the glass is heated uniformly to about 1300°F and then rapidly cooled by blowing air uniformly on both surfaces at the same time. This process locks the outer surfaces of the glass in a state of high compression, while the core is locked in opposing tension.

The properties of the glass that change with tempering are the tensile strength, bending strength, and fragment type. Tempered glass is about four times stronger than annealed glass in tensile and bending strength. When tempered glass breaks, it fragments into thousands of small fragments that are roughly cubical in shape. This qualifies it as a safety glazing material. However, because of the stresses built into the glass during the tempering process, a tempered glass sheet cannot be cut, drilled, or edged. Attempts to do so will result in the entire sheet breaking into thousands of little pieces.

When solar collectors first came into production in the middle 1970s and tempered glass was chosen as one of the most practical glazing materials, three sizes of tempered glass were generally available: 34 × 76 in., 34 × 96 in., and 46 × 96 in. Collectors were designed around these size availabilities, because it is not practical to manufacture a specific size of tempered glass unless there is a very large market for it. Therefore, most early solar collectors are about 36 in. wide and range from just under 7 ft to just over 8 ft in length. For economic reasons the trend today is toward 4 × 10-ft and larger collectors. By increasing glass thickness to $\frac{5}{32}$ in., one-piece glazing sheets up to 48 × 108 in. may be used. Most large collectors use several small glazing sheets to reduce weight and simplify handling.

Polycarbonate and acrylic thermoplastic glazing materials are also commonly available in 36 × 96-in. and larger sizes.

The Glazing Jig

Figure 14.1 shows a plan for a glazing jig built to handle 76-in.-long tempered glass and thermosplastic sheets as well as absorber plates. The details of the jig's construction are as follows.

1. A 4 × 8-ft sheet of plywood was cut down to 39 × 96 in.
2. A $1\frac{1}{2}$ × $1\frac{1}{2}$-in. timber was glued and screwed to the right end and to the bottom edge of the plywood sheet to form an L shape.
3. Two $1\frac{1}{2}$ × $1\frac{1}{2}$-in. skids were glued and screwed to the rear side of the plywood 8 in. in from the outer edges of the sheet. Three cross members were added as stiffeners.
4. Four $\frac{3}{4}$ × $1\frac{1}{4}$-in. cross members were run across the sheet at the points shown on the drawing. These were glued and screwed in place *with the heads of the screws carefully countersunk below the upper surface.*
5. On the left end, two $\frac{3}{4}$ × $1\frac{1}{4}$-in. members were screwed to the plywood 1 ft in from and parallel to each edge.
6. One 1-ft-long, $\frac{3}{4}$ × $1\frac{1}{4}$-in. member was screwed to the upper edge of the plywood sheet and another was screwed on next to the $1\frac{1}{2}$ × $1\frac{1}{2}$-in. timber.
7. A $\frac{3}{4}$ × $2\frac{1}{2}$-in. piece was screwed on top of these members parallel to the edge of the plywood.

Steps 5 to 7 produce a stiffened jig end that has good hand holds and will allow a rope sling to be attached for lifting the jig onto a roof.

8. A $\frac{3}{4}$ × $2\frac{1}{2}$-in. piece was cut $94\frac{1}{2}$ in. long and placed on the top edge over the members already screwed to the plywood. C clamps were used to hold the piece in place for the next two steps.
9. Four $\frac{3}{4}$ × $1\frac{1}{4}$-in. pieces were cut 39 in. long and placed on top of

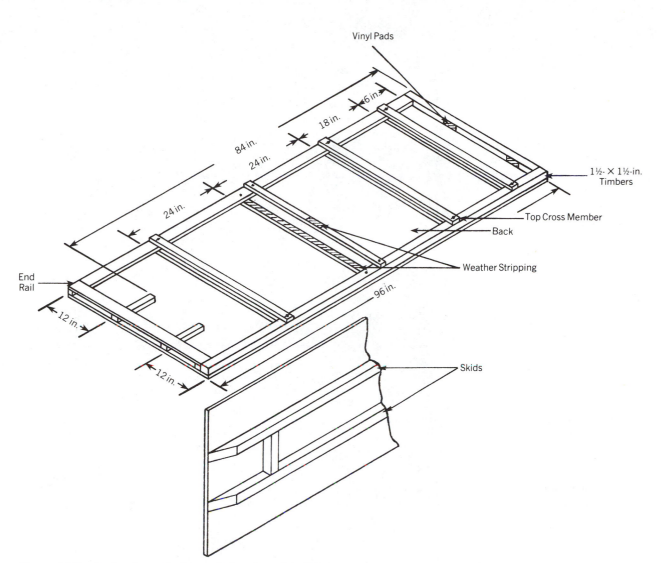

Figure 14.1 A plan for a glazing and absorber plate jig.

the cross members shown in step 4. C clamps were also used to hold these in place for the next step.

10. At the end of each of the four cross members, $\frac{1}{4}$-in. holes were drilled from the top cross members down through the other members and out the back of the plywood. Then $\frac{1}{4} \times 3$-in. bolts with washers and wing nuts were used to hold the jig together, and the C clamps were removed.

11. A foam-filled door weather stripping was secured to the four lower cross members with countersunk carpet tacks.

12. Vinyl door sill was cut into small strips and mounted with contact cement at strategic points where the edges of the glass would lie and at the center of the four top cross members.

Figure 14.2 is a photograph of the completed jig. Figure 14.3 shows the jig with the cross braces and top member removed, ready to receive the glazing or the absorber plate.

In Figure 14.4 there is a close-up of the top cross braces. Figure 14.5 shows the lifting end details. Figure 14.6 shows the details of the opposite end. Figure 14.7 is a close-up of the bottom rail, and Figure 14.8

Figure 14.2 The completed and assembled jig.

Figure 14.3 The jig with the top member and cross braces removed.

Figure 14.4 A close-up of the top cross braces.

shows the bolt and weather stripping at the top edge. Finally, Figure 14.9 shows the skids mounted on the rear with stiffeners between them.

The jig can readily be rope slung either horizontally or vertically. To make a jig for a larger sheet of glazing or absorber plate, use a 10-ft sheet of plywood and add one additional set of cross members. The use of this jig will allow you to transport and replace glazings and absorber plates safely, without damaging them.

Figure 14.5 The lifting-end details.

Figure 14.6 The details of the glass-bearing end.

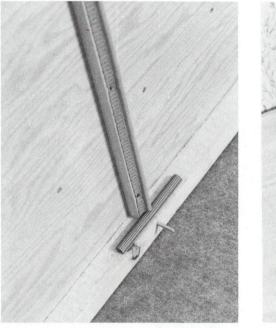

Figure 14.7 A close-up of the bottom rail.

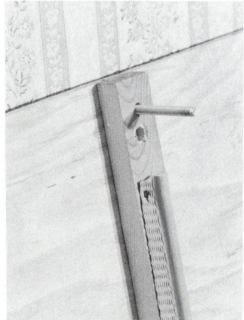

Figure 14.8 The top edge bolt and weather-stripping details.

Figure 14.9 The rear skids and stiffeners.

Collector Disassembly

The collector that will be taken apart has a single tempered glass glazing, a copper tube in strip, selective surface absorber plate, fiberglass back and edge insulation, and is housed in an all-aluminum housing. The headers are designed as internal manifolds and protrude from the sides of the collector close to the top and bottom of the housing.

Figure 14.10 shows the back of the collector standing against a wall. The back is an aluminum sheet held in place with stainless steel screws.

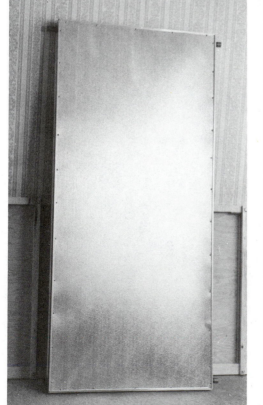

Figure 14.10 The back of the collector.

Figure 14.11 A close-up of the back screws.

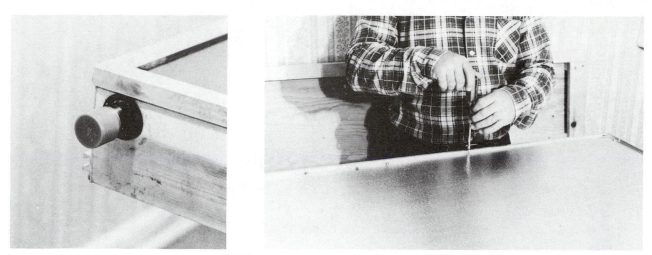

Figure 14.12 A close-up of the manifold end.

Figure 14.13 Removing the back screws.

Figure 14.11 shows a close-up of the screws, and Figure 14.12 shows a close-up of the end of one manifold.

The collector was placed, glazing down, on a pair of sawhorses. (It is best to put two 2 × 4-in. pieces across the horses first. These were not used here because they would be in the way of taking the photographs.) The screws were then removed with a Phillips screwdriver (Figure 14.13), and the back was slipped off (Figure 14.14).

Figure 14.15 shows the back insulation rolled up into one end of the collector. Figure 14.16 is a shot of the back of the absorber plate with the insulation removed and set aside. Notice in Figure 14.16 that the edge insulation is compressed into the sides of the outer box and held in place by the absorber plate, which pinches it against the sides. The absorber plate itself is screwed to a series of standoffs, with a thermal break mounted between the standoff and the absorber plate. The arrow in Figure 14.17 points to one of these standoffs.

Figure 14.14 Slipping the back off.

Figure 14.15 The insulation rolled into one end of the collector.

Figure 14.16 The back of the absorber plate. Notice the way the edge insulation is held in the collector.

Figure 14.17 The absorber plate is mounted to standoffs with a thermal break between them.

Figure 14.18 Removing the glazing frame screws.

Figure 14.19 The third screw up from the bottom holds the glazing frame on the collector box.

In Figure 14.18 the collector has been turned face up on the sawhorses, and the screws that hold down the glazing frame are being removed. It is best to loosen all the screws slightly before removing them. This relieves any tension on the glazing frame.

Figure 14.19 shows the end of the collector. In this particular model there are four screws holding each corner together. The bottom two screws hold the collector box together. The third screw up from the bottom, which is shown being removed, holds the glazing frame on the collector, and the top screw holds the glazing frame together.

The four corner screws holding the glazing frame onto the collector were removed and the corners of the glazing frame were gently pried upward to loosen them by using a screwdriver at the edge of the frame leveraged against the manifold (see Figure 14.20). The sides and ends were then hand loosened and, as shown in Figures 14.21 and 14.22, the

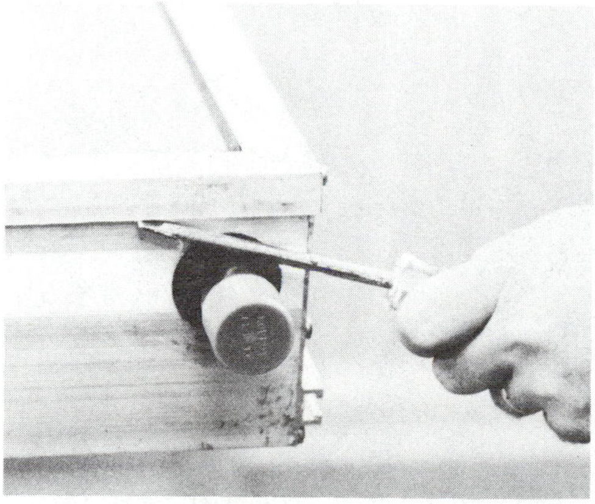

Figure 14.20 Pry the corners up carefully to loosen the glazing frame.

Figure 14.21 Hand loosen the sides of the glazing frame.

Figure 14.22 Slide the glazing frame off the collector.

glazing frame was slid over the side and placed in a horizontal position against the collector.

The glazing jig, as shown in Figures 14.23 and 14.24, was placed on the top of the collector and the glazing frame was placed into the jig. Then the four screws holding the glazing frame together were removed. You can see one of these screws in Figure 14.24.

Figure 14.23 Place the glazing frame in the jig.

Figure 14.24 Remove the corner screws from the glazing frame.

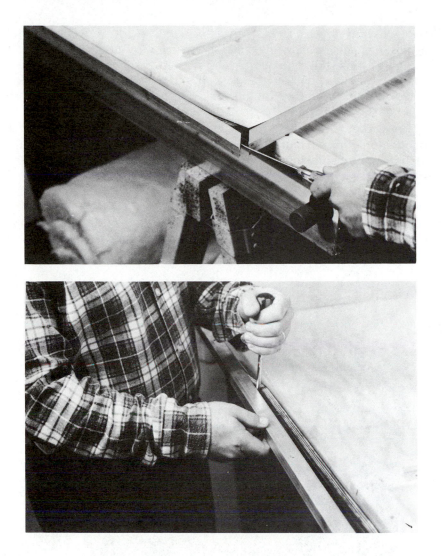

Figure 14.25 A small screwdriver can be used to pry the corner of the glazing frame carefully apart.

Figure 14.26 A smooth object such as a large screwdriver can be inserted between the gasket and the frame to allow the frame to be pulled off by hand.

To remove the glass from the glazing frame (see Figures 14.25 and 14.26), a small screwdriver was placed in one corner joint and the joint was pried open. *It is very important not to exert pressure on the edge of the glass.* Once the four corners had been loosened, the sides and ends of the frame were carefully pulled off using a large, smooth screwdriver shank between the gasket and the glass to allow the aluminum strips to

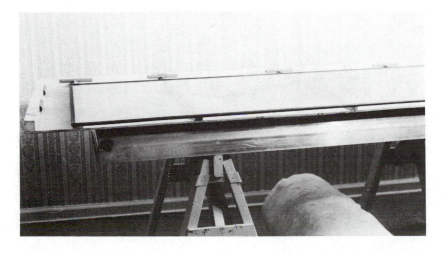

Figure 14.27 Here is the glazing with the gasket still in place resting on the jig.

be pulled away. *Again, it is very important not to exert high pressure against the glass or allow the metal screwdriver to rest directly against it.*

Figure 14.27 shows the glass, with the gasket still attached, resting on the jig. From this point on, the glass should only be handled on the long sides and not picked up by the end. (Do not pick the glass up in the manner shown in Figure 14.28.)

The gasket is not glued to the glass and can readily be removed by hand in one piece (Figure 14.29). Once the gasket is removed, the glazing is slipped tightly into the jig and the jig is reassembled. Figure 14.30 shows the assembled jig containing the glass.

Figures 14.31, 14.32, and 14.33 show the absorber plate being removed. The screws holding the plate to the standoffs are removed (see Figure 12.31), the two lower screws at the corners of the enclosure are removed (see Figure 14.32), and one side of the box is slipped off the manifold ends (see Figure 14.33).

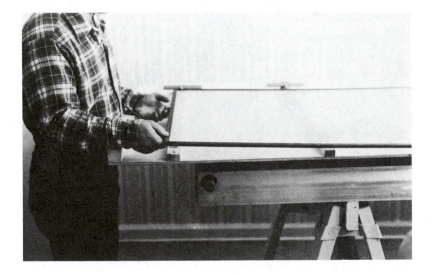

Figure 14.28 Do not pick up the glass by the ends after the frame is removed.

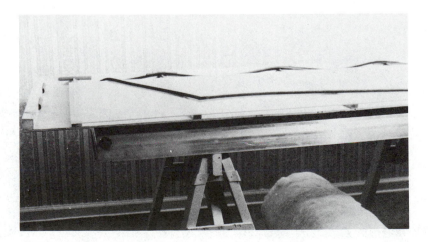

Figure 14.29 Remove the gasket by hand in one piece.

Figure 14.30 The assembled jig with the glazing stored in it.

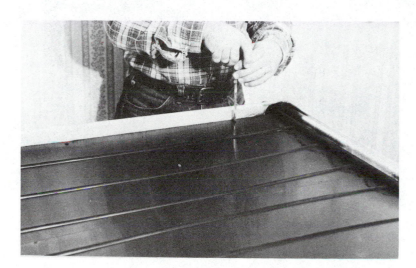

Figure 14.31 Remove the screws holding the absorber plate to the standoffs.

Figure 14.32 Remove the two lower screws left at the corners on one side.

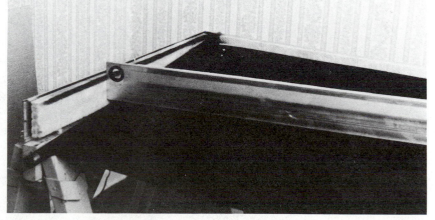

Figure 14.33 Slip the side rail off the manifold ends.

Repair the Collector **273**

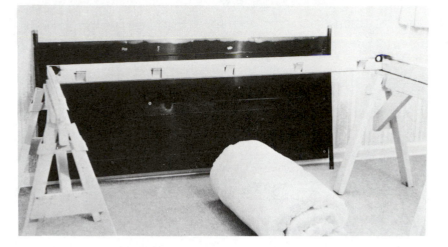

Figure 14.34 Slip the absorber plate out of the other side rail.

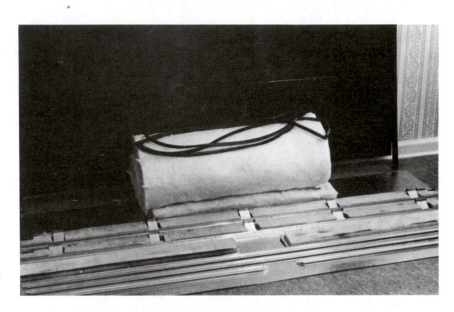

Figure 14.35 Here are all the parts except the glazing.

Figure 14.36 The edge insulation and the manifold grommet can now be removed from the side rails.

Figure 14.37 A close-up of the edge insulation, standoff, and manifold grommet.

The manifold ends were then slipped out of the other side; the absorber plate was removed and stood against the wall (see Figure 14.34). The four screws holding the rest of the collector box together were removed from the corners and disassembly was completed. All the parts except the glazing are shown in Figure 14.35.

Look at Figure 14.36. This shows the edge insulation and the manifold grommet removed from the side rail. Figure 14.37 shows a close-up of the insulation, the standoff with its thermal insulator, and the manifold grommet.

Some Other Collector Designs

Not all collectors disassemble this easily. Wedge locks may be substituted for screws. A wet silicone sealant may have been applied during construction for a primary weather seal. The back of the collector may not be easily removed. Finally, the glazing may not be mounted in an independent frame.

These different constructions can be dealt with fairly easily. Figure 14.38 shows a collector assembled with wedge locks and wet silicone sealants. First, look at the back of the collector. When the collector was assembled, a bead of silicone sealant, represented by the dotted area, was placed in the rail's lower channel. The back was placed down over the sealant, and wedge lock angles were inserted into the side rail channels to hold it tight. It is impractical to remove this collector's back.

Now look at the way the absorber plate is held in the collector. It lies on top of the insulation and is held away from the enclosure by a thermal strip. Only the manifold grommets hold this collector plate in place. It is not screwed down.

The glazing was installed by placing a gasket around the glass and setting the glazing into a channel on the side and end rails. Then an ell-shaped extruded aluminum top cap was placed over the glazing gasket, and wedge locks were slipped at intervals into the top side rail channel to lock down the top cap. The wedge locks are shown in solid black on

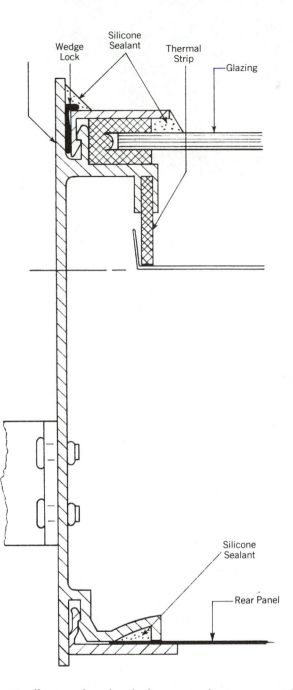

Figure 14.38 A cross section of a collector assembled with wedge locks and wet silicone sealants.

Figure 14.38. Finally, two beads of silicone sealant were applied to seal the unit. One bead covers the wedge lock channels, and the other bead seals the glass to the top cap.

To remove the glazing in this unit, cut away the silicone sealant in the wedge lock channel with a sharp knife to expose the wedge locks. With a pair of pliers, draw the wedge locks out and remove the glazing still attached to the frame. Place the glazing with its frame on the jig. Cut away the sealant between the glass and top cap. Remove the top cap and the gasket.

To remove the absorber plate in this unit, disassemble one side by removing the corner screws and prying the rear cover out. Then slip the absorber plate out. This exposes the insulation for easy removal.

When reassembling the collector, you will have to replace the wet sealants with one-part silicone sealant, which is readily available at most hardware stores. Silicone sealants mend easily, and the new sealant will stick to the old sealant with no problem.

Figure 14.39 shows a construction where the back is inserted in a channel and riveted to the side rails. Again, this back is not removable. The absorber plate rests on a solid foamed insulation and is held in place by the manifold or header pipes where they exit the housing. The glazing, with its gasket, rests on the top of the side rails and is held in place with a top cap that is screwed tightly to the top rail.

To remove the glazing, take off the top cap and transfer the glazing, with its gasket, to the jig. Disassemble one side by removing the corner screws to slip out the absorber plate. The insulation can be replaced once the absorber plate is removed. Stay with factory-approved insulation because it must support the absorber plate.

Figure 14.40 shows a collector where the corners of the enclosure are welded together and cannot be disassembled. The back plate of the collector is an asphalt impregnated sheet attached to a solid, closed-cell, urethane insulation. A 1-in.-thick fiberglass insulation is placed on top of this insulated back.

In this collector all disassembly proceeds from the face of the collector. Remove the top cap screws and transfer the top cap, with the glazing attached, to the jig. Carefully cut away the silicone sealant and remove the top cap. The gasket can now be removed and the glass slipped into place in the jig.

The absorber plate rests on the insulation and is held in place by the header pipes. These pass out the collector through notched openings in the enclosure. A notch grommet is used to hold these pipes. To disassemble, merely carefully lift the absorber plate out of the box along with the grommets.

The closed-cell urethane insulation would rarely ever need replacing. With the absorber plate removed, the fiberglass insulation can easily be replaced.

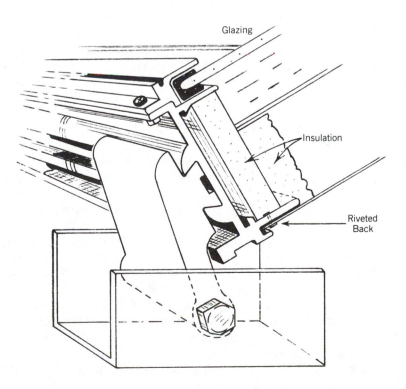

Figure 14.39 A collector with a riveted back that cannot be removed.

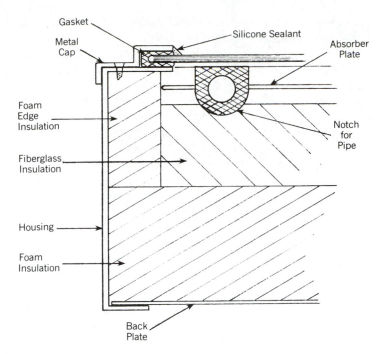

Figure 14.40 A collector box that cannot be disassembled except at the top cover.

When reassembling this style collector, carefully replace the silicone sealant with new one-part silicone sealant.

Collector Repairs

Servicepeople must be aware of what can be repaired and what must be replaced in a collector. They need to know where the materials can be obtained and what materials may be substituted if the manufacturer no longer supplies them or cannot be located.

Glazings

Tempered glass glazings cannot be repaired once damaged. In most cases they will be shattered into thousands of tiny pieces that must be carefully removed from the collector. A shop vacuum cleaner can be helpful in this operation, but one with a paper vacuum collection bag should not be used.

The replacement of glass glazings is relatively simple. Tempered glass is readily available from local suppliers as well as from solar manufacturers. Since the sizes are patio door sizes, the glass is often stocked locally. However, there is a difference in light transmission from one grade of tempered glass to another, and care must be taken to replace the glazing with the same grade and thickness if the original performance is to be restored.

We recommend the use of Solite from AFG Industries, Inc., in Kingsport, Tennessee. Solite is supplied in $\frac{1}{8}$-in., $\frac{5}{32}$-in., and $\frac{3}{16}$-in. thicknesses.

The replacement glass must be the same thickness or thicker than the glass being replaced. The collector designer very carefully chose a certain thickness based on the physical strength of the glass as well as other qualities. Thickness can be determined by examining the gasket originally used with the glass. If you increase thickness, be certain that you obtain the proper gasket.

If you have problems obtaining gaskets locally or from the original manufacturer, one of the leading solar glazing gasket manufacturers is the Pawling Rubber Corporation, 157 Maple Boulevard, Pawling, New York 12564; telephone (914) 855-1000.

Several different types of thermoplastic glazings are used in flat plate solar collectors. The most popular ones are rigid glass-reinforced polyester sheeting (Filon, Vistron Corporation); polycarbonate sheeting (Lexan, General Electric); and polyvinyl fluoride flexible sheeting (Tedlar, E. I. duPont). Some polymethyl methacrylate rigid sheeting (Plexiglas, Rohm & Haas) is also used.

These materials vary greatly in their properties from grade to grade; the collector manufacturer should be consulted before reglazing with a different material.

Glazing Gaskets It is wisest to replace the glazing gasket when replacing the glazing if the collectors are more than a few years old. Rubbers become brittle and lose their resiliency in the solar environment. It is hard to get a good seal with an old gasket. Generally, either EPDM or silicone elastomers are used.

Wet Glazing Sealants A silicone sealant should be the only glazing sealant used on solar collector glazings. The silicone sealant weathers well and has a long life; it also adheres well to glass. The glass and the metal must be clean, dry, and free of oxide coatings to achieve the best results. Silicone sealants are self-healing and, if the surface is fresh and clean, the new sealant will bond extremely well to the old sealant.

Absorber Plates

If the absorber plate is leaking from its fluid passages, it should be repaired or replaced. The most common cause of absorber plate leaks in nonheat-exchanged systems is freeze damage caused by water freezing in water tubes and bursting one or more tubes or causing solder joints to fail.

Freeze-damage cracks can often be repaired with silver solder or 95/5, tin/antimony solder. It is even possible to cut out a damaged riser section and replace it with a short section of copper tube.

Before soldering a burst collector riser tube or a leaking header tee joint, clean the area to be soldered with a power-driven wire wheel. The copper should have a highly polished shine before any heat or flux is applied. Dirt in the cracked tubing or in the factory-soldered joint may keep solder from adhering. If the wire wheel does not leave a clean open gap for solder to flow into, a hacksaw or jab saw may be used to create a shallow depression in the metal over the leak that solder will fill by capillary action. 50/50 solder is generally not strong enough for permanently filling the gaps in a burst tube. 95/5 solder or silver solder are recommended in most cases. Wet rags can be used to prevent softening of surrounding factory-soldered joints.

Extensive collector corrosion is generally caused by acidic water, degraded heat transfer fluid, or excessive fluid velocity. In heat-exchanged systems, assuming no freeze damage, most leaks in fluid passages are caused by internal corrosion. A leak in one place generally indicates that leaks will appear shortly in other locations. The serviceperson should try to determine the cause of the corrosion before proceeding with repairs so that future problems will not arise.

To determine whether corrosion is localized or general, inspect the header or riser tube at several locations. If evidence of generalized corrosion is not seen, proceed to repair the leaks instead of replacing the absorber plate.

If the fins have pulled away from the fluid tubes, repairs can often be effected. This problem is generally not seen in well-manufactured plates but may be seen in older systems. If any soldering is done in making these repairs, be certain that the fluid passages have been drained, that there is no pressure in the system, and that tin/antimony solder is used. It is best to be very cautious about using silver-soldering techniques, because the absorber plate was generally manufactured by silver-solder brazing. The heat needed for repairs could loosen the joints between the risers and the headers. A mechanical repair could also be made in many systems by using pop rivets and banding to reattach the risers to the fins. Place a heat transfer cement between the tube and the fin if necessary to increase the heat transfer across the repaired section.

Absorber plates are usually standard items. Generally, a new plate can be obtained that will fit the old configuration. Two of the more active manufacturers in the absorber plate business who can supply replacement absorbers are as follows.

- Terra Light Corp.
  54 Cherry Hill Drive
  Danvers, Massachusetts 01923
  (617) 744-5452
- Thermatool Corporation
  280 Fairchild Avenue
  Stamford, Connecticut 06902
  (203) 357-1555

If the surface coating has peeled away from the fins, a quick repair can be effected using standard spray paint. Use a selective or flat black, high-temperature paint recommended by the absorber plate manufacturers.

Selective surfaces cannot be repaired on the job. Loss of color, leaving a copper-colored absorber plate, would be indicative of such damage. If the damage does not warrant removing the collector or replacing the absorber, spray paint the damaged areas.

Damage to manifolds and to headers is often mechanical and can be repaired using standard plumbing techniques. Make certain that the system is drained and all pressure relieved before doing any soldering on the collector. Use tin/antimony solder or low-temperature silver solder, and be careful about loosening other joints in the collector.

Insulation

Damage to insulation will be obvious and consist of movement, slumping, or wetness. If the insulation must be replaced, use high-temperature, low-binder-content fiberglass materials or foil-faced isocyanurate. Information on fiberglass materials is available from Owens-Corning Fiberglass Corporation, Fiberglass Tower, Toledo, Ohio 43659; telephone (419) 248-8000. Information on isocyanurate materials is available from Celotex Corporation, Building Products Division, Insulation Group, Box 22602, Tampa, Florida 33622; telephone (813) 871-4811.

Collector Box

Collector boxes that screw together in the corners usually require a sealant on the inside of the corner to prevent leakage. Silicone one-part sealants are generally used. The metal should be carefully cleaned to obtain good adhesion. The sealant tubes come with a long nozzle that allows the sealant to be applied easily to the entire length of the joint.

Most collector boxes are designed to breathe. When collector boxes cannot breathe readily, a sudden thermal shock, such as cold rain on a hot collector, can quickly cool the collector and cause a vacuum within it. This could place high stress on the glazing and break it. The collector that was disassembled breathed because the back of the collector was not assembled using a sealant. The other collectors discussed contained small weep holes in the bottom rail down close to the back of the collector. Check the collector that you are working on and be certain that it can breathe. Do not make any new holes to enhance breathability unless it is apparent that the collector glazing shattered because of thermal shock. If you add weep holes, be certain to locate them where rain or snow cannot drip or wick into the collector.

If the collector box is unanodized aluminum and if the air at the location contains corrosive materials, the enclosure may become badly pitted with time. Little can be done to stop this pitting. Paints will not adhere well to corroded aluminum and, in many instances, the pitting will continue under the paint. A thin, transparent, protective coating of the type sold for spraying pitted aluminum doors and storm windows may help arrest further pitting without peeling off.

If the pitting is excessive, carefully check the mountings to make certain that the collector will remain safely fastened down. If new mountings are indicated, it is usually best to attach suitable mounting brackets with stainless steel pop rivets. Stay well away from the edges of the glazing and the absorber plate. If you accidentally drill into the edge of the glazing, it may shatter. Drilling into the absorber plate may cause it to leak.

There may be instances where you have to cut into the back of the collector box to replace insulation or mount an internal controller sensor. A 0.032-in.-thick aluminum sheet, riveted in place and lap sealed with silicone sealant, will generally effect a good repair.

CHAPTER
15

Replace the Heat Exchanger and Transfer Fluid

In general, a damaged heat exchanger cannot be permanently repaired at the job site and must be factory rebuilt or replaced. On the other hand, heat transfer fluids can generally be tested on site and replaced if necessary.

Replacing Heat Exchangers

The seven classes of heat exchangers that exist in residential solar water heaters are plate around tank, tank around tank, cascade over tank, internal-flued tank, coil in tank, external separate, and air collector-to-liquid storage exchangers.

When the Exchanger Is Part of the Tank

It is usually necessary to replace both the tank and the exchanger when the exchanger is part of the tank. This is a major repair job. It should be preceded by an analysis as to whether the system was correctly matched to the collector array and to the load. If it was, an exact replacement from the same manufacturer will be easiest to install.

If the system is one of the early double-wall plate-around-tank or tank-around-tank systems, or if it is a cascade-over-tank or internal-flued tank system, do not attempt to substitute an identical exchanger/tank system. The product may be ineffective, generally unavailable, not easily obtained in your area, or more expensive than an easily obtained alternative.

Two characteristics must be checked before making an exchange; the heat-exchanger pressure limits and the heat-exchanger/heat transfer fluid compatability. The new system must be able to withstand the system design pressures, and the metals must be compatible with the heat transfer fluid.

The two alternative systems that will work as substitutes in almost all systems are the coil-in-tank system or the external separate heat-exchanger system. Generally, you should consider the coil-in-tank system for existing one-pump systems and the external separate heat-exchanger system for existing two-pump systems.

Coil-in-tank replacements are available nationwide from suppliers carrying Vaughn, Ford, and A. O. Smith water heaters. Separate heat exchangers and standard glass-lined storage tanks are generally available from almost all plumbing and air-conditioning suppliers.

Plate-Around-Tank Substitution   There are two types of plate-around-tank systems: single-circuit systems with one pump and double-circuit systems with two pumps.

Use a coil-in-tank replacement for the single-circuit plate-around-tank system, because this system was generally inefficient. A properly chosen coil-in-tank replacement should increase system heat-exchanger efficiency.

Several sizes of both tank and exchanger coil are available. Choose a tank that has between 1.2 and 2 gal of storage per square foot of collector. Follow the manufacturer's recommendations in sizing the heat exchanger.

If the exchanger/tank system being replaced was double circuited with two pumps, a separate heat exchanger and if necessary new tank can be substituted with few modifications.

Choose a tank that has between 1.2 and 2 gal of storage per square foot of collector. Choose a heat exchanger that has an effectiveness of at least 50% as determined by checking the manufacturer's specifications against the general effectiveness formula or by working with the manufacturer's engineers or distributor personnel to arrive at the proper size and rating.

Before making the final choice, check the specifications for the pumps and be certain that the present pumps will handle the desired flow rate.

You can substitute a separate exchanger/tank system in a single-circuited system by installing an additional pump. The pump will go in the exchanger/storage loop. It should be electrically connected in parallel with the collector/storage pump so that both turn on and off at the same time. Check the controller specifications to make certain that the 120-V controller relay will handle the additional load.

In all substitutions check to see if a double-wall exchanger is required. Use the heat transfer fluid guidelines developed in Chapter 10 to determine this.

Tank-Around-Tank Substitution   Tank-around-tank systems will almost always be single circuited. Therefore, the coil-in-tank system should be considered first. Substitution guidelines are the same as for a plate-around-tank system.

Cascade-over-Tank Substitution   The cascade-over-tank system was withdrawn from the market shortly after its introduction because of excessive corrosion problems. Very few exist. The original cascade-over-tank system used a steel tank and aluminum collectors. No substitutions should be made until the collectors are checked to see if they contain aluminum water passages. If the collector water passages are aluminum, consult the collector manufacturer before making any substitutions.

The bulk of all aluminum absorber plates was supplied by The Reynolds Metals Company of Richmond, Virginia, and the Olin Company, Olin Brass Division, of East St. Louis, Illinois. Reynolds supplied the entire collector; Olin sold absorber plates to collector manufacturers. Both manufacturers have discontinued their aluminum absorber plate lines.

The cascade-over-tank system is not a closed system. It is a drainback system, and this must be taken into account when substituting. Again, the best substitutions are the coil-in-tank and the separate exchanger/tank systems. However, the pump is usually not sized correctly for these substitutions and may also have to be changed.

Internal-Flued Tank Systems The internal-flued tank system has also been withdrawn from the market. Either go with a good draindown tank or with a heat-exchanged system, depending on how the collector array is designed.

Some internal-flued tank systems used steel collectors. Copper heat exchangers may not be suitable for these systems. They may cause corrosion within the passages of the steel collectors. Consult with the collector manufacturer before substituting a copper heat exchanger.

Coil-in-Tank Systems Replacement of the current coil-in-tank system with an equivalent-sized coil-in-tank system is usually best. A separate exchanger/tank system and a second pump can also be substituted.

External Separate Heat-Exchanger Systems It is unusual for both the tank and the exchanger to go bad in an external separate heat-exchanger system. Generally, the exchanger is copper and the tank is glass-lined steel. With time, the tank will develop corrosion without hurting the exchanger.

Glass-lined tank replacements are satisfactory. The size should be in the range of 1.2 to 2 gal/ft^2 of collector. A premium-grade tank with a large magnesium anode should be chosen. The piping inlets and outlets should be in roughly similar locations to minimize any piping changes.

In the case of a drainback system with a separate reservoir for the collector fluid, also inspect the reservoir tank for corrosion and replace as needed.

In all cases, if a steel exchanger was used, it may be necessary to replace it with another steel exchanger. Before switching to a copper exchanger, evaluate the other metals contained in the collectors and the storage tank. A copper exchanger will speed the corrosion of both aluminum and iron.

In all cases of exchanger replacement, carefully evaluate the need for a double-walled exchanger as indicated by the choice of heat transfer fluid.

Air Collector-to-Liquid Storage System Again, these systems generally contain copper water passages in the exchanger and glass-lined steel tanks. The tank is usually the replacement item. Handle this in the same manner as the replacement tank in the separate exchanger system.

If the exchanger needs replacement, use a similar exchanger or one that is rated for the same or better effectiveness.

Heat Transfer Fluid Replacement

Chapter 10 discussed the various heat transfer fluids. It should have become clear from that discussion that *one heat transfer fluid is not substituted for another without very carefully considering the effects of the fluid on the entire system.*

Chapter 10 gave detailed coverage of fluid evaluation, inspection, and testing. It also covered methods of draining and filling the loop when replacement was necessary.

Proceed to replace the heat transfer fluid by following the guidelines established in Chapter 10.

CHAPTER
16

Repair
the Pump

The easiest way to learn about pump repair is to take some pumps apart. This chapter will show how a magnetic drive pump and a canned wet-rotor pump are disassembled and will present a troubleshooting chart that will help to isolate any pump problems quickly.

The Magnetic Drive Pump

Figure 16.1 shows a small magnetic drive pump of the type normally found on the heat exchanger/storage loop. Four long screws, shown by the arrows in Figure 16.2, hold the pump body to the motor. When the screws are removed, the pump can be separated into two parts, the motor and the pump. Figure 16.2 shows the pump; Figure 16.3 shows the motor complete with its magnet. Note the motor cooling fins on the rear of the circular magnet.

With the screws removed, the pump impeller unit can be removed from the brass pump body, as shown in Figure 16.4. There is an O-ring seal between the two that is seen laying on top of the pump body in Figure 16.5.

Figure 16.6 shows the impeller being removed from the impeller housing; Figure 16.7 shows the impeller and its attached magnet laid beside the impeller housing. The impeller and its magnet ride on the small shaft in the center of the housing. In Figure 16.8 the impeller housing has been turned over to display the interior of the housing where the motor magnet rides. Figure 16.9 shows all the parts of the disassembled pump.

The Canned Wet-Rotor Pump

Figure 16.10 shows a typical canned wet-rotor pump. The pump body is on the right, and the pump motor is on the left. The arrow on the pump body points to the direction of fluid flow. The electrical connections and the starting capacitor are located in the small box at the lower left on the underside of the motor.

The pump body is mounted to the motor frame with four bolts. The arrow in Figure 16.11 points to one of these bolts.

Figure 16.1 A small magnetic-drive pump.

Figure 16.2 The pump case with the motor removed. The arrows point to the four screws that hold the motor to the pump volute.

Figure 16.3 The motor with its drive magnet. Notice the motor cooling fins at the rear of the magnet.

Figure 16.4 The pump impeller unit is removed from the pump volute.

Figure 16.5 An O ring seals the impeller unit to the pump volute.

Figure 16.6 The impeller being removed from the impeller housing.

Figure 16.7 The impeller and its attached magnet.

Figure 16.8 This is the interior of the housing. The motor's drive magnet rides in here.

Figure 16.9 All the parts of the pump laid out in order.

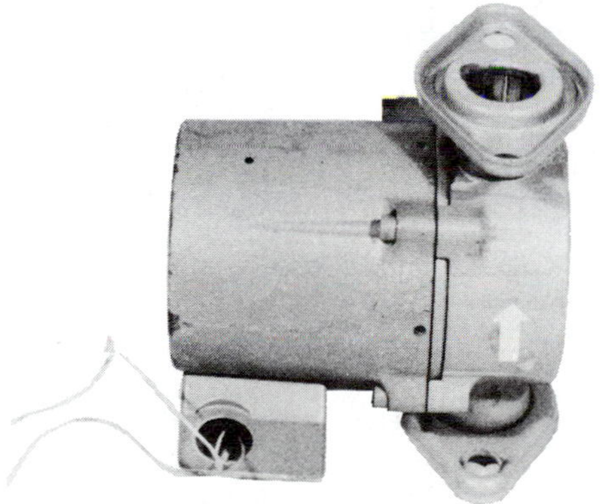

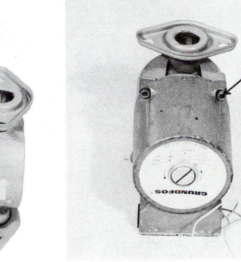

Figure 16.10 This is a canned wet-rotor pump.

Figure 16.11 The motor is attached to the pump volute with four bolts. The arrow points to one of them.

Figure 16.12 shows the bolts being removed with an Allen wrench. Once they are removed, the pump body frame can be pulled off, as shown in Figure 16.13, and the pump volute can be separated from the motor frame. In Figure 16.14 the pump volute is the piece in the center with the two flanges.

The arrow in Figure 16.14 points to the pump impeller, which is pressed into the motor rotor. This unit can now be easily slid out of the motor body. Figure 16.15 shows the rotor unit next to the pump body. In Figure 16.15 the arrow points to the gasket, which seals the pump volute to the motor body.

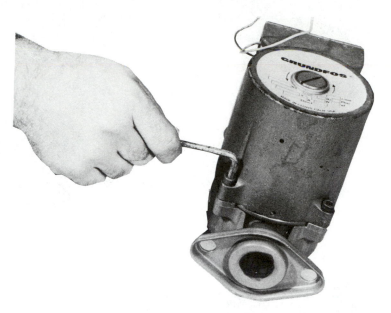

Figure 16.12 An Allen wrench is used to remove the bolts so that the pump can be disassembled.

Figure 16.13 The pump body frame has been removed.

Figure 16.14 The pump volute has been separated from the motor frame. The arrow points to the impeller.

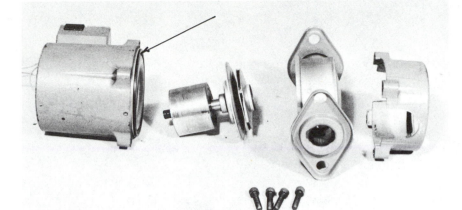

Figure 16.15 Here are all the parts of the pump. The arrow points to the gasket that seals the volute against the motor body.

Figure 16.16 The motor case turned 90° to show the end bearing for the motor shaft.

Figure 16.17 A bearing inspection plug, sealed with an O ring, is located on the other end of the motor case.

Figure 16.18 A plain flange with a pipe thread. This type of flange has no shutoff valve.

In Figure 16.16 the motor case has been turned 90° and propped up on a pump flange so that you can see the end bearing for the rotor unit's shaft. In Figure 16.17 the other end of the motor case is shown with the bearing inspection port plug removed. Note the O ring that is used to seal the plug.

Figures 16.18 and 16.19 show two different types of pipe flanges that

Figure 16.19 An isolation flange. This type of flange contains a ball valve that may be closed by turning it with a screwdriver where the arrow is shown.

Figure 16.20 Here the ball valve is open and fluid can flow.

Figure 16.21 Here the ball has been almost closed to demonstrate its action.

are commonly used with this type of pump. Figure 16.18 shows a plain flange with a pipe thread. The flange is sealed to the pump frame by placing a rubber gasket between them.

Figure 16.19 shows a flange containing an isolation valve. The flange is mounted to the frame in the same manner. However, the flange contains a ball valve that can be closed or opened by using a screwdriver in the slot at the arrow.

In Figure 16.20 the ball valve is open and a clear path exists for fluid flow; in Figure 16.21 the ball valve has been almost closed to demonstrate its action.

Troubleshooting

Figure 16.22 represents a simplified wiring diagram of the pump circuit. The electric service provides a fused 120 V at point 5. One wire goes directly to the pump and the other wire goes through the relay switch of the controller, which is shown located between points 3 and 4. (In some cases, the relay switch may open both wires.) The wiring then goes to the pump motor at point 2.

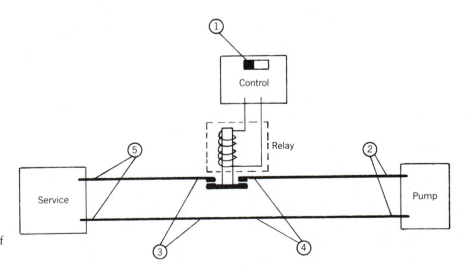

Figure 16.22 A simplified wiring diagram of the pump electrical circuit.

Figure 16.23 is a troubleshooting chart for this circuit. Proceed to troubleshoot the pump as follows.

1. Set the controller switch to *manual.*
2. If the pump turns on and fluid flows, the pump is okay. Switch the system to automatic and troubleshoot the control system.
3. If the pump turns on but no fluid flow is developed, listen for motor hum. If the motor does not hum or heat up, go to step 7.
4. If the motor hums or heats up, there is a mechanical problem inside the pump volute or the piping. The pump must be repaired or replaced, or a blockage in the piping must be eliminated.

If the pump is a large mechanical seal pump where the disassembly of the pump motor, shaft, and impeller assembly from pump volute is complex, check for piping problems first. If disassembly is straightforward, proceed as follows.

- Close the pump isolation valves.
- Switch off the power to the pump.

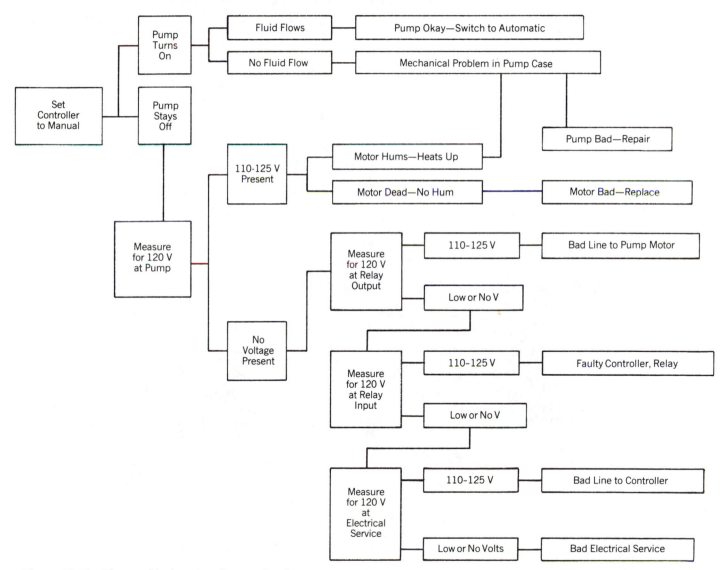

Figure 16.23 The troubleshooting diagram for the pump circuit.

- Take the pump motor with its attached impeller out of the pump volute. Place the motor on a dry, insulated surface such as a chair seat.
- Momentarily apply power to the pump and check to see if the pump impeller is turning freely.

5. If the impeller will not spin freely, repair or replace the pump motor.

6. If the impeller spins freely and no probable cause for the lack of fluid flow is found, reassemble the pump motor, impeller, and volute. Open the pump isolation valves and purge air from the pump as needed. Recheck the piping for the following:

- Closed valves.
- Air plugs in the piping's high points.
- Crushed pipes.
- Foreign matter in the strainer or the piping.
- Gate or check valves "cold welded" shut by corrosion or foreign matter.

7. (Proceeding from step 3.) Turn the controller to off, open the electrical box on the pump, and place either a neon line voltage tester or a voltmeter between the two wires. Figure 16.24 shows a voltmeter connected in this manner. *Take care that the two wires do not touch each other in this step.* With the voltmeter or the neon line voltage tester attached, turn the controller switch to manual and observe the light or measure the voltage.

8. If the neon light glows or if the voltmeter shows 120 V, perform steps 4 to 6.

9. If no voltage is present, the voltage across the line at the output of the relay must be measured with the voltmeter or the neon line voltage tester. The relay is located in the controller. Figure 16.25 shows the voltmeter attached across a typical relay output.

10. If 100 to 125 V are present with the controller switched on manual, the line running between the pump and the controller relay output is bad and should be repaired or replaced.

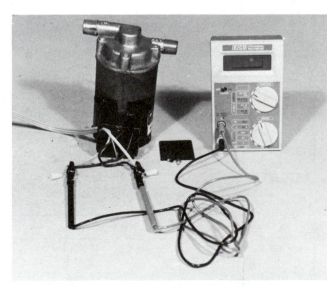

Figure 16.24 How the voltmeter or neon line voltage tester connects across the pump circuit at the pump's electrical box.

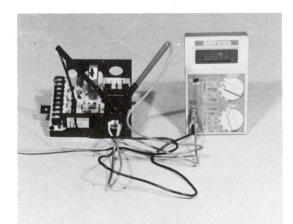

Figure 16.25 How the voltmeter or the neon line voltage tester connects across the pump circuit at the relay's output terminals.

11. If just stray voltages (a few volts) or no voltage is measured at the relay output, the voltage across the lines at the input to the relay should be measured. This measurement is made where the wires from the electrical service enter the relay circuit. In Figure 16.25 these would be the two terminals just above the relay output wires. Normally, these will be marked *ac input*.

12. If the voltage at the relay input measures 110 to 125 V, the controller or the relay is faulty. Proceed to troubleshoot the control system.

13. If there is no voltage at the relay input, measure the voltage across the electrical service or at the plug in the cases of a plug-in system (point 5).

14. If 110 to 125 V are measured across the electrical service or plug, the wires to the controller relay are faulty and should be repaired or replaced.

15. If there is no voltage across the electrical service or the plug, the electrical service must be repaired.

This completes the checkout of the pump.

Repairing or Replacing the Pump and Motor

The repairs of the pump and motor differ with the style and type of pump. Most manufacturers will supply repair information upon request. Generally, there are three major classes of pump problems to consider.

Noisy Pump Operation

1. **Air trapped in pump.** Open the vent port to bleed air. Check the automatic air vents for proper operation.

2. **Loose impeller.** Tighten the impeller bolt or replace the impeller/shaft assembly.

3. **Foreign matter or corrosion in volute (impeller housing).** Open and clean the volute and impeller. Check the clearance between the volute and impeller. Replace the impeller assembly (and motor in a canned-rotor pump) and/or the volute as needed for proper clearance. If the pump is cast iron and shows corrosion, consider replacing it with a bronze or stainless steel pump.

4. **Dry or worn bearings or seals.** With the electric power off, check that the impeller can be rotated freely by hand. Lubricate the bearings if possible. Replace bearings or seals as needed in the mechanical shaft seal pump if it is still noisy. Replace the motor and impeller assembly in a canned wet-rotor pump if necessary.

If it seems that the pump has been run dry, eliminate the cause before repairing the pump. Recheck for air trapped in the pump.

Circuit Breaker Trips on Start-Up

1. **Short circuit in controller wiring.** Disconnect the pump from the supply wires at point 2 and cap each wire with a wire nut. Check to see if the circuit breaker still trips when the controller is in manual *On* or automatic position. If so, troubleshoot the controller and supply wiring.

2. **Short circuit in pump. Disconnect the pump from the power circuits for safety before making these checks.** Pump wiring or windings are shorted or grounded. Check for signs of broken insulation and replace any questionable wiring. Check the wiring inside the pump electrical box.

   Finally, check for an internal pump winding short as follows with a voltohmmeter (VOM) set to the lowest-resistance scale.

   - Connect the VOM across motor supply leads coming from the pump.
   - If resistance is 0, the pump is shorted. Replace the pump motor. (An infinite resistance would indicate that the winding was open, but this condition would not trip the circuit breaker.)
   - A resistance reading of 4 to 100 Ω may be normal, depending on the pump under test. See the manufacturer's information or compare with a new motor. If the resistance is low, replace the pump motor.

3. **Pump winding shorted to ground. Disconnect the pump from the power circuits for safety before making these checks.** Set the VOM to the highest-resistance scale. Connect one VOM lead to one motor supply lead and connect the other lead to the body at the frame grounding screw or some other bare metal location. The resistance should be infinite. Repeat the test using the other motor lead and ground. Again, the resistance should be infinite. Replace the motor if the resistance reads other than infinity.

Circuit Breaker Trips After Short Delay Following Start-Up

1. **High-resistance short to ground.** Delayed tripping may indicate a high-resistance short to ground that may not have appeared when you tested for grounded windings. Repeat the test. As a last resort, replace the motor and see if the problem stops.
2. **Arcing in pump motor.** Place a clamp-on ammeter on the pump lead. Start up the pump and measure the amperage. Compare the results to rated amperage. Replace the motor if amperage is excessive. As a last resort, replace the pump and see if the problem stops.
3. **Circuit overload.** The pump circuit or the building circuits may be overloaded. If the circuit breaker continues to trip:

   - Test the pump on a separate circuit.
   - Test the controller on a separate circuit.

   If both the pump and the controller work on a separate circuit, have an electrician check the building's wiring.

If the decision is made to replace the pump, make certain that the pumping curves for the two pumps are compared and that the new pump will handle the required fluid flow. If a pumping curve for the old pump is unavailable, perform the flow rate and head loss calculations shown in Chapter 11 before making a final replacement pump choice.

CHAPTER 17

Repair the Controls

The solar water-heater system's controls consist of a solid-state electronic differential temperature controller, one or more 120-V relays, and two or more temperature sensors.

Chapter 11 explained what the system controls do, the controller's logic, how the controller works, what the controller system consists of, and control system installation considerations.

This chapter provides a troubleshooting guide for the control system and explains how to use it.

The Pump/Control System Troubleshooting Procedure

The control system and the devices that it controls, such as the pump and any valves, must be treated as a unit in troubleshooting. Here is a sequential order of troubleshooting that usually works best.

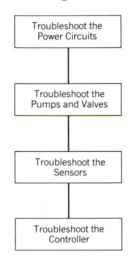

Troubleshooting the pump/control system requires that the serviceperson know how to

- Use a VOM or multimeter.

Figure 17.1 A VOM that uses an ammeter dial for read out.

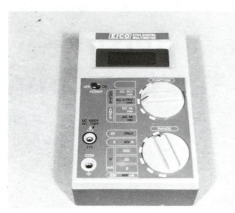

Figure 17.2 A multimeter that uses a digital display for read out.

- Measure sensor resistance and convert the resistance to temperature.
- Check out controllers with a sensor substitution box.

The Multimeter or Voltohmmeter

In Chapter 16 you used a **voltohmmeter,** or VOM, to check the voltage in the pump circuit. A voltohmmeter is an instrument that will measure the voltage, resistance, and current in an electrical circuit.

There are two basic types of voltohmmeters. The older type displays the results by moving a pointer over a calibrated dial. Figure 17.1 shows such an instrument. It is very satisfactory for use in evaluating solar water-heater controller circuits.

The newer type of voltohmmeter converts the results into a digital display much like a calculator. The results are easier to read, so the user is less liable to make a reading error. Figure 17.2 shows a digital display multimeter.

Both types operate in the same way. The entire difference is in how the results are displayed.

How to Use the Digital Multimeter

The meter in Figure 17.2 is the EICO 274A digital multimeter, a typical instrument that can be used to check out the control system. Figure 17.3 is a drawing of the face of this multimeter.

Heliotrope General, one of the major manufacturers of controllers, supplies a control system test kit for servicepeople that contains this EICO multimeter. Since it is a typical digital multimeter, it will be used to explain multimeter use.

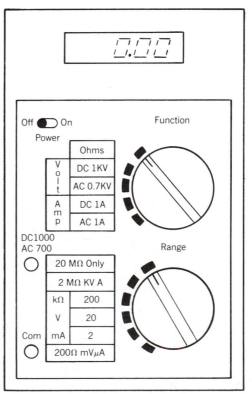

Figure 17.3 The face of the EICO 274A digital multimeter.

In the upper left corner of the meter's control board there is a power switch labeled off/on. To operate the meter, the power switch must be on.

Turn the meter over (see Figure 17.4) and remove the cover. Inside the compartment there are two fuses. The lower fuse is mounted in the circuit, and the upper fuse is a spare. *Do not operate the meter unless it contains the proper size (1 A) fuse or you may melt the electronic circuits.*

The compartment also contains a 9-V battery that operates the circuits. The meter will not operate unless the battery is good. There is room in the compartment to carry a spare battery.

Replace the cover and turn the meter back over. Push the power switch on. If the battery is bad, the digital display will read *BT :*, and the meter cannot be used until the battery is replaced.

If the battery is good, the meter will read either *00.0* or *1 .*, depending on how the function and range switches are set. The decimal point may appear in a different position in the 00.0 display.

Turn the power switch off to conserve the battery.

On the right side of the control board there are two switches. The upper switch is labeled **function,** and the lower switch is labeled **range.** The function switch sets up the instrument to measure resistance (ohms), potential (volts), or current (amps) and indicates the largest value that the meter can measure. The range switch is used to set a specific intermediate range of values to work in because the meter is most accurate when the value being measured does not fall at the lowest or highest ends of the meter's sensitivity range.

On the lower left side of the control board there are two terminals. One is labeled *DC 1000, AC 700, (+)*; the other is labeled *COM*, which stands for common or ground. The *COM* terminal is also known as the *(−)* terminal.

The meter comes supplied with two measuring probes that fit into these two terminals. The red lead mounts in the + terminal, and the black lead mounts in the *COM* terminal.

The meter is used by placing these two probes across the circuit being measured. Figure 17.5 shows the meter with the two leads inserted. Note the alligator clips lying next to the probes. These can be slipped over the tips of the probes to allow the probes to be clipped to the measuring points.

Figure 17.4 The rear of the EICO 274A digital multimeter showing the battery and fuse compartments.

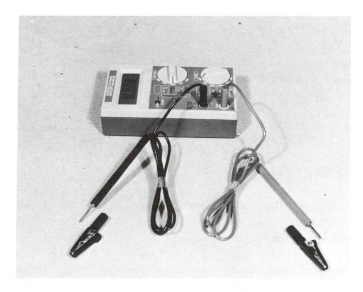

Figure 17.5 The meter with the probes inserted. Alligator clips can be slipped over the probe ends.

Multimeters come complete with instruction manuals, so the general operation of a multimeter will not be repeated here. Instead, a chart of function and range switch settings will be given for different types of measurements. If a different multimeter is used, the ranges provided by the meter may differ from this chart.

MULTIMETER SETTINGS FOR TYPICAL MEASUREMENTS

TO MEASURE	IN THE RANGE OF		SET FUNCTION SWITCH AT	SET RANGE SWITCH AT	MULTIPLY READING BY
	MINIMUM TO MAXIMUM				
The circuits must be dead with the electricity off to measure resistance.					
Resistance	0	199	Ohms	200	1
Resistance	200	1,999	Ohms	2 K	1,000
Resistance	2,000	19,999	Ohms	20 K	1,000
Resistance	20,000	199,999	Ohms	200 K	1,000
Resistance	200,000	1,999,999	Ohms	2 M	1,000,000
Resistance	2,000,000	19,999,999	Ohms	20 M	1,000,000
Never measure circuits containing over 700 ac volts. Always place probes across the lines. Never insert probes in series with the lines. If the voltage is unknown, start with the highest range.					
ac Voltage	0	0.19	ac 0.7 kV	200 mV	0.01
ac Voltage	0.2	1.9	ac 0.7 kV	2 V	1
ac Voltage	2	19.9	ac 0.7 kV	20 V	1
ac Voltage	20	199.9	ac 0.7 kV	200 V	1
ac Voltage	200	700.0	ac 0.7 kV	2 kV	1,000
Never measure circuits containing over 1,000 dc volts. Always measure across the lines. Never insert probes in series with the circuit. If the voltage is unknown, start with the highest range.					
dc Voltage	0	0.19	dc 1 kV	200 mV	0.01
dc Voltage	0.2	1.9	dc 1 kV	2 V	1
dc Voltage	2	19.9	dc 1 kV	20 V	1
dc Voltage	20	199.9	dc 1 kV	200 V	1
dc Voltage	200	1,000.0	dc 1 kV	2 kV	1,000

Amperage measurements are not needed in the troubleshooting of controllers. This multitester only measures up to 1 A of current safely. Most pump circuits use more than 1 A. Do not use this tester as an ammeter placed in series with the pump or valves. Same expensive multimeters have extended 10 A current ranges, but even these may be damaged if the circuit has a low-resistance short.

The multimeter's condition must be checked before use. To check out the condition of the meter, proceed as follows.

- Turn the *power switch* to *ON*. If the display reads *BT :*, the battery is low and must be replaced. If *BT :* does not appear, proceed with the checkout.
- Set the *function switch* to *OHMS*. Set the *range switch* to *20 K*. Connect the tips of the two probes. The meter should read about *0.00*. If the meter does not read at all, the fuse is blown and must be replaced.

Testing Sensors

Chapter 11 taught that most control systems contain two types of sensors, the temperature-dependent, variable-resistance thermistor and the temperature-dependent switching sensor.

Sensors are the most vulnerable part of the control system because they are located away from the controller box and are connected to it by long wires. Therefore, the first step in learning to troubleshoot control systems is learning to measure the resistance of sensors.

Sensors can also play another troubleshooting role for the serviceperson. In evaluating systems it is necessary to take temperature readings at different locations in the system. The multimeter and the temperature-dependent, variable-resistance thermistor can be used together to take accurate temperature readings when evaluating the system.

Testing the Thermistor

Although a number of different thermistors are used in solar water-heater control systems, the 3000-Ω (3-k) and the 10,000-Ω (10-k) negative temperature coefficient thermistors are the most common.

The following table shows the resistance of the 3-k and 10-k thermistors used by Heliotrope General and Independent Energy, Inc. at various temperatures.

THERMISTOR RESISTANCE TABLE

TEMPERATURE (°F)	3,000-Ω THERMISTOR	10,000-Ω THERMISTOR
(Thermistor open)	Infinite	(Infinite)
(Thermistor shorted)	0	0
30	10,400	
32	9,790	32,660
34	9,260	30,864
36	8,700	29,179
38	8,280	27,597
40	7,830	26,109
42	7,410	24,712
44	7,020	23,399
46	6,650	22,163
48	6,300	21,000
50	5,970	19,906
52	5,660	18,876
54	5,370	17,905
56	5,100	16,990
58	4,840	16,128
60	4,590	15,315
62	4,360	14,548
64	4,150	13,823
66	3,940	13,140
68	3,750	12,494
70	3,570	11,885
72	3,390	11,308
74	3,230	10,764
76	3,080	10,248
77	**3,000**	**10,000**
78	2,930	9,760
80	2,790	9,299
82	2,660	8,862
84	2,530	8,449
86	2,420	8,057
88	2,310	7,685
90	2,200	7,333
92	2,100	7,000
94	2,010	6,683
96	1,920	6,383
98	1,830	6,098
100	1,750	5,827
102	1,670	5,570
104	1,600	5,326
106	1,530	5,094

THERMISTOR RESISTANCE TABLE (Continued)

TEMPERATURE (°F)	3,000-Ω THERMISTOR	10,000-Ω THERMISTOR
108	1,460	4,873
110	1,400	4,663
112	1,340	4,464
114	1,280	4,274
116	1,230	4,094
118	1,180	3,922
120	1,130	3,758
124	1,040	3,453
128	953	3,177
132	877	2,925
136	809	2,697
140	746	2,488
144	689	2,298
148	637	2,124
152	589	1,966
156	546	1,820
160	506	1,688
165	461	1,537
170	420	1,402
175	383	1,280
180	351	1,170
185	321	1,071
190	294	982
195	270	901
200	248	828
210	210	702

Note: The values at which the products are rated are indicated in boldface type.

The thermistor must be removed from the circuit to measure its resistance. This is accomplished by removing the thermistor wires from the controller terminals. Figure 17.6 is a picture of a typical terminal board and shows where the sensor wires attach.

Figure 17.7 shows a 10K thermistor taped tightly to a pipe and hooked to the multitester with the function switch set on *ohms* and the range switch set on *20-k*. With the power switch on, the meter reads 10.08, or 10,080 Ω. Referring back to the thermistor resistance table, the temperature of the pipe is 77°F.

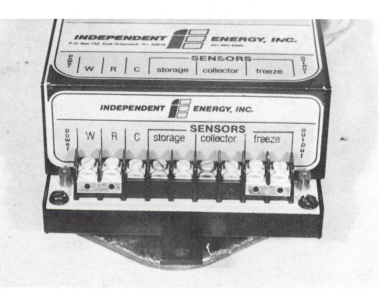

Figure 17.6 A typical controller terminal board showing where the sensor wires attach.

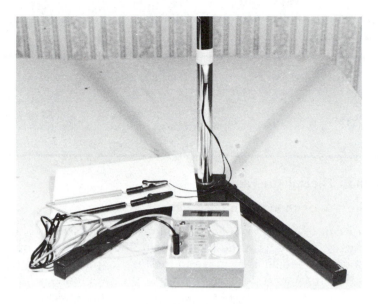

Figure 17.7 A 10K thermister taped to a pipe for making temperature measurements with the multitester.

In Figure 17.8 a polystyrene coffee cup has been slit down the side, slipped over the thermistor, and taped back together. The cup was stuffed with fiberglass insulation. A simple setup like this allows the temperature of any pipe to be read accurately. Masking tape was used for this room temperature setup, but a tape suited for high temperatures would be better in the field.

Another method that works well in the field is to attach the sensor and cover it with a piece of foam pipe insulation. In both instances, use thermal grease between the sensor and the pipe if available.

For testing thermistor sensors in the system, just disconnect the leads and clip one multimeter probe to each one. Set the meter's function switch to *ohms* and the range switch to *20-k* (assuming a 3-k to 10-k

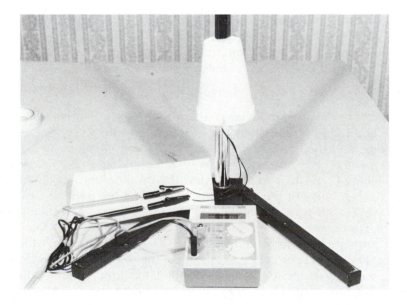

Figure 17.8 For accurate measurements, cover the thermistor.

thermistor) and read the meter. The following results should be seen with a 10-k thermistor.

THERMISTOR CONDITION	METER READING	REMARKS
Open circuit	1 .	Infinite resistance
Shorted circuit	00.0	Zero resistance
Hot pipe (78–210°F)	0.7–9.76	700–9760 Ω
Cold pipe (32–76°F)	10.2–32.7	10,250–32,700 Ω

Common Thermistor Faults

Here are six common thermistor faults.

Open Circuits   If the circuit is open, either the thermistor is bad or one of the wires leading from the thermistor to the controller is open. The best procedure usually is to move to the location of the thermistor and test the thermistor at its lead-in wires. This will quickly determine whether the thermistor must be replaced or whether new wires must be run from the thermistor to the controller.

High-Resistance Circuits   A very high, out-of-range but not infinite resistance also indicates a faulty sensor or sensor lead-in wire. Disconnect the sensor from its wires and evaluate its resistance. Twist the two lead-in sensor wires together at one end and read the resistance at the other ends. If the resistance exceeds 10 Ω, look for a damaged sensor lead-in wire or replace the wiring.

Shorted Circuits   If the circuit is shorted, the short can be in the thermistor or in the wires leading to the controller. Move to the location of the thermistor. Disconnect the lead-in wires and test both the wires and the thermistor separately. This will determine whether the thermistor or the wiring is faulty.

Low-Resistance Circuits   Sensor circuits can have partial shorts, giving a very low but not zero resistance. When a 3-k sensor measures 100 Ω or less or a 10-k sensor measures 300 Ω or less, either the sensor or the wiring is most likely faulty. Check the sensor first. If it tests good, test the wiring in the following manner.

- Disconnect the wires from the sensor
- Insulate the wires from each other
- Disconnect the wires from the controller
- Connect the VOM across the wires at the controller end.

The VOM should show an infinite resistance.

Thermistors That Test in Range   A thermistor may test within the correct range and still not be operating correctly. Thermistors can drift in value on aging, and their resistance characteristics can change if moisture or corrosion enter the case. The only way to be positive that the thermistor is operating correctly is to place a new thermistor alongside it and measure the resistance of the two at several different temperatures.

However, this type of malfunction is relatively rare, particularly if the manufacturer has preaged the thermistor and if the sensor installation has been protected properly. Generally, it is best to assume that the thermistor is operating correctly and check out the rest of the control system. If *one* sensor wire is inadvertently grounded, the ground may confuse the controller even though the sensor gives a correct temperature reading at the controller. Replace the sensor lead-in wires if this condition is suspected.

Inadvertent Thermistor Temperature Changing   During the testing, the thermistor may change in value if left exposed to the air or if the pump is cycled on and off. Always be certain that the service conditions have not affected thermistor temperature.

Measuring Temperature-Actuated Switching Sensors

The temperature-actuated switching sensor is either open or closed, depending on the temperature. Therefore, its resistance should either measure infinite *(1 .)*, indicating an open switch, or shorted *(0.00)*, indicating a closed switch. Any resistance other than infinite or zero indicates a defective sensor *unless* a temperature-measuring sensor is also located in the circuit. Always test the sensors disconnected from the circuit before making a final decision to change them.

To test this switching sensor, three things must be known.

- Whether the switch is normally open (n.o.) or normally closed (n.c.).
- The temperature at which the switch opens or closes.
- The temperature of the sensor.

The schematic of the controller is used to determine if the sensor is expected to be an n.c. or n.o. switch. Usually this schematic is shown in the manufacturer's literature or on the inside cover of the controller. Figure 17.9 shows a typical inside cover schematic.

The temperature at which the switch opens or closes can be determined by the function of the sensor. A freeze-protection sensor will operate at about 40 to 44°F. A high-limit sensor will operate at about 160 to 180°F. There are also some special-purpose switching sensors that open at other temperatures. Know what function you are testing.

The temperature of a freeze-protection sensor can be determined by measuring the temperature of the air surrounding the sensor or the temperature of the pipe to which the sensor is attached. The temperature of the high-limit sensor can be checked by measuring the temperature of the pipe or surface to which it is attached.

Care must be exercised in determining whether the switch is operating

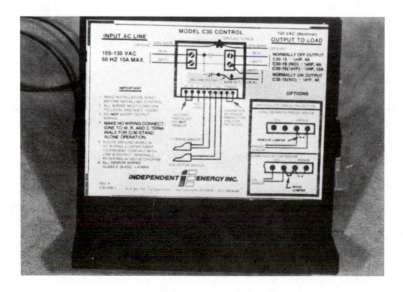

Figure 17.9   A typical inside cover controller schematic.

correctly, because the switch resistance is zero or infinite. Obviously, a shorted lead-in wire or open lead-in wire would also measure zero or infinite. Therefore, the lead-in wire and the sensor must be checked separately by disconnecting the lead-in wire from both the controller and the sensor. With the lead-in wire disconnected, proceed as follows.

- Separate the lead-in wires at one end so that they do not contact each other. Place the multimeter across the other end of the leads with the function switch on *ohms* and the range switch on *20 M*. The meter should read *1 .*, indicating infinite resistance (open circuit).
- Twist the bare ends of the lead-in wires together at one end and connect the multimeter at the other end with the range switch set to 200 Ω. The meter should read from *00.0* to *3.0*, indicating a shorted circuit.
- Clip the multimeter across the leads of the sensor with the function switch on *ohms* and the range switch on *20 K*. The meter should read either *1 .* or *0.00*, depending on whether the switch is *n.o.* or *n.c.* at the current temperature. Any other reading indicates leakage across the switch and requires sensor replacement.

These measurements tell whether the sensor *at rest* is in the right condition. They do not tell whether the switching action is occurring. The switch must be cycled through the switch action temperature to determine its action. The freeze switch must be cooled down to below 40°F and the high-limit switch must be heated to above 180°F to check switch action.

Switching sensors can be checked on the job by heating and cooling them. Do not use water immersion to check switching sensors (water inside the sensor capsule is a common cause of sensor failure). If it is not convenient to check the sensors on the job, check out the wiring. If the wiring is good, just replace the sensor when the service problem is one of tank overheating or system freeze-up. (*Quick test tip:* Freon spray can be used to simulate freezing.)

Checking Out the Controller

The controller is checked out with the sensors disconnected and replaced by resistors of known values. The exact procedure used will depend on the type of controller and how it is hooked into the system.

The Sensor Substitution Box

A *sensor substitution box* is used to check out the controller. A sensor substitution box is a test box containing resistors of known value that can be substituted in the circuit for the sensors.

Figure 17.10 is a schematic of a substitution box built to handle 3K thermistors. It contains a 0- to 1000-Ω variable resistor to replace the collector sensor and a 1000-Ω fixed resistor to replace the storage sensor.

To use this box, the thermistors are disconnected from the controller and the substitution box is wired into the circuit. Figure 17.11 shows a Model TDT-1 tester from Heliotrope General connected to a typical controller. Figure 17.12 shows the face of the substitution box. Note that the dial for the variable-resistor knob is calibrated in degrees and runs from 0 to 180°F.

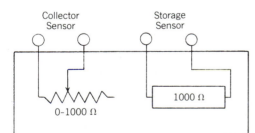

Figure 17.10 A schematic of the circuits in a 3-k thermistor sensor substitution box. This is the Model TDT-1 tester from Heliotrope General.

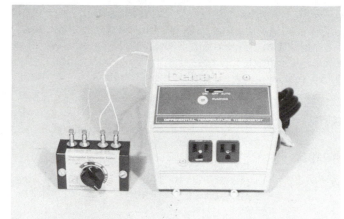

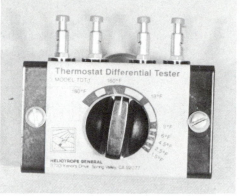

Figure 17.12 This is the face of the TDT-1 tester. Note how the dial is calibrated.

Figure 17.11 The TDT-1 tester connected to a Heliotrope General controller.

The 1000-Ω fixed resistor is connected across the storage tank sensor leads. At 1000-Ω resistance, the controller thinks it is looking at a 125°F tank temperature.

The 0- to 1000-Ω variable resistor is connected across the collector tank sensor leads. When the knob is turned fully clockwise so that the dial reads 0°F, 1000-Ω is placed across the collector leads and the controller thinks that it is seeing a 125°F collector temperature, or a difference of 0°F between the tank and the collectors. Under these conditions, the pump would remain off.

Turning the knob counterclockwise lowers the resistance placed across the collectors. The dial is calibrated at 1.5°F, 2.5°F, 4.5°F, 6°F, 9°F, and 18°F hotter than the storage resistor. For instance, when the dial reads 9°F, about 840-Ω are across the collector leads and the controller thinks it is looking at a 134°F collector, a temperature differential of 9°F.

If the controller were set to turn on with a 9°F temperature difference, the pump would start up when the knob reached this point.

Continuing to turn the knob counterclockwise lowers the resistance across the collector sensor terminals more and more and tests the controller with hotter and hotter collectors.

The dial is also calibrated at 160°F and 180°F. These two calibration points are set at 506 Ω and 351 Ω, which correspond to a 3K thermistor's resistance at 160°F and 180°F. The two points are used to simulate a 3K thermistor used as a high-limit switch.

A 10K thermistor can be checked in the same way, but the substitution box should be manufactured with 3000-Ω fixed and variable resistances.

Figure 17.13 is a schematic of a substitution box that can be used for both 3K and 10K thermistors. The substitution box contains both 1000-Ω and 3000-Ω variable and fixed resistors. Because 3000 Ω is not a standard carbon resistor value, a 1.2KΩ and a 1.8KΩ are connected in series to make up 3000 Ω.) A double-pole, single-throw switch is connected to the collector and the storage terminals and either 1000- or 3000-Ω fixed and variable resistors can be connected to the terminals.

Figure 17.14 shows an additional circuit that is useful in a sensor substitution box.

The circuit contains two double-pole, single-throw switches and four resistors. This circuit can be built to check either freeze thaw action (FT) or high-limit action (HL) for either 3K or 10K thermistors.

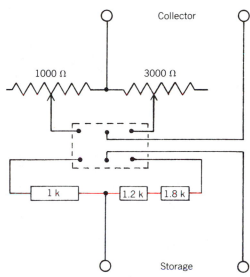

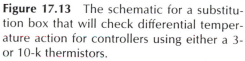

Figure 17.13 The schematic for a substitution box that will check differential temperature action for controllers using either a 3- or 10-k thermistors.

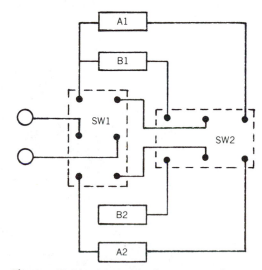

Figure 17.14 A circuit for testing freeze thaw (FT) or high-limit (HL) action for both 3- and 10-k thermistors.

Switch SW2 selects the resistors. When SW2 is closed to the left, resistors B1 and B2 are selected. When the switch is closed to the right, resistors A1 and A2 are placed in the circuit. Switch SW1 selects whether resistor 1 or resistor 2 is connected across the thermistor circuit.

Here are the values that would be used in these circuits.

HIGH-LIMIT CIRCUIT

RESISTOR	THERMISTOR	TEMPERATURE (°F)	VALUE (Ω)
A1	10 K	185	1.2K
A2	10 K	155	1.8K
B1	3 K	185	330
B2	3 K	155	560

FREEZE THAW CIRCUIT

RESISTOR	THERMISTOR	TEMPERATURE (°F)	VALUE (Ω)
A1	10 K	38	27 K
A2	10 K	54	18 K
B1	3 K	38	8.2 K
B2	3 K	54	5.6 K

A Versatile Sensor Substitution Box A sensor substitution box can be purchased for a particular manufacturer's controllers when working with only one or two types of controllers. In that case the manufacturer will supply complete instructions on how to check out the controller.

A sensor substitution box that can be used with almost all controllers can be built by using the two circuits shown previously.

Figure 17.15 shows the face of this sensor substitution box. The box contains two variable-resistance circuits for checking out differential turn on and turn off and high-limit thermistor action, and two fixed-point resistance circuits for testing high-limit and freeze thaw action separately.

The top of the box has a set of terminals for the collector sensor substitution and a set of terminals for the storage sensor substitution. On the left side of the box is a set of terminals for high-limit (HL) checking; on the right side of the box is a set of terminals for freeze thaw (FT) checking.

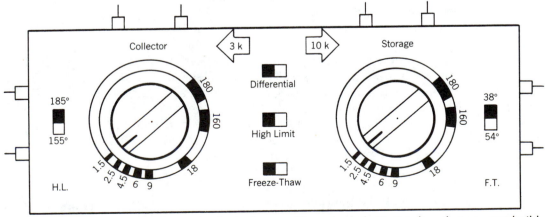

Figure 17.15 The face of a versatile sensor substitution box that you can build.

The center of the box contains three switches that set the terminals for either 3K or 10K thermistors. To the left of these switches is a calibrated dial for checking the differential action of a 3K thermistor controller; on the right of the switches is a calibrated dial for checking differential action of a 10K thermistor controller.

A fixed-point HL switch is located at the far left. With the switch up, a 185°F resistance value is placed across the controller. With the switch down, a 155°F resistance value is placed across the controller.

A fixed-point FT switch is located at the far right and places either a 38°F or 54°F resistance value across the freeze thaw controller circuit.

Figures 17.16 and 17.17 show the inside of the box, so the view is reversed left to right. Figure 17.16 shows the wiring for the two variable resistors. This corresponds to the schematic in Figure 17.13. The wiring for HL and FT is not shown. Figure 17.17 shows the wiring for the HL and FT circuits. These circuits correspond to the circuit shown in Figure 17.14. The location of the resistors has been changed slightly to accommodate an easy circuit layout. Use 5% tolerance (gold band), $\frac{1}{2}$-W resistors throughout the circuit. It is fine to use $\frac{1}{4}$-W resistors if desired but, mechanically, they are hard to work with. Resistors with 10% and 20% tolerances could be used if they were first *premeasured* with the multimeter and then selected to be close to the desired value.

A simple way to build this sensor substitution box is to mount the collector and storage insulated terminals, the variable resistors, and the five switches on the L-shaped cover of a metal chassis box while mounting the HL and FT terminals on the sides of the box, as shown in Figure 17.18.

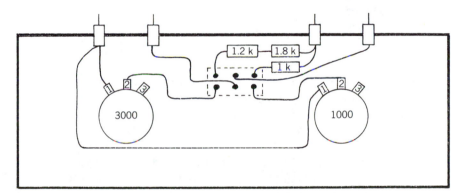

Figure 17.16 This is the wiring diagram for the differential temperature control circuits.

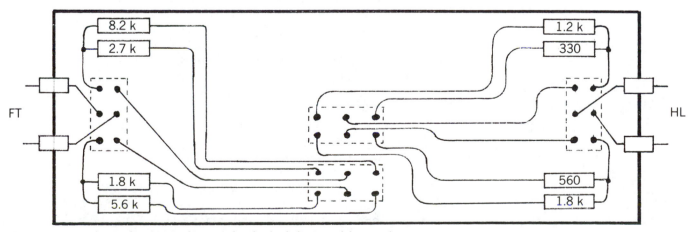

Figure 17.17 This is the wiring diagram for the high-limit and freeze-thaw control circuits.

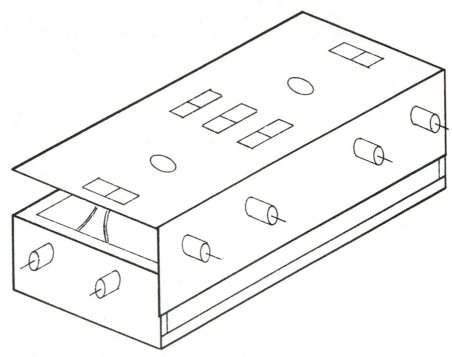

Figure 17.18 A simple method of construction is to use an L-shaped metal chassis box.

The entire circuit is then built on the inside of the cover, and leads from the center poles of the HL and FT switches are soldered to the HL and FT terminal posts.

The variable-resistor dials are calibrated with the multimeter after the circuit is completed in the following manner.

- Switch the differential switch to the 3 K setting. Place the multimeter across the storage terminals with the function switch set on ohms and the range switch set on 20 K. The tester should read between 950 and 1050 Ω with a 5% resistor. Record this reading.
- Place the multimeter across the collector terminals with the function switch on ohms and the range switch on 20 K.
- Turn the 1 K variable-resistor shaft fully clockwise. The multimeter should read about 1000 Ω. If it reads down close to 0 Ω, turn the panel over, remove the wire from the variable resistance terminal marked 1, and solder to the terminal marked 3. The multimeter should now read between 950 and 1050 Ω.
- Place the knob on the resistor shaft and turn it so that the pointer points to about eight o'clock. Tighten the knob on the shaft and mark the pointer's location on the panel.
- Subtract 27 Ω from the resistance of the fixed resistor measured in the first step. Set the knob pointer so that the multimeter reads that value. For example, if the fixed resistor measured 982 Ω, 982 − 27 = 955 Ω. Mark that point as 1.5°F on the dial.
- Subtract 47 Ω from the fixed resistor. Turn the pointer to that location and mark the location 2.5°F.
- Subtract 87 Ω, measure, and mark the location 4.5°F.
- Subtract 133 Ω, measure, and mark the location 6°F.
- Subtract 168 Ω, measure, and mark the location 9°F.
- Subtract 307 Ω, measure, and mark the location 18°F.
- Turn the knob until the meter reads 506 Ω. Mark the dial 160°F at this location.
- Turn the knob until the meter reads 351 Ω. Mark the dial 180°F at this location.

This completes the calibration of the 3K thermistor dial. The 10K thermistor dial is calibrated in the same manner using the following resistance readings.

POINTER	RESISTANCE READING
Fixed resistor	2850 to 3150 Ω
1.5°F	Storage — 90 Ω
2.5°F	Storage — 149 Ω
4.5°F	Storage — 261 Ω
6°F	Storage — 343 Ω
9°F	Storage — 498 Ω
18°F	Storage — 903 Ω
160°F	1688 Ω
180°F	1170 Ω

You should build a set of four or more 24- to 36-in. leads for use with this box. Use different-colored wires and solder an alligator clip to each end of each lead.

Checking Out the Differential Temperature Controller Action

The differential temperature controller action of almost all controllers can be checked by removing the collector and storage sensors from the circuit and replacing them with the proper substitution resistor circuits. Figure 17.19 gives a typical checkout procedure using the custom substitution box previously described. Mark the location of each sensor wire before disconnecting it.

The operation of the dump valves is not specified after the first step because, in some systems, the valves are set to dump the fluid in the collectors at high limit while in other systems, the valves do not open at high limit. The serviceperson should determine how the controller is set up from the owner's manual and check to see that the dump valves are also working properly.

Checking Out the Freeze-Protection Controller Action

The freeze thaw sensor can be a thermistor or a switching sensor. Both 3K and 10K thermistors are used. The switch is an n.o. switch in some circuits and an n.c. switch in other circuits. Proceed to test freeze-protection action as shown in Figure 17.20 using the custom sensor substitution box.

Checking Out High-Limit Controller Action

The high-limit sensor can also be a thermistor or a switching sensor. Both 3K and 10K thermistors are used. The thermistor can be a separate thermistor or the storage thermistor may be used. The switching sensor is separate from the storage thermistor and both n.o. and n.c. switches are used. Proceed to test high-limit action with the custom substitution box as shown in Figure 17.21. Remember that two different controller strategies, "boil and bake," can be called for in draindown systems.

The Types of Faults The six major controller faults that these test procedures show are

1. Pump runs with collectors and storage at the same temperature.
2. Pump does not turn on at turn-on threshold temperature.
3. Pump does not turn off when the tank reaches 160 to 180°F.

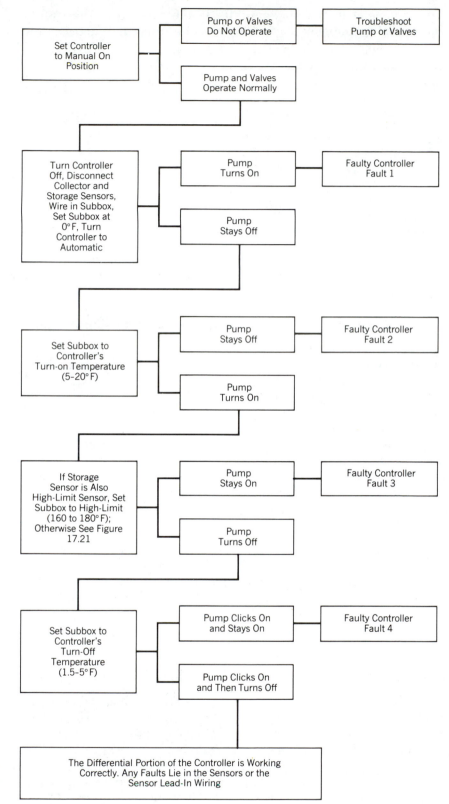

Figure 17.19 A troubleshooting chart for checking differential temperature controller action.

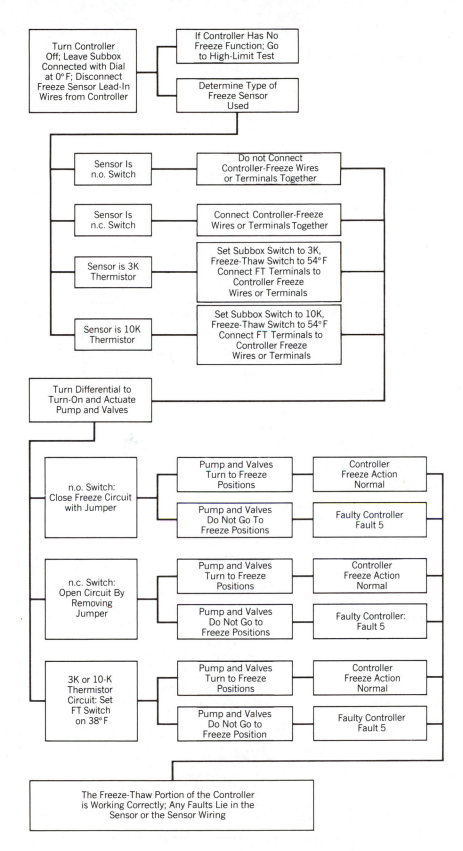

Figure 17.20 A troubleshooting chart for checking freeze-thaw controller action.

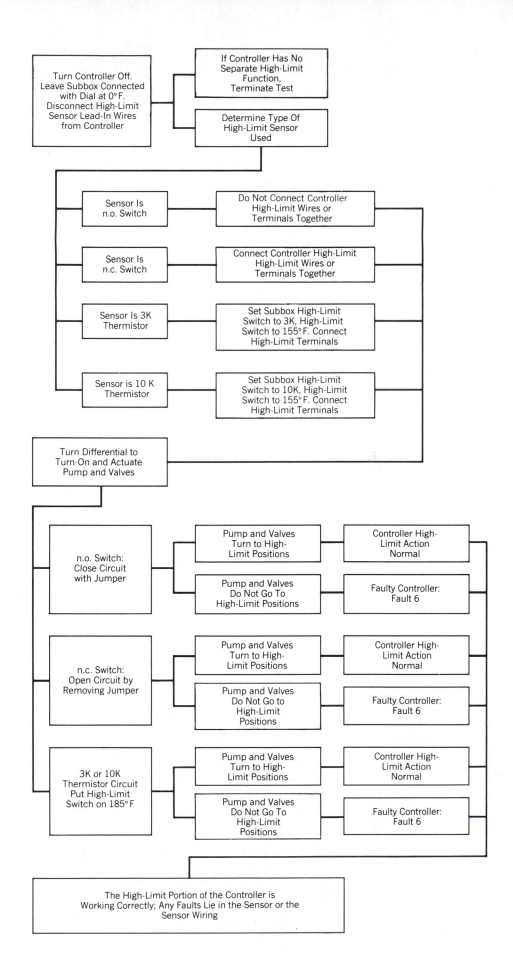

4. Pump does not turn off when collectors cool down to the tank temperature.
5. Freeze thaw operation is not occurring.
6. High-limit action is not occurring.

Controllers Can Be Confusing to Service

The serviceperson must keep in mind that the normal operation of the controller requires an interplay between differential temperature sensors, freeze sensors, high-limit sensors, and sometimes factory-installed jumpers.

If a normally closed freeze-sensing snap switch wire is broken, the controller will receive a freeze signal. No amount of controller or resistance substitution box manipulation will turn the pump or the valves on. The same is true if an n.c. freeze sensor is removed from the circuit during the test unless a jumper wire is clipped across the terminals to replace it. This is one reason why it is wise to make a preliminary check of the sensors before fully checking out the controller.

It is just as important to make certain that the manufacturer's jumpers are properly installed as per the instructions for the desired control strategy. Refer to the manufacturer's manuals for specific troubleshooting procedures whenever possible. Use a jumper wire to simulate a normally closed switch and an open circuit to simulate a normally open switch when testing differential temperature controls.

The principle that you must keep in mind is that both a freeze signal and a high-limit signal will always override the normal action of the differential temperature controller that runs the pump.

Effecting Repairs

Most controller manufacturers do not want their controllers field repaired. Therefore, they do not supply service information for the electronic circuits. Some controllers, like those from Independent Energy, are constructed so that a new electronic board may be substituted easily. Others, like those from Heliotrope General, require the replacement of the entire controller.

Sensor repair is not possible either. Faulty sensors must be replaced with new ones having similiar specifications.

Figure 17.21 A troubleshooting chart for checking high-limit controller action.

CHAPTER 18

Outfit the Toolbox

Solar water-heater servicing is a multitrade occupation. The solar serviceperson must bring the skills of the electrician, plumber, carpenter, and sheet-metal worker to the job site. Consequently, the solar toolbox contains a composite of the tools and supplies used in all four trades.

Additionally, solar servicepeople must deal with some unique needs, such as the need to evaluate, troubleshoot, and repair solid-state control systems, to evaluate collector array orientation, tilt, and shade; and to transport, lift, and replace glass glazings and often entire collectors.

Invest in the Best

Reliable tools are a key factor in supplying quality servicing at reasonable costs. "Flea market specials" may be cheap to purchase, but they are expensive to work with.

The professional serviceperson must make a substantial investment in tools and training before making solar service calls. Of course, if the serviceperson's background is in one of the previously mentioned trades, he or she will already own many of the listed tools.

In the service business time is money and reliable tools are time-savers. It is hard to justify a 1-hour trip to replace a cheap tool. Do not get caught short on the job. Invest only in the best.

How the Tool Lists Are Laid Out

Power tools and accessories are shown first. A list of general hand tools follows. Then the rest of the tools are listed by trade. In some cases specific tools have been called out by manufacturer's name and/or number. These are the tools that experience shows are superior for the job. Seriously consider buying those models unless you already own a tool in those categories.

Power Tools

- **A reversible, variable-speed, 500-rpm, $\frac{1}{2}$-in. power drill** This heavy-duty, slow-speed drill is essential for penetrating masonry, driving collector mounting screws and lag bolts, and general-purpose, heavy-duty drilling.

- **A circular saw with a $7\frac{1}{4}$-in. blade.** This saw should be capable of cutting lumber blocks, plywood, sheet metal, and masonry. A heavy-duty saw is recommended but not critical.
- **A reversible, variable-speed, 2500-rpm, right-angle head, $\frac{1}{4}$- or $\frac{3}{8}$-in. drill.** This drill is needed for installing fasteners and making holes in tight locations where access is restricted.
- **A sabre saw or a reciprocating saw (optional).** This tool is only needed occasionally.
- **A power hammer (optional).** This tool, which is only needed occasionally, should be capable of driving up to $2\frac{1}{2}$ in. heat-treated fasteners into concrete to fasten lumber, electrical boxes, and similiar items to concrete and concrete block walls. Carry an assortment of fasteners ranging from $\frac{3}{4}$ to $2\frac{1}{2}$ in.
- **A 50- to 100-ft, 12-gauge, three-wire, grounded outdoor extension cord.** This heavy-duty cord is used to power the electric tools. A 12-gauge wire, 50-ft cord will carry 15 A; a 12-gauge wire, 100-ft cord will carry 13 A. The tools suggested here use 5 to 13 A.
- **A grounded, caged trouble light and a 50-ft, grounded, 18-gauge extension cord.** This will be needed for inspection and troubleshooting indoors.
- **Plug-in adaptor assortments.** Three-wire to two-wire plug adapters, light socket-to-plug adapters, and three-in-one outlet taps should be in the toolbox to allow obtaining power from other-than-normal sources.

Power Tool Accessories

- A $7\frac{1}{4}$-in. combination circular saw blade.
- A $7\frac{1}{4}$-in. fine tooth plywood circular saw blade (optional).
- A $7\frac{1}{4}$-in. masonry circular saw blade.
- A $7\frac{1}{4}$-in. metal cutting circular saw blade.
- Metal and wood blades for the sabre or reciprocating saw.
- Set of high-speed, carbon steel twist drills, $\frac{1}{32}$ to $\frac{1}{2}$ in.
- Set of masonry drills, $\frac{3}{16}$ to 1 in.
- Set of spade bits for cutting wood, $\frac{1}{4}$ to $1\frac{1}{4}$ in.
- Set of holesaws and mandrels, $\frac{7}{8}$ to 2 in.
- A 12- $\times$ $\frac{1}{4}$-in. electrician's bit.
- An 18-in. drill bit extension for spade bits (optional).
- Socket drive adapters, $\frac{1}{4}$ and $\frac{3}{8}$ in.
- Magnetic power nut drivers, $\frac{1}{4}$ to $\frac{1}{2}$ in. (optional).
- Power screwdriver bits (optional).
- Drill and saw blade sharpeners (optional).

General Hand Tools

- A 24-ft extension ladder.
- An 8-ft stepladder.
- A 50- to 100-ft safety line or rope.
- Carpenter's hammer.
- Prybar.
- Utility knife.
- An 8-ft tape measure.

- 25-ft $\times$ 1-in. tape measure.
- Torpedo level.
- Combination square with scribe awl and level.
- Ball of string for laying out piping runs.
- Chalkline for laying out straight lines.
- Flashlight.
- A $\frac{1}{2}$-in. and a 1-in. steel-handled wood chisel.
- Caulking gun.
- Screwdriver assortment:

  Phillips screwdrivers, No. 0 to 2 plus No. 2 stubby.
  Slotted (standard) screwdrivers, $\frac{1}{8}$ to $\frac{1}{4}$ in., plus $\frac{1}{4}$-in. stubby.
  Set of refrigerator nut screwdrivers (optional).
  Set of jeweler's screwdrivers.
  Set of Phillips and slotted ratcheted offset screwdrivers.

- Saw assortment:

  Hacksaw with assortment of blades (Sandvik 225).
  Keyhole saws for wood and metal.
  Jab saw with metal cutting blade (Milwaukee) (optional).
  Crosscut hand saw (optional).

- Mechanic's hand tool assortment:

  $\frac{3}{8}$-in. hardened steel impact socket set.
  $\frac{3}{8}$-in. socket wrench set with ratchet and 12-in. extension (breaker bar—optional).
  $\frac{1}{4}$-in. socket wrench set with ratchet, T bar, and 6-in. extension.
  Adjustable wrenches, 6 and 12 in.
  Allen wrench set.
  Pliers, 6 in.
  Visegrips, 10 in.
  Set of open- and closed-box combination wrenches, $\frac{1}{4}$ to $1\frac{1}{4}$ in.

Sheet-Metal Worker's Tools

- Aviation metal shears.
- Long nose pop rivet gun.
- Hand or power nibbler.

Electrician's Tools

- 8-in. lineman's pliers.
- 7-in. long nose pliers.
- 6-in. diagonal cutters.
- Wire stripper, 20 to 10 gauge.
- Crimp-on connector pliers.
- Color-coding tape in four colors.
- 1-ft, 12-gauge bare copper wire (stud locator).
- Romex stripping tool.
- Yankee screwdriver with screwdriver tips, drill bits, and $\frac{1}{4}$-in socket drive adapter (optional).

Plumber's Tools

- Propane torch (Turbotorch STK-9).
- Spark lighter with spare flints.
- Ratchet tubing cutter (Reed TC1Q).
- "Knuckle" tubing cutter.
- Pipe wrenches, 12, 14, and 18 in.
- Pump pliers, 12 in.
- Rat-tail files.
- Flat file assortment.
- Fitting brushes for $\frac{1}{2}$-, $\frac{3}{4}$-, and 1-in. fittings.
- Wire brush, small.
- Windex spray bottle (for cooling pipes after soldering).
- 14-in. chain wrench (optional).
- Portable or truck-mounted pipe vise (optional).
- Small sheet of aluminum flashing for heat shields.

Solar-Oriented Tools

- Carbon steel kitchen knife for cutting insulation.
- Toothbrush for cleaning circuit boards and fine parts.
- Inspection mirror.
- Safety matches.
- 24-in. $\times$ $\frac{1}{2}$-in., water hose with female fitting on one end.
- 24-in. $\times$ $\frac{1}{2}$-in., water hose with female fitting on both ends.
- Female-to-female hose connector.
- MIP-to-male hose connector.
- 24-in. of vinyl tubing for siphoning water out of pipe traps.
- Empty 2-lb coffee can.
- Rubber dishpan.
- Sponge assortment.
- Hose shutoff valve.
- Indelible felt pen.
- Transfer pump for heat transfer fluid or 5-gal Extrol tank fitted with shutoff valve.
- Small air compressor or hand pump.
- Insulation miter tubes. These are shop-fabricated 12-in.-long from $1\frac{1}{2}$-in. and 2-in. PVC pipes with a 45° angle cut on one end and a $22\frac{1}{2}$° angle cut on the other end. They are used to produce precision angle miters on pipe insulation.
- Service manuals from the various solar water-heater equipment manufacturers.

Safety-Oriented Equipment

- Work gloves.
- Welder's gloves (optional).
- Nonskid safety shoes.
- Bump cap or hard hat.
- Safety goggles.
- Dust mask.
- Fire extinguisher.

Supplies

The choice of supplies must be dictated by the types of heaters that the serviceperson will encounter so, except for a few items, only a general description will be given.

- Copper and brass fittings and pipe, $\frac{1}{2}$, $\frac{3}{4}$, and 1 in.
- Dielectric unions, $\frac{1}{2}$ and $\frac{3}{4}$ in.
- Valve assortment, $\frac{1}{2}$, $\frac{3}{4}$, and 1 in:

  Automatic air vents.
  Boiler drain valves.
  Temperature and pressure relief valves.
  Pressure relief valves (20 to 30 psi).
  Gate, balancing, and ball valves.
  Spring check valves.
  Swing (flapper) check valves.
  Electrically actuated valves and spare parts.
  Sunspool valves and spare parts.
  Tempering valves.
  Vacuum relief valves.
  Wye strainers.

- Soldering and brazing materials.

  50/50 tin/lead solder with flux (for nonsolar loop use).
  95/5 tin/antimony solder with flux (for solar loop use).
  Silver solder with flux (for minor, light-duty component brazing repairs).
  Assortment of rags.

- Compounds, sealants, and glues.

  Dow Corning 111 compound.
  Silicone one-part sealant (for potting).
  Insulation glue.
  WD-40 penetrating lubricant.
  Thread compound and Teflon tape.
  Heat sink compound

- Control system parts.

  Spare controllers (optional).
  Spare pumps (optional).
  Thermistor sensors (3K and 10K).
  Snap switch sensors (high limit and freeze).
  Sensor hookup wire.
  12-V and 24-V transformers (optional).
  Pump gaskets, seals, and O rings.
  Dielectric union gaskets.

- Assorted pipe insulation.
- Assorted hangers and fasteners.
- Plastic cable ties.
- Electrical fittings and connectors.
- Romex, flexible conduit, and No. 14 type TW wire.
- Pressure-sensitive labels; flow direction arrows, valve and control labels.

CHAPTER
19

Purchase the Test Equipment

Instruments are needed to troubleshoot solar water heaters properly. The serviceperson must be able to measure liquid temperatures, pressures, and flow rates and electrical resistance, current, and voltage. The control system operation must be tested. Measurements of collector array orientation, tilt, and shading may be required. Finally, the properties of heat transfer fluids, such as pH and concentration, may have to be evaluated.

Again, Invest in the Best

A wide variety of instrumentation is available. Some of it is well suited for solar, some is not, but there are only two levels of quality: reliable and unreliable. Test equipment is used to make decisions. Those decisions lead to action that results in solving service problems. False or improper readings result in time-consuming and expensive actions that may not be needed. Therefore, unreliable test equipment cannot be tolerated.

The test equipment must be reliable, it must be suited for the job, and it must be durable and last for many years, so it pays to invest only in the best test equipment.

How the Instrument Lists Are Laid Out

The instrument lists are laid out by the type of testing that needs to be performed. The first list will cover the instruments required to test temperature, pressure, and flow rates. The second list covers measuring electrical resistance, current, and voltage. The third list will show what is needed to make control system sensor substitution measurements. The fourth list will detail what is needed to measure collector array orientation, tilt, and shading. The fifth list will contain the instruments needed for testing heat transfer fluids.

In many cases manufacturer's names and model numbers will be suggested. The instruments suggested have been proven in use as reliable.

In a few cases the fabrication of a special test instrument will be suggested. This fabrication should be undertaken when a commercial instrument is unavailable or unwieldy to use.

Instruments for Measuring Temperature, Pressure, and Flow Rate

- **Temperature measurements** Temperature measurements are most important in testing solar water-heater operation. The basic decisions shown in the troubleshooting charts in this book are based on taking accurate temperature measurements.

  The Taylor pocket dial thermometer represents the least expensive instrument. It is accurate to ±2°F when properly wetted with heat transfer paste and protected from ambient air temperatures. A minimum of two will be needed.

  The electronic thermometer costs more but has higher accuracy, gives faster readings, and is more convenient to use. This thermometer uses built-in probes or sensors for measurement and reports the results on a digital readout in degrees Fahrenheit. An Omega 450, an Electromedics SH-66, or their equivalent is suggested.

  A voltohmmeter coupled with thermistor sensors can be used as an electronic thermometer, as shown in Chapter 17. However, a chart that converts the instruments's resistance reading to degrees Fahrenheit is required.

- **Pressure measurements** Two types of pressure measurements are made in evaluating solar water heaters: service-water pressure measurements and closed collector loop pressure measurements.

  A water-pressure gauge that measures from 0 to 150 psi should be used to measure city water pressure. For convenience, the gauge should be mounted to a female hose coupling that will allow pressure to be measured at any hose bib in the system. Any high-quality gauge may be used.

  A water-pressure gauge measuring from 0 to 30 psi is very useful for evaluating pressures in a closed-loop heat-exchange system. Again, mounting it to a female hose bib allows it to be connected to the drain and fill valves.

  An air pressure gauge that measures from 0 to 50 psi is used for measuring the pressure on Extrol tank diaphrams and for monitoring the tightness of systems when they are being pressure tested after leak repair. The gauge should be equipped with a female fitting that will screw to a Schraeder air valve. (A Schraeder air valve is a valve like the air valve on an automobile tire.)

- **Flow measurements** A portable flow meter is useful for making approximate flow rate measurements in pumped loops, as shown in Chapter 6. The Blue/White or the Hedland in-line 1- to 10-gpm flow meter or its equivalent is recommended. The flow meter is attached as shown in Figure 6.20.

  Standard hose fittings will not withstand the pressures and temperatures found in many solar systems. Have a hydraulic equipment supplier make up two heavy-duty $\frac{3}{4}$-in. inside diameter hoses with female hose fittings on either end. Attach these permanently to adapters connected to the flow meter.

Instruments for Electrical Measurements

- **Resistance measurements** Resistance measurements are made with a multimeter or a voltohmmeter (VOM), as discussed in detail in Chapter 17. The digital readout VOM is highly recommended for ease of operation, speed, and accuracy.

Specifications can be found in Chapter 17. Invest in a shock-resistant design or carry the instrument in a foam-cushioned case. Anticipate that it will be dropped several times during its life. The Beckman HD-110, the EICO 274A, or an equivalent VOM are recommended.

- **Voltage measurements** The VOM can be used for making the voltage measurements as outlined in Chapter 17, or a neon line voltage tester may be used. The Wiggy neon line voltage tester or its equivalent is recommended because these units can distinguish between 110-V and 220-V circuits. This distinction can be the difference between life or death for a solar component or the serviceperson.

  A 3-ft safety electrical line cord is also a useful tool. A grounded cord should be used. This safety line cord consists of a 3-ft, 110-V power cord with a male plug on one end and a set of three insulated alligator clips on the other end. It is used to bring 110 V directly to a component such as a pump without removing the component from its location.

Caution Tests perfomed using the safety line cord require special caution, as does all work with live electrical circuits. The regular circuit must be disconnected first. Always attach the alligator clips to the component before plugging in the cord. Do not allow the alligator clips to touch each other or any grounded surface. Disconnect the cord plug before removing the alligator clips from the component.

- **Current measurements** A clamp-on ammeter such as the Amprobe RS-1 or its equivalent works best. This instrument clamps on the outside of the wire conductor over the conductor's insulation. The strength of the magnetic field passing through the wire is converted to a dial measurement of amperage. It is used to identify overloaded circuits and damaged pumps.

Caution Some digital multimeters may be damaged if used to measure currents of the magnitude found in solar water-heating systems. Read instructions carefully.

Instruments for Control System Measurements

- **Sensor measurement** The VOM is used for making sensor measurements. Full details on making these measurements are located in Chapter 17.
- **Sensor substitution measurements** The sensor substitution box is used to measure differential control turn-on/turn-off points, to measure high-limit points, and to measure freeze-protection points. The measurements are made by substituting known test resistances and switches in the sensors' circuits. Chapter 17 covers the routines for making these measurements in depth.

  Chapter 17 also gives the plans for constructing a universal sensor substitution box. This route can be followed, or the serviceperson can purchase sensor substitution boxes from the major control suppliers. Boxes for Rho Sigma, Heliotrope General, and Independent Energy controllers will cover most systems.

  Keep firmly in mind when using sensor substitution boxes that these are not designed as precision instruments. Variable resistors set with a digital VOM are best used for making more precise measurements.

- **Precision resistance measurements** A 10,000-Ω (10K) and a 20,000-Ω (20K) variable resistor are handy to have in the test kit. These can be set to precise resistances with the digital VOM. The variable resistors are wired with two 24-in. leads. One lead goes on the center lug and the other goes on either of the two side lugs. The other ends of the leads terminate in alligator clips. The resistance required is dialed in by turning the shaft of the resistor.

  The resistors are used by substituting them for the sensors in the circuits under test and slowly dialing them to the point where the desired circuit action takes place. Then the resistor is carefully removed and its resistance measured with the VOM. When the resistance is compared to the correct thermistor chart, the exact temperature of operation is found.

  If the proper sensor substitution box is not available, a pair of 10K or 20K variable resistors can be used in its place.
- **Sensor lead-in wire continuity measurements** The sensor substitution box does not test the continuity of the lines running between the controller and the sensor. A 50- to 100-ft, two-conductor, low-voltage wire, such as is used to connect telephones, is handy for such testing. One end of the wire should be prepared with insulation-piercing alligator clips, which may be purchased in any electronics store; the other ends can be prepared with standard small alligator clips for attaching to the controller. The voltages running through these wires should be very low. Special precautions are generally not needed, but it is wise to check the voltage across the terminals before proceeding in case there is an internal short circuit within the controller.

  Four or five 2-ft lengths of color-coded, low-voltage wire with insulated alligator clips soldered to both ends are useful in checking out valve control lines or sensor connections within the mechanical room.

Instruments for Evaluating Collector Array Orientation, Tilt, and Shading

- **Orientation measurement** A compass is the fastest way to measure collector array orientation. Obtain a compass that has a fluid-damped movement and a rotatable bezel or backing marked off in degrees. The fluid-damped movement will give a quick response, and the rotatable bezel allows the magnetic deviation from true south to be set ahead of time.
- **Tilt angle measurement** The collector tilt angle is quickly and easily checked with an inclinometer, which is a weighted, rotating pointer set into a transparent, fluid-filled dial that is calibrated in angular degrees. When the inclinometer is placed on the collector glazing, the slope of the tilted surface with respect to the horizontal may be read directly from the instrument. Most hardware stores stock reasonably accurate inclinometers. Select a shock-resistant model that will survive being dropped from the roof.
- **Shading measurement** It is very hard to estimate properly by eye the shading patterns caused by nearby trees or buildings. A solar siting instrument is required. The Solar Site Selector or the Solar Pathfinder are two relatively inexpensive instruments that will allow these measurements to be made with a degree of accuracy.

Both instruments are used in the same way. The instrument is placed in a level position and faced geographic (true) south. Printed or etched lines on a transparent surface indicate the sun's path at any time of the year. Sight through the range finder across the sun's path lines and determine if any shading occurs. Very often, the use of these instruments will explain otherwise unexplained decreases in solar collection.

Instruments for Measuring Heat Transfer Fluid Properties

As explained in Chapter 10, heat transfer fluids vary from inhibited water to silicone fluids. Most of these fluids require analysis by a laboratory to spot major problems, with the exception of the glycol/water mixtures. Therefore, the following instruments are restricted to measuring the properties of water/glycol systems.

- **Specific-gravity measurements.** The concentration of ethylene glycol in ethylene glycol/water systems can be measured by measuring the mixture's specific gravity. This is done with a hydrometer, which is essentially a glass tube containing float balls of different densities. Depending on the ratio of water to glycol, some or all of the balls float in the solution. This hydrometer may be purchased at any automotive supply store.

Caution Propylene glycol concentrations cannot be accurately determined with specific-gravity measurements. Instead, the refractive index must be measured.

- **Refractive index measurements.** The concentration of propylene glycol in a propylene glycol/water mixture is determined by measuring the refractive index of the mixture with a refractometer. American Optical Company refractometer model 7391 is recommended. A sample of the fluid is placed in the instrument, and the operator peers into the eyepiece to read the refractive index. The refractive index is then compared to a chart showing concentration versus refractive index obtained from the fluid manufacturer. Dow Chemical Bulletin 173-1190-82 contains such a chart for its Dowfrost and a picture of the refractometer is shown in Bulletin 173-1112-81. The refractometer may also be obtained from Dow Chemical Company or one of its local distributors.
- **Acidity/alkalinity measurements** As explained in Chapter 10, glycol systems contain inhibitors to neutralize the acids that are formed when the glycol breaks down under heat. The condition of these inhibitors is measured by measuring the pH of the fluid.
PH paper may be obtained from a local chemical supply house or from some drugstores for measuring this "reserve alkalinity."

GLOSSARY

Active solar water heater A system that requires outside mechanical or electrical energy for proper operation. Although the solar industry calls thermosiphon and ICS systems passive solar water heaters, this is technically incorrect as they require mechanical energy in the form of water pressure for proper operation.

Air eliminator A mechanical device that separates air from liquid by passing the stream of liquid over a baffle. Commonly seen in liquid heat-exchanged systems.

Air handler A mechanical package containing an air blower or fan, appropriate directional dampers, and controls to circulate, direct, and control the flow of the air in an air heat-exchanged system.

Air heat-exchanged system Solar water heater system that uses air for the collector fluid and an air-to-water heat exchanger to transfer the air's heat to the storage water.

Air purge valve See *automatic air vent valve.*

Air separator See *air eliminator.*

Air-to-water heat exchanger See *finned tube heat exchanger.*

Air vent valve See *automatic air vent valve.*

Ambient air sensor A controller sensor used to measure the temperature of the outdoor air around the collector array.

Ammeter An instrument used to measure the flow of electrical current.

Amperes A measure of the rate at which an electrical current is flowing.

Anode rod A rod made of magnesium and suspended in a storage tank to slow tank corrosion.

Anodic More electrically positive. The opposite of cathodic.

Autoignition point The temperature at which a fluid will ignite without an outside flame being applied to it.

Automatic air vent valve A float valve that automatically opens to vent air from the system when the liquid level of the system drops. Usually placed at the highest point on the collector array and, in liquid closed-loop, heat-exchanged systems, on the air eliminator. May also be placed at other troublesome points in the system where air tends to gather, such as on the top of storage tanks in draindown systems.

Auxiliary heater A fossil-fuel or electric heater that is placed in series with a solar preheater to raise the solar-heated water to a higher delivery temperature. Also see *solar water preheater.*

Backflush valve A valve in the solar system that can be opened to flush liquid through the system's collectors in the opposite direction to normal flow. Used to clean out collector's liquid passages.

Bake option A control strategy for draindown systems where, under high-limit conditions, the pump is shut down and the collectors are drained.

Ball valve. A shutoff valve with a moving part consisting of a round ball with a port through it. Can be used as a flow control or throttling valve. Both two-way and three-way ball valves are available.

Batch heater See *breadbox solar water heater.*

Bellows connector A type of connector for handling differential expansion between collectors and external manifolds. It utilizes a pipe formed into a bellows.

Boiling point The temperature at which a liquid changes to a gas.

Boil option A control strategy for draindown systems where, under high-limit conditions, the circulator pump is shut down and the collectors are left full of water.

Breadbox solar water heater A simple ICS solar water heater that consists of a black-painted tank held within an insolated enclosure with a glazed cover. The enclosure often resembles a breadbox.

Canned wet-rotor pump A type of centrifugal pump where the motor rotor runs in the liquid being pumped.

Catastrophic failure Catastrophic failure is failure that results in physical damage to the system.

Cathodic More electrically negative. The opposite of anodic.

Centrifugal pump A type of pump that has blades that rotate and sling the fluid off the blade tips into a discharge port. The fluid is thrown outward by centrifugal force.

Check valve A valve that only allows the flow of fluid in one direction.

Closed drainback system Drainback systems that do not allow the introduction of fresh oxygen during drainback. They use an air vent line instead of a vacuum relief valve.

Closed loop A piping system that is both closed to the atmosphere and separated from the service water. Typically connects the collectors with the heat exchanger in a liquid-exchanged system.

Coiled pipe connectors A type of connector for handling differential expansion between collectors and external manifolds. Uses a 360° coil of soft copper tubing for the connector.

Collector fluid See *heat transfer fluid.*

Collector sensor The differential thermostat controller sensor located at the collector.

Compatibility, fluid See *fluid compatibility.*

Compressed air drainback system A drainback solar water-heating system that creates storage for the collector water by pumping high-pressure air into the storage tank to drive down the water level. We do not recommend the use of air-pressurized drainback systems in freezing weather.

Continuous service (electrical service) A situation where the equipment being controlled draws electrical current continuously throughout its use cycle. Requires a controller and wiring rated for continuous service.

Controller A device that automatically turns the pump off and on and opens and closes the proper valves as conditions within the system signal for these actions.

Controller logic The path of reasoning that the controller travels in making its decisions. Controller logic varies from system type to system type.

Counterflow A heat-exchanger configuration where the two fluids flow in opposing directions relative to the exchanger's heat transfer wall.

Density A measure of weight per unit of volume. A gallon of water weighs 8.33 lb/gal.

Differential temperature controller A controller that measures the difference in temperature between the collectors and the storage tank and controls the system based on this temperature differential.

Differential thermal expansion Expansion and contraction that take place at different rates or in different directions.

Differential thermostat See *differential temperature controller.*

Diode A semiconductor device that will only pass electrical current in one direction.

Double-walled heat exchanger A heat exchanger that has two separate walls between the heat transfer fluid and the storage water. Such exchangers usually incorporate a leak path to the environment so that leaks can be quickly detected.

Downcomer The pipe in a pumped-fluid solar heating system that runs from the top of the collector array to the storage tank. Also, the pipe in a thermosiphon system that runs from the bottom of the storage tank to the bottom of the collector array.

Drainback system A solar water-heater system that is designed with air space in the heat exchanger or in the

storage tank so that the water drains out of the collector array when the pump is not operating.

Draindown system A solar water-heater system designed so that the water in the collectors and the outdoor piping is automatically drained when the temperature drops to 40 to 42°F.

Drain valve Globe or gate valves used in solar water-heater systems on tanks and in the piping to give drainage. Usually, a boiler drain valve is used.

Dribble valve A thermostatic valve that opens at 40 to 42°F to allow water to circulate slowly through the collectors.

Dump position The open position assumed by a draindown valve when the power is removed. This is a fail-safe position in that the system automatically assumes a freeze-protected configuration if power is lost.

Dynamic head The pressure drop in a pumped system caused by the friction created when the fluid is moved through the pipes. Usually measured in feet of water or pounds per square inch.

Environmental protection The action of properly protecting the collectors and the associated piping and wiring from the weather.

Equivalent lengths of pipe A method of stating pressure drop across fittings, valves, and equipment. The pressure drop is said to be equivalent to the pressure drop in a certain number of feet of pipe.

Evaporator/condenser collector A collector that contains a heat exchanger to boil the liquid (the evaporator) and a second heat exchanger to condense the vapor (the condenser).

Expansion tank A small, air-filled tank in a closed loop that has room for the fluid when it expands. Usually, the fluid is separated from the air by a rubber bladder in solar system applications.

External manifold A manifold in a collector array that does not pass through the collector box.

External motor mechanical shaft seal pump A type of centrifugal pump that is characterized by an external motor

attached to the pump through a mechanical shaft seal. This type of pump is rarely used in small, residential solar systems. It is typically designed only to pump water, not organic liquids.

Fastening The action of properly bolting or lagging the collectors and the collector racks to a solid support.

Fill valve Globe or gate valve used in solar water-heater systems to fill the system with water or heat transfer fluids. Typically, a boiler drain valve is chosen.

Finned tube heat exchanger A heat exchanger consisting of a series of copper tubes passing through a bank of closely spaced metal fins. Air passes over the fins while liquid is pumped through the tubes. An automobile radiator is a typical finned tube heat exchanger.

Fire point The temperature at which a fluid will continue to burn if it is ignited by an external flame and the flame is then removed.

Flash point The temperature at which a fluid will burn when an external flame is applied to it but will self-extinguish when the external flame is removed.

Float vent See *automatic air vent valve.*

Flow control valve See *throttle valve.*

Fluid compatability The determination of whether the fluid is suitable for use in a particular system. This is a complex determination involving many factors.

Freeze dump circuit The portion of the system controller circuit that opens the dump valves in a draindown system and stops the pump.

Freeze dump sensors The sensors in the freeze dump circuit that signal the onset of freezing temperatures. The sensors are usually located at critical spots on the collector array.

Function switch The switch on the voltohmmeter that sets the meter to measuring resistance, potential, or current. Most voltohmmeters are not suitable for measuring high currents.

Gallons per minute (gpm) A common way of describing the flow of liquid through a pipe or pump.

Gallons per minute per square foot of collector (gpm/ft²) A common way of describing the flow rate through collectors.

Gate valve A positive shutoff manual valve that uses a gate to open and close the valve. Commonly used where nonrestrictive flow is required.

Gosselin rating A rating of fluid toxicity in humans based on animal feeding studies. See also *lethal exposure test*.

Halocarbon fluids Fluorocarbon chemicals that are commonly used throughout industry as refrigerants and aerosol propellents. A common trade name is Freon.

Headers The large liquid passages running across the top and bottom of the collector absorber plate. The finned riser tubes are normally connected to a top header and a bottom header.

Heat exchanger A device, usually made of metal, that transfers heat across an internal wall that separates two gases, two liquids, or a liquid and a gas.

Heat-exchanger effectiveness test A performance test that measures how much heat an exchanger actually transfers versus how much heat it could theoretically transfer if it were completely (100%) effective.

Heat of fusion The heat required to melt 1 lb of a solid substance. The temperature of the substance is not changed.

Heat of vaporization The heat required to change 1 lb of a substance from a liquid to a gas. The temperature of the substance is not changed.

Heat rejection option A control strategy for draindown systems where, under high limit conditions, the system is engineered to dissipate heat actively.

Heat transfer fluid The fluid that circulates through the collectors and the heat exchanger in a liquid heat-exchanged system. A fluid that freezes at subzero temperatures and has a high boiling point is generally chosen.

High-limit sensor A controller sensor that overrides the differential turn-on/

turn-off circuit when the storage tank has reached a preset temperature in the range of 160 to 180°F. Commonly mounted near the top of the solar storage tank.

Horizontal tank thermosiphon solar water heater A thermosiphon solar water heater that uses a storage tank lying on its side in a horizontal position.

Hose connectors A type of connector between collectors and manifolds that handles differential thermal expansion by using a weather-resistant, high-temperature rubber hose as the connector.

Hydrometer An instrument used to determine the specific gravity and the freeze point of (poly)ethylene glycol/water mixtures.

Integral collector-storage system (ICS) A solar collector that also contains solar storage. See also *breadbox solar water heater*.

Internal manifold A collector array manifold that runs inside the collector box.

Isogonic chart A map that shows deviation of magnetic north from true north for all geographic locations.

Latent heat of vaporization. Heat accompanied by a phase change from the liquid state to the gaseous state. There is no change in the fluid's temperature during phase change.

Lethal exposure test A fluid toxicity test based on animal feeding studies. See also *Gosselin rating*.

Liquid heat-exchanged system A solar water-heating system that uses a collector fluid other than water in a closed loop and a heat exchanger to transfer the solar heat to the storage water.

Load-side heat-exchanger coil A large coil of tubing, usually copper, that is inserted into the drainback tank to provide heat transfer from the collector fluid to the service water. The tubing contains the service water, and the tank holds the heat transfer fluid.

Magnetic-drive pump A pump that contains a drive magnet sealed within the body of the pump. The pump is driven by a second magnet located on the pump shaft external to the pump body.

Manifold The piping that feeds and removes the fluid from the collector when the pump is running. A collector array contains a top and a bottom manifold. The top and bottom collector headers may be used as the manifolds, in which case the system is said to be internally manifolded, or the system may have separate manifolds external to the collectors.

Melting point The temperature at which a substance changes from a solid to a liquid.

Motorized valve A type of electrically actuated valve that uses a gear motor to open and close the water passages.

Multimeter See *voltohmmeter*.

Multiple-point freeze protection The use of more than one freeze sensor to activate the freeze cycle of the system controller.

Negative coefficient thermistor A thermistor sensor that decreases in resistance as the temperature increases.

Night sky radiation The term used for the nighttime radiation of energy from a warm collector to a cold night sky. On a very clear night the sky can have a radiant temperature as low as −50°F. This radiant effect can cause standing water in a solar collector to freeze and damage the collector tubes.

Nonpotable A fluid that is not suitable for human consumption.

Nonrestrictive flow valve A flow valve that, when fully opened, does not restrict the flow through a pipe. Ball valves and gage valves are typically nonrestrictive flow valves.

Normally closed (n.c.) Used to describe the operating position of an electrically operated switch or valve when it is not energized. A normally closed valve is shut when it is not energized. The opposite of normally open.

Normally open (n.o.) Used to describe the operating position of an electrically operated switch or valve when it is not energized. A normally open valve is open when it is not energized. The opposite of normally closed.

Ohm's law A formula describing the relationship of current, resistance, and voltage in an electrical circuit. $E = IR$, where E = volts, I = amperes, and R = ohms.

Open drainback system Drainback systems that allow fresh oxygen to enter the system through a vacuum relief valve during drainback.

Orientation The compass direction that a solar collector array faces.

Parallel-connected collectors Collectors that are connected so that the flow is divided across them. The bottom collector manifolds are connected to the cold riser, and the top collector manifolds are connected to the hot downcomer.

Passive solar water heater A system that does not require outside mechanical or electrical energy for proper operation. The solar industry calls thermosiphon and ICS systems passive solar water heaters. This is technically incorrect as they require water pressure for proper operation.

pH A value that represents the acidity or alkalinity of a liquid system.

Phase-change heat-exchanged solar water-heating systems Solar water-heater systems that use a boiling liquid as the heat transfer fluid.

Piping The action of interconnecting the collectors, storage tanks, and other components together so that liquid flow can be established among them.

Positive coefficient thermistor A thermistor that increases in resistance as the temperature increases.

Positive freeze protection A feature of a solar water-heating system that indicates that the system is designed to protect itself against freezing even if the system loses external power.

Pot The action of encapsulating a wire or a component in an insulating compound. Potting prevents water from entering an electrical splice or connection.

Potable A descriptive term for a fluid that is fit for human consumption. Potable liquids are acceptable for drinking. Water is a common potable liquid.

Preheater See *solar water preheater*.

Pressure drop The resistance to flow encountered in pipes and other fluid transport system components. Usually expressed in feet of liquid or pounds per square inch.

Pressure gauge A gauge that measures liquid or gaseous pressure. Usually used in closed-loop solar water-heating systems on the collector loop.

Pressure relief valve A valve that automatically opens when a predetermined pressure is reached. Commonly seen on collector arrays and in closed-loop liquid heat-exchanged systems.

Proportional controller A controller that delivers electricity to the pump in varying amounts proportional to the temperature difference between the collectors and the storage tank. This causes the pump to run faster when the temperature difference gets greater.

Pump A mechanical device designed to circulate liquid through a system.

Pumped secondary loop drainback system A drainback system that uses a load-side heat-exchanger tank and a pump between the exchanger tank and the solar storage tank to circulate the storage water through the exchanger coil.

Pumped service-water solar system A solar water-heating system that pumps the service water through the collectors to heat it but does not drain down nor drain back for freeze protection.

Pump off/on circuit The controller circuit that manages the operation of the system's pumps.

Range switch The switch on a voltohmmeter that determines the operating range of the meter. The range may be changed to fit the conditions being measured.

Recirculation freeze-protection system A pumped service-water solar water heater that recirculates warm water through the collectors when the temperature drops below 40 to 44°F to prevent the system from freezing.

Refractometer An instrument that checks the refractive index of a fluid.

Reverse-return parallel connection A method of connecting collectors that

equalizes liquid flow, temperature, and efficiency throughout the system.

Reverse thermosiphoning An event where the heat collected in the storage tank passes back into the collectors and is dissipated out to the ambient air.

Right angle-collector tilt angle The tilt angle formed when the collector is titled so that its face is perpendicular to the sun's rays. Also known as the solar normal-collector tilt angle.

Riser piping The pipe in a pumped fluid solar heating system that runs from the bottom of the storage tank to the bottom of the collector array. Also the pipe in a thermosiphon system that runs from the top of the collector array to the top of the storage tank.

Riser tubes The tubes in the collector that connect the top and bottom manifolds. Usually have heat transfer fins attached to them.

Rubber hose connector See *hose connector*.

Safety glazings Collector glazings that meet the Consumer Product Safety Commission's "Safety Standard for Architectural Glazing Materials." The willful violation of this standard is considered a criminal offense.

Sensible heat Heat accompanied by a change in the fluid's temperature.

Sensor A device used in the collector circuits to determine the temperature at a point remote from the controller.

Separate condenser gravity-return system A phase change heat-exchanged solar water heater in which the condenser is separate from the evaporator but located higher than the evaporator so that the condensed collector liquid can return to the lower manifold of the collector by gravity.

Separate condenser pumped-return system A phase change heat-exchanged solar water heater in which the condenser is separate from the evaporator but located lower than the evaporator so that the condensed collector liquid cannot return to the lower manifold of the collector by gravity but must be pumped up with a liquid refrigerant-type pump.

Series connections Connections where the inlet of one device is connected to the outlet of the device it follows so that only one path is established for liquid flow. Also used in electrical circuits to describe how the current flow is established.

Solenoid valve A type of electrically actuated valve that uses an electromagnet to open and close the water passages.

Shading The existence of objects that prevent the sun's rays from striking the solar collectors.

Silver solder A brazing compound used to join metals under high-heat conditions. Usually used in shop fabrication instead of on the job.

Site dependent A phrase describing the dependency of the situation on the geographic site. The orientation of a collector is site dependent.

Solar module A descriptive name for a package of preplumbed valves, controls, and pumps mounted together and supplied as a single unit. May also contain the heat exchanger.

Solar water preheater A solar water heater connected in series with an auxiliary fossil-fueled or electric heater so that the cold water that would normally flow to the auxiliary tank is warmed by solar heat first.

Specific gravity The weight of a given volume of a material as compared to the weight of the same volume of water. Water has a specific gravity of 1. A material that weighed 9 lb/gal would have a specific gravity of 1.080 (9/8.33 = 1.080).

Specific heat or **specific heat capacity** A measure of how much energy it takes to heat a given mass of a substance by 1°F. In the English system of measurement it is reported as the number of British thermal units it takes to heat or cool 1 lb of the substance by 1°F. Water has a specific heat of 1. It takes 1 Btu to raise the temperature of 1 lb of water by 1°F.

Spool-type valve A special type of draindown valve built specially for draindown systems. Replaces electrically actuated valves.

Stagnation or **stagnating** A condition where the system is at rest and the fluid is not circulating. Can result in

temperatures as high as 300 to 350°F in some collector arrays.

Standoff bracket A collector bracket that holds the collector up off the roof so that water, ice, and snow are free to drain from under the collectors.

Static head See *vertical pumping head*.

Stone-lined tank A water-heater tank that has been manufactured with a spun-concrete lining. Such tanks are free of corrosion and have very long lives.

Storage tank sensor See *tank sensor*.

Stratification The layering of water within a tank into hot and cold sections.

Substitution box A test device containing fixed and variable resistors that can be used to check out the operation of a controller. The resistors in the box are substituted for the sensors in the system.

Summer bypass ductwork Air ductwork in an air heat-exchanged solar system that allows the solar heated air to bypass the house and the rock storage bed during the summer months.

Swage-lock fitting A quick connect/disconnect fitting for piping.

Switching relay An electrical relay that allows a low-powered control circuit to energize a high-powered system device such as a pump or motorized valve.

System dependent A descriptive term that expresses the dependency of a device on the design of the system. Collector flow rate, for example, is system dependent.

System pressure drop The total resistance to flow found in a solar waterheater system. Generally expressed in feet of water or pounds per square inch. The pump must be able to provide the proper rate of flow through the system at the system pressure drop encountered.

Tank sensor The differential thermostat controller sensor located low on the storage tank wall.

T-drilled manifold An external manifold made with a pipe-drilling technique where a special T drill is used to

drill a hole in the manifold pipe. T drilling is normally a shop fabrication procedure and not a job-site procedure.

Temperature and pressure relief valve A valve installed on water heaters that automatically opens when the temperature rises above 210°F or the tank pressure reaches 125 to 150 psi.

Temperature-variable resistor See *thermistor*.

Temperature/viscosity slope A valve showing the rate of change in the viscosity of a liquid with a change in the temperature of that liquid.

Tempered glass A heat-treated glass used in collector glazings. Meets the standard test for safety glazing.

Tempering valve A valve that automatically adds cold water to the hot water stream when it is hotter than 120 to 140°F.

Thermal conductivity A measure of the rate at which heat is conducted through a 1-in. thickness of any given material.

Thermistor An electronic (or "solid-state") sensor whose resistance changes with temperature at a predetermined rate.

Thermosiphon The convective circulation of a fluid occurring in a closed system wherein the denser cold fluid sinks and displaces the less dense, warm fluid upwards.

Thermosiphon solar water heater A solar water heater that works on the thermosiphon principle. Thermosiphon systems do not contain pumps or controllers.

Thermostatic three-way valve A valve that automatically diverts the flow of fluid to one of two outlets depending upon the fluid's temperature.

Three-way electrically actuated valve A solenoid or motorized valve with one inlet and two outlets. One outlet is closed when the other is open.

Throttle valve A valve placed downstream from the collector-to-storage pump that can be adjusted to increase or lower the flow of liquid through the collectors. Normally a ball or butterfly valve is used.

Tilt The angle between the collector face and a horizontal line.

Toxic An adjective meaning poisonous.

Toxicity A noun used to describe how poisonous a substance is to humans.

Two-way electrically actuated valve A solenoid or motorized valve that has one inlet and one outlet. The valve is either open or closed.

U-tube connector A collector-to-manifold connector that uses a pipe formed in the shape of a U to cope with differential thermal expansion.

Vacuum relief valve A solar system valve used to relieve any vacuum in a solar system by admitting ambient air to the system. Normally located at the highest point in the collector array.

Vertical pumping head The resistance to flow in a system caused by the need to pump water up a vertical pipe where no compensating downleg exists to counterbalance the weight of the water in the pipe.

Vertical tank thermosiphon solar water heater A thermosiphon water heater that contains a vertically mounted storage tank.

Viscosity A measurement of a fluid's resistance to flow under shear.

Voltage drop A measurement of the change in voltage across a resistive electrical device.

Voltohmmeter (VOM) An instrument used to measure the voltage and the resistance in electrical circuits.

INDEX